INTERPRETING NEWTON

This collection of specially commissioned essays by leading scholars presents new research on Isaac Newton and his main philosophical interlocutors and critics. The essays analyze Newton's relation to his contemporaries, especially Barrow, Descartes, Leibniz, and Locke, and discuss the ways in which a broad range of figures, including Hume, MacLaurin, Maupertuis, and Kant, reacted to his thought. The wide range of topics discussed includes the laws of nature, the notion of force, the relation of mathematics to nature, Newton's argument for universal gravitation, his attitude toward philosophical empiricism, his use of "fluxions," his approach toward measurement problems, and his concept of absolute motion, together with new interpretations of Newton's matter theory. The volume concludes with an extended essay that analyzes the changes in physics wrought by Newton's *Principia*. A substantial introduction and bibliography provide essential reference guides.

ANDREW JANIAK is Creed C. Black Associate Professor of Philosophy at Duke University. He is the editor of *Newton: Philosophical Writings* (Cambridge, 2004) and author of *Newton as Philosopher* (Cambridge, 2008).

ERIC SCHLIESSER is BOF Research Professor, Philosophy and Moral Sciences at Ghent University. He has published widely on Newton, Huygens, and their eighteenth-century reception (especially by Hume and Adam Smith) as well as in philosophy of economics.

INTERPRETING NEWTON

Critical Essays

Edited by

ANDREW JANIAK AND ERIC SCHLIESSER

CAMBRIDGE
UNIVERSITY PRESS

CAMBRIDGE UNIVERSITY PRESS
Cambridge, New York, Melbourne, Madrid, Cape Town,
Singapore, São Paulo, Delhi, Tokyo, Mexico City

Cambridge University Press
The Edinburgh Building, Cambridge CB2 8RU, UK

Published in the United States of America by Cambridge University Press, New York

www.cambridge.org
Information on this title: www.cambridge.org/9780521766180

First published 2012

Printed in the United Kingdom at the University Press, Cambridge

A catalog record for this publication is available from the British Library

ISBN 978-0-521-76618-0 Hardback

CONTENTS

NOTES ON CONTRIBUTORS

ORI BELKIND is Assistant Professor of Philosophy at the University of Richmond. He is the author of *Physical Systems: Conceptual Pathways between Spacetime and Matter* (forthcoming), and has also published on Leibniz and Newton.

ZVI BIENER is Assistant Professor of Philosophy at the University of Cincinnati. He works primarily on early modern conceptions of the unity of science and the large-scale structure of fields of knowledge, and has published on Galileo, Descartes, and Newton.

KATHERINE BRADING is William J. and Dorothy K. O'Neill Collegiate Professor of Philosophy at the University of Notre Dame. In addition to writing on seventeenth-century natural philosophy, she has published on symmetry in modern physics and on the history of early twentieth-century physics.

GRACIELA DE PIERRIS is Associate Professor of Philosophy at Stanford University. She has published widely in historical topics in epistemology, especially Locke, Hume, and Kant. She has also written on Frege, Wittgenstein, and Quine.

MARY DOMSKI is Assistant Professor of Philosophy at the University of New Mexico. She has published widely on early modern natural philosophy and mathematics as well as on Kant. She is the co-editor of *Discourse on a New Method: Reinvigorating the Marriage of History and Philosophy of Science* (2010).

LISA DOWNING is Professor of Philosophy at Ohio State University. She has published widely on Locke, mechanist conceptions of body and their justification, debates surrounding gravity/attraction, and changing views of scientific explanation in the early modern period.

KATHERINE DUNLOP is Assistant Professor of Philosophy at Brown University. She is primarily interested in theories of exact science and of perception, especially in Kant. Her research extends to Kant's early modern predecessors and to the reception of his views (and theirs) in twentieth-century analytic philosophy.

MICHAEL FRIEDMAN is Frederick P. Rehmus Family Professor of Humanities at Stanford University. He is the author of *Foundations of Space-Time Theories: Relativistic Physics and Philosophy of Science* (1983), *Kant and the Exact Sciences* (1992), *Reconsidering Logical Positivism* (Cambridge, 1999), *A Parting of the Ways: Carnap, Cassirer, and Heidegger* (2000), and *Dynamics of Reason: The 1999 Kant Lectures at Stanford University* (2001).

DANIEL GARBER is Stuart Professor of Philosophy at Princeton University. He is the author of *Descartes' Metaphysical Physics* (1992), *Descartes Embodied* (2001), and *Leibniz: Body, Substance, Monad* (2009).

WILLIAM HARPER is Professor of Philosophy at the University of Western Ontario. He has published widely in philosophy of science, including decision theory, Kant, and Newton. He is the author of *Isaac Newton's Scientific Method: Turning Data into Evidence about Gravity and Cosmology* (forthcoming).

NICHOLAS HUGGETT is Professor of Philosophy at the University of Illinois at Chicago. He has edited *Space from Zeno to Einstein* (1999) and co-authored *Physics Meets Philosophy at the Planck Scale* (Cambridge, 2001). His most recent book is *Everywhere and Everywhen* (2010).

ANDREW JANIAK is Creed C. Black Associate Professor of Philosophy at Duke University. He is the editor of *Newton: Philosophical Writings* (Cambridge, 2004) and author of *Newton as Philosopher* (Cambridge, 2008). His most recent article is "Substance and Action in Descartes and Newton," *The Monist* 93 (October 2010). He writes on early modern natural philosophy and on Kant.

LYNN JOY is Professor of Philosophy at the University of Notre Dame and Fellow of the Reilly Center for Science, Technology, and Values. Her publications include *Gassendi the Atomist: Advocate of History in an Age of Science* and "Scientific Explanation from Formal Causes to Laws of Nature," in *The Cambridge History of Early Modern Science.* She is currently working on a study of the concept of dispositions in contemporary philosophy of mind, entitled *Dispositions and Intentionality in the Humean Tradition.*

MARCO PANZA is Research Director in History and Philosophy of Sciences, Institut d'Histoire et de Philosophie des Sciences et des Techniques, UMR 8590 of the CNRS and the University of Paris 1. He has published widely in the history and philosophy of mathematics, and is the author of *Isaac Newton*

(2003), *Newton et l'origine de l'analyse, 1664–1666* (2005), *Nombres. Eléments de mathématiques pour philosophes* (2007).

ERIC SCHLIESSER is BOF Research Professor of Philosophy at Ghent University. He has published widely on Newton, Huygens, and their eighteenth-century reception (especially Hume and Adam Smith) as well as in philosophy of economics. He is the co-editor of the forthcoming *Oxford Handbook on Isaac Newton.*

CHRIS SMEENK is Associate Professor of Philosophy at the University of Western Ontario. He has published on the philosophy and history of modern cosmology and has co-authored works on Newton with Zvi Biener, George Smith, and Eric Schliesser.

GEORGE SMITH is Professor of Philosophy at Tufts University. His research has focused heavily on the development of evidence in the exact sciences with special emphasis on Newton's *Principia* and its aftermath. He is the co-editor of the *Cambridge Companion to Newton* (2002) and of the forthcoming *Oxford Handbook on Isaac Newton.*

ACKNOWLEDGEMENTS

First and foremost, I wish to thank the contributors to this volume for their outstanding work and impressive patience. I would also like to thank Hilary Gaskin for supporting this project from its inception, and for her wise counsel, as always. Finally, and perhaps most importantly, I thank my co-editor, Eric Schliesser, for seeing the manuscript through its many transformations over the past few years. Eric, I still owe you a drink!
Andrew Janiak

This volume would not have been possible without the generosity shown toward me by Syracuse University, Leiden University, Netherlands Organisation for Scientific Research (which supported me with a four year grant to study Newton and his impact on philosophy), and Ghent University. Along the way I incurred many debts of gratitude toward flexible Deans and Department Chairs, especially the late Stewart Thau and Cathy Newton at Syracuse, Pauline Kleingeld at Leiden, and Freddy Mortier as well as Erik Weber at Ghent. I am grateful to all the contributors to this volume for sharing such wonderful work with Andrew and me; they are terrific philosophers and scholars as well as an inspiring intellectual community. I thank the folk at Cambridge, especially Hilary Gaskin, for believing in this project. Andrew, you are a swell, selfless editor, and you owe me a drink! Finally, my share of this volume is dedicated in love to Sarit and our Avi.
Eric Schliesser

The idea of a volume was born at a workshop on "Newton and/as Philosophy" hosted at Leiden University in the Summer of 2007. Half the papers collected here were presented there.
Andrew Janiak and Eric Schliesser

Introduction

ANDREW JANIAK AND ERIC SCHLIESSER

It may be anachronistic to say that Isaac Newton and his *Principia* decisively changed physics and philosophy, because separate fields of physics and philosophy did not yet exist. But the notion of decisive change captures something significant about the continuing relevance of studying Newton. What has been aptly termed "Newton's new way of inquiry" (Harper and Smith 1995) was baffling for even his most sophisticated contemporaries, and it took Europe's brightest astronomers and mathematically inclined natural philosophers almost a century in order to evaluate and assimilate the *Principia*. But for reasons that need not detain us here, few of these figures (e.g., Clairaut, Euler, Laplace), who were fully immersed in Newton's work, really offered a definitive account of the methodology of the *Principia*. Of course, many scholars from Newton's day onward have offered interpretations of Newton's explicit methodological claims, but surprisingly few have combined this approach with detailed knowledge of Newton's technical practice. As is well known, by the time physics became enshrined as the leading part of the disciplinary structure of science, its attitude toward its own history did not encourage close scrutiny of past practices. In this volume, the three chapters on methodology by George Smith, William Harper, and Ori Belkind all capture important aspects of Newton's new way of inquiry.

Newton also changed philosophy in two important ways. First, the body of work eventually known as "Newtonian mechanics" became a privileged form of knowledge that had to be dealt with somehow within metaphysics and epistemology. Second, it initiated a slow process in which philosophy defined itself in terms that often contrasted with – or were modeled on – Newtonian success. But as a consequence, in philosophy's evolving self-conception Newton stopped being central to the history of philosophy. Somewhat surprisingly, philosophical interest in Newton revived at the beginning of the twentieth century, precisely when his physical theory was called into question by Einstein's revolutionary work. Most of the papers in this volume engage with Newton's place within the history of philosophy. Before we turn to a detailed description of the chapters collected here, we offer a brief introduction to the scholarship that in many ways forms the shared background of recent philosophically motivated work on Isaac Newton.

The philosophical study of Newton's thought has undergone a series of revolutionary developments during the past century.[1] During the first half of the twentieth century, in the context of Einstein's transformative science, much of the key work was accomplished by European historians of science such as Hélène Metzger and Alexandre Koyré, but E. A. Burtt's *The Metaphysical Foundations of Natural Science* (1924) also had a lasting influence.[2] Burtt's text is remarkable for its insistence on treating canonical modern philosophers – such as Descartes and Hobbes – in tandem with canonical natural philosophers – such as Boyle and Newton – in the same fashion, thereby expressing a more expansive conception of philosophy's history than one might have encountered elsewhere.[3] For her part, Metzger wrote insightfully about Newton's place within the history of chemistry and also about a number of substantive philosophical and theological issues that some of Newton's interpreters took to be raised by his work. At the time of her tragic death in Auschwitz, Metzger was writing another work that traced developments in chemistry and optics from Newton through to later figures.[4] Koyré's prodigious scholarship concerned a host of philosophical and historical issues, encompassing a range of figures from Galileo to Newton[5] – his influence was felt not merely in his own scholarship, but through the magisterial critical edition of the *Principia* that he undertook with I. B. Cohen (see below). Any historian or philosopher wishing

1 This introduction must be brief, and will therefore inevitably leave out discussions of important scholars who have grappled with Newton during the past hundred years. Even a volume-length introduction could not provide a comprehensive treatment of scholarly developments during that time frame, let alone a standard introduction. We deal here with those authors who seem, according to the editors of this volume, to have been the most significant twentieth-century figures from the perspective of the philosophical engagement with Newton. Discussions of mathematics, physics, alchemy, optics, politics, etc., would obviously focus on other scholarly figures, texts and traditions. Finally, in what follows, when we cite and discuss the scholarship of important figures who have worked on Newton, we focus solely or principally on their main works concerning Newton and his influence (many historians and philosophers have written on various topics over the years).

2 Pierre Duhem's remarks about Newton in (1906) – which was translated as *The Aim and Structure of Physical Theory* by Phillip Wiener in 1954 (reprinted 1982) – were also a significant aspect of the reception of Newton in the pre-war period. Duhem argues in particular that Newton's "deduction" of the principle of universal gravity from Kepler's Laws is fundamentally flawed (see Duhem 1982, pp. 190–195). For a critical engagement with Duhem's criticism, see Smith (2007b).

3 Serious philosophical engagement with Newton's work even in the 1970s would still involve a citation or discussion of Burtt's work – see, e.g., Westfall (1971).

4 Metzger was the author, *inter alia*, of (1930) and (1938); for an extensive discussion of her life and work, see Freudenthal (1990).

5 The scope of Koyré's scholarship in the history of science and in the history of philosophy, to the extent that they can be distinguished, was immense. By the time of the beginning of the Second World War in 1939, he had already published several monographs, a collection of essays entitled *Études Galiléennes*, and a translation and commentary on Copernicus's magnum opus. For details of Koyré's life and scholarship, see Herivel (1965b).

to engage with Newton in, say, the 1950s or 1960s would have felt it necessary to begin with – although not necessarily to end with – the work of these figures.[6] The sale of many key Newton manuscripts in 1936 – in which John Maynard Keynes played a crucial role[7] – enabled many scholars to edit and publish texts in the post-war period. These texts now form an essential component of our understanding of Newton's life and thought.[8]

During the immediate post-war period, Newton scholarship underwent another major revolution at the hands of two towering figures in Britain and the United States, I. B. Cohen and Sam Westfall.[9] Having received the first American Ph.D. in the history of science (1947), Cohen probably did more than any single figure in the past century to make Newton's texts available to scholars and to the general public. In 1972, Cohen published *Isaac Newton's Philosophiae Naturalis Principia Mathematica*, an edition co-edited with Koyré, whose untimely death prevented him from seeing it through to completion. In 1958, Cohen had already edited *Isaac Newton's Papers and Letters*, which expertly collected a number of key Newtonian texts, including his optical papers from the 1670s[10] and his correspondence with Bentley (first published in the mid-eighteenth century), and many years of work with Anne Whitman would eventually lead (in 1999) to the first fully new English translation of the *Principia* in two centuries.[11] Cohen's outstanding editorial work was matched

6 In the post-war period, the French tradition of Newton scholarship was continued by a number of important figures, including Michel Blay (1995), and Francois de Gandt (1995). The work of Léon Bloch (1908) was an early component of twentieth-century French scholarship.

7 John Maynard Keynes' 1946 lecture "Newton the Man," characterized Newton's interest in alchemy, "with one foot in the Middle Ages and one foot treading a path for modern science," see: www-groups.mcs.st-andrews.ac.uk/~history/Extras/Keynes_Newton.html

8 Perhaps the most significant post-war investigation of Newton's alchemical manuscripts was presented in Betty Jo Teeter Dobbs's (1975). There has been a tremendous amount of interest in Newton's alchemy in the past two decades – see, for instance, Figala's assessment in (2002) and, more recently, Newman (2006).

9 In addition to *Force in Newton's Physics* and *Never at Rest*, Westfall was also the author of (1958), (1971b) and numerous articles on Newton and the history of science. Cohen's works include (1956), (1971), and (1980). Before his untimely death in 1996, Westfall was slated to edit the *Cambridge Companion to Newton*, which ultimately became a significant institutional signal that Newton's work was of continuing importance for philosophers working in the English-speaking world. The volume, which eventually appeared in 2002, was published under the editorship of I. B. Cohen and of George Smith.

10 During the past twenty years, the most significant research into Newton's optics, and the most important work on the scholarly editions of Newton's work in optics, has been published by Alan Shapiro (Newton 1984).

11 Cohen's editorial and scholarly work on Newton's manuscripts and on the *Principia* has certainly been matched by the immense, decades-long project represented by D. T. Whiteside's *The Mathematical Papers of Isaac Newton*, and by H. W. Turnbull *et al.*'s crucial project, *The Correspondence of Isaac Newton*. These are indispensable scholarly editions.

by his immense range of publications on many aspects of Newton's life and thought.

After finishing a Ph.D. in 1958 at Yale on science and religion in seventeenth-century England, Westfall began to focus almost exclusively on Newton for the next few decades, culminating in his two great works: *Force in Newton's Physics* (1971) and his unsurpassed biography, *Never at Rest* (1980). Although Westfall held a Ph.D. in history, he taught in the History and Philosophy of Science Department at Indiana University (founded in 1960 by the philosopher Norwood Russell Hanson), and he showed a remarkable capacity for combining a subtle understanding of historical detail with an insightful analysis of philosophical issues and problems. Hence *Force in Newton's Physics* is a contribution not only to our understanding of Newton's physical theory, but also to our conception of how Descartes's and Leibniz's work in dynamics intersects with their broader concerns and preoccupations. Even today, every scholar must grapple with the enormously important work of Cohen and Westfall, which have reshaped our understanding of Newton in numerous ways. Indeed, their contributions may never be surpassed.

Beginning in the 1950s and 1960s, Marie Boas (later Marie Boas Hall) and A. Rupert Hall made available for the first time a series of Newton's manuscripts – most notably "De gravitatione et Aequipondio Fluidorum," a crucial unpublished anti-Cartesian tract that has garnered enormous attention in recent years – that have been central to the scholarly understanding of his life and work ever since.[12] In addition to their editorial and archival work, Hall and Hall published a number of articles and books that deal partly or centrally with Newton's thought, including Marie Boas's classic monograph, "The Establishment of the Mechanical Philosophy," and A. R. Hall's numerous books about Newton and his milieu.[13] During the 1960s and 1970s, key contributions to the philosophical understanding of Newton were made by Howard Stein, J. E. McGuire and Ernan McMullin.[14] Stein's most influential paper, "Newtonian Space-Time," was presented in 1967 and then published

12 Hall and Hall's collection is Newton (1962). An updated translation of "De Gravitatione" by Christian Johnson (with the assistance of Andrew Janiak) is available in Newton (2004).

13 See Marie Boas's monograph-length article on the mechanical philosophy, Boas (1952). Much of her subsequent work concerned Boyle and also the history of the scientific revolution, including: Hall (1958), (1962), and (1991). A. Rupert Hall wrote, inter alia, the following influential works: (1963), (1980), and (1992). Together, Hall and Hall also edited Henry Oldenburg's correspondence (Hall and Hall 1965), which is obviously crucial for understanding the history of the Royal Society of London, as Oldenburg was its secretary for many years. Many of Newton's published and unpublished writings, on a wide variety of topics, are now available via The Newton Project: www.newtonproject.sussex.ac.uk.

14 Although Max Jammer did not write specifically about Newton during this period, his famous trilogy in the history of science contained substantial engagement with Newton's ideas: Jammer (1954), (1957) and (1961). At least two of the concepts Jammer discussed, force and mass, are given their canonical modern formulation by Newton in 1687.

in 1970 – it served to introduce a generation of philosophers, many of whom worked not on Newton but on more general issues concerning philosophy of science, to Newton's thinking about space, time and motion.[15] The originality and power of Stein's contribution were felt for many decades. In the ensuing years, Stein has continued to influence many philosophers – including Michael Friedman and Robert DiSalle – seeking to understand Newton's place within the early modern tradition of natural philosophy and the modern tradition of philosophy of physics. J. E. McGuire's life-long engagement with Newton's philosophical work began in the 1960s with a remarkable series of papers concerning the then-neglected alchemical aspects of Newton's unpublished oeuvre. During the ensuing decades, McGuire also made seminal contributions to the study of key Newtonian concepts such as space, time, and force, connecting them to various philosophical and scientific traditions of the late Renaissance and early modernity.[16] For his part, McMullin's wide ranging scholarship on the history and philosophy of modern physics included a crucial early monograph entitled *Newton on Matter and Activity* (1978). McMullin's text was one of the only systematic treatments of Newton's philosophical views to have been written in the post-war period, and its influence is still evident in contemporary scholarship.

***

The contributions to this volume build on the influential work in the twentieth century discussed above, and they often see Newton through the various lenses provided by that work. Indeed, contemporary philosophical engagement with Newton must not only react to the myriad published and unpublished works that form the known Newtonian corpus, they must also respond – both sympathetically and sometimes critically – to the vast field of twentieth-century scholarship on Newton and his influence. The editors have divided the fifteen contributed papers in this volume into three sections: (1) Newton and his contemporaries; (2) Philosophical themes in Newton; (3) the reception of Newton. Such a division is a bit arbitrary, of course, because there is considerable overlap among the papers in different sections. In this introduction we call attention to five broad themes that break new ground in Newton studies and that are shared by a number of contributions.

First, the study of Newton's methodology has long been the focus of George Smith's groundbreaking research.[17] This volume concludes with a new major

15 See Stein (1970a), (1970b), (1990b), and (2002), which presents and expands upon many classic themes from Stein's forty-year engagement with Newton.

16 Many of McGuire's papers are collected in his (1995). Together with Martin Tamny, McGuire edited Newton (1983), an edition of the notes Newton kept as an undergraduate at Trinity College, Cambridge in the 1660s. McGuire's influence is also felt through the many students he trained at Leeds and at Pittsburgh.

17 At present, Smith is probably the most influential English-speaking philosopher working on Newton. His now renowned course on Newton at Tufts University has introduced at

study of his that attempts to characterize Newton's conception of inquiry in the process of articulating nine different ways in which Newton changed physics. Smith details how physical research was predicated on the theory of gravity. Smith's study will be of interest not merely to historians and methodologists, but also to those exercised by the nature of scientific knowledge of gravity before and after Einstein's revolution. Smith's methodological researches have influenced many other contributions to this volume.

Smith's sometime co-author, William Harper, contributes a paper in which he contrasts Newton's methodology of successive approximations with the methodological views of Newton's greatest scientific contemporary, Christian Huygens, who articulates a view of methodology characteristic of the hypothetico-deductive approach. Harper focuses on Newton's richer ideal of empirical success. In particular, Harper calls attention to the importance within Newton's method of accurate theory-mediated measurement of the parameters of the model which explain the predicted phenomena. In line with Smith's approach to Newton, according to Harper's reconstruction a major feature of Newton's philosophy of science is the acceptance of theoretical propositions as guides to research in which empirical deviations from the model count as new theory-mediated phenomena to be exploited as carrying information to aid in developing a more accurate successor.

Ori Belkind shares in Smith's and Harper's rejection of attributing to Newton the hypothetico-deductive method. And like Harper, Belkind calls attention to the importance of Newton's strategy of contingently accepting certain (what Belkind calls) "structural assumptions." In his study of Newton's argument for universal gravity, Belkind calls attention to the importance in Newton's thought of the composition of the quantity of motion and the compositional nature of the gravitational force. By showing that such composition is legitimate, it becomes possible to treat measurement as a way of answering theoretically interesting questions.

A second major theme in which the volume breaks new ground is in its focus on Newton's matter theory, which is the subject of four papers. Zvi Biener and Chris Smeenk use the queries of Roger Cotes, the very able editor of the *Principia*'s second edition (1713), to highlight linked tensions in Newton's matter theory and empiricist methodology, and to stress their development in Newton's thought. Following Cotes, Biener and Smeenk identify two competing views on the nature of matter in Newton. On what they call the "dynamical conception of matter," quantity of matter is measured through a body's response to impressed force. They argue that this conception is dominant in the *Principia* and is justified by a quantitative empiricist method that

least an entire generation of students *and faculty* to Smith's powerful approach to Newton's work. In addition, see the following: Smith (1999), (2001b), (2002a), and (2002b). He also co-wrote an important article with Bill Harper (Harper and Smith 1995).

relies on theory-mediated measurement of parameters that play a role in the laws of motion, as articulated by Smith and Harper. On what they call the "geometrical conception of matter," quantity of matter is measured by the volume a body impenetrably fills. Biener and Smeenk argue that this Cartesian conception is dominant in *De Gravitatione* and is justified by an essentially qualitative empiricist method. They show that the tension between these two conceptions threatens to undermine the argument for Universal Gravitation. It is in response to this threat (as outlined by Cotes's queries) that Newton more decisively endorses a dynamical conception and casts off the vestiges of *De Gravitatione*'s Cartesianism.

Katherine Brading approaches Newton's account of bodies by way of a comparison between Descartes and Newton. She argues that Newton offers a law-constitutive solution to the problem of bodies, according to which the definition of bodies is incomplete prior to the specification of the laws of nature, and completed by those laws. She argues that according to Newton, it is a necessary condition for the individuation and identity of physical bodies that they satisfy the three laws of motion. She then spells out how Newton can be seen as generating a research program of identifying the laws that can account for the necessary and sufficient conditions for the individuation and identity of physical bodies.

Lynn Joy investigates Newton's treatment of body by comparing Boyle and Newton on dispositional properties. She claims that the very idea of a disposition itself underwent a major conceptual change between Boyle and Newton. She argues that Newton turned Boyle's philosophical theory of dispositions on its head by showing that mass could be conceived as an exclusively dispositional property of bodies without requiring that mass be causally grounded in the categorical properties of Boyle's matter. Joy also calls attention to the open-ended nature of Newton's science and philosophy; they were open to the revolutionary possibility that the disposition of mass, when conceived of as a natural force acting according to certain mathematical laws, constitutes an existence more fundamental than that of Boyle's matter.

Daniel Garber's paper compares Leibniz's and Newton's views on the nature of force. Garber spells out some of their most fundamental differences in terms of their different approaches toward thinking about the natural world. Garber sees Leibniz as inheriting a program in natural philosophy from Descartes that provides an account of bodies as such, one grounded in an understanding of their true causes. Garber sees Newton as inheriting a Galilean project that offers a quantitative account of the world, one that favors mathematical description over an account of ultimate first causes. Garber also argues that whereas Leibniz's interest in force is a means to illuminate the nature of body, Newton's account of force is allowed to remain explanatorily basic.

Strictly speaking, Nick Huggett's piece is not on Newton's matter theory, but on Newton's views on space and motion. Nevertheless, it reinforces and refines some of the other contributors' conclusions on how to think of Newton's "dynamics" and the status of the laws of motion. Huggett develops and then challenges Howard Stein's and Robert DiSalle's influential readings of Newton's Scholium on space. Huggett builds his case on the observation that Newton does not introduce "true motion" and "absolute motion" as synonyms; "absolute motion" connotes change of absolute place, while "true motion" connotes a special privileged sense of motion. More specifically, the latter concept gets meaning from the laws of mechanics – it is the concept of motion implicit in the laws. In other words, according to Huggett, in "true motion" Newton consciously held an extremely sophisticated conception of motion. The theoretical part of the concept is that of contemporary "dynamical" interpretations, which also hold motion to be that which the laws refer to as motion in the frames in which the laws hold. On Huggett's interpretation, Newton cannot be said to have advocated a purely dynamical view in the Scholium, but rather the view that motion with respect to absolute space satisfied the dynamical concept.

Two papers, one by Katherine Dunlop and the other by Marco Panza, focus on Newton's mathematics. They illuminate the relationship between mathematics and the science of motion in Newton, which is the third broad theme. Katherine Dunlop relates Newton's views to those of his teacher, Isaac Barrow, emphasizing continuities between teacher and pupil in order to call attention to Newton's departures. She explains the significance of Newton's Preface to the *Principia*, with its focus on postulates as the link between geometry and mechanics. By building on the methodological work of Smith and Harper, she explains the way in which geometry's first principles secure physical significance for the conclusions of theory-mediated measurement. The main point of Marco Panza's investigation of Newton's development of his theory of fluxions is to locate a crucial step in the origins of analysis, conceived as an autonomous mathematical theory. By closely analyzing Newton's *De Analysis* and *De Methodis* as well as Newton's reaction to Roberval's method of tangents, Panza argues that fluxions were conceived by Newton as abstract quantities related to other abstract quantities, called "fluents." By contrast, that which Newton called (in his notes of 1665–66) 'motion', 'determination of motion' or 'velocity,' was understood as (a scalar component of) punctual speeds of motions generating particular geometric magnitudes, typically segments. Panza's interpretation helps explain, in part, why in the *Principia* Newton did not rely on fluxions, but instead turned to geometry.

In the fourth broad theme, ever since the French Enlightenment, Locke and Newton have been considered intellectual fellow-travelers; in this picture Newton is seen as providing the physics and Locke the metaphysics for the new sciences. In much recent scholarship, what are often called the "empiricist" similarities between Locke and Newton have also been emphasized. Building on previous work by Howard Stein, three papers force a reconsideration of the relationship between Locke and Newton. Graciela De Pierris brings out some crucial differences between Locke and Newton, in part by comparing Hume's empiricism with Locke's. She argues that Locke remains wedded to the demonstrative ideal of the mechanical philosophy and lacks Newton's understanding of fruitful inductive generalization. She then reads Hume as simultaneously articulating many of the fruits of the Newtonian method while also offering a skeptical challenge to it.

Mary Domski reconsiders the famous "master-builders" and "under-Labourer" passage in the *Essay*. She argues that in the fourth edition of the *Essay*, Locke emphasizes Newton's success as a mathematician, but not as a mathematical natural philosopher. She also shows that in Locke's other writings from the 1690s, Newton is praised for his application of mathematics to a very specific domain of nature, namely, the motions of planetary bodies. According to Domski, then, Locke took Newton's work to be emblematic, not of a general physics, but of a sub-discipline of natural philosophy dealing only with the forces and motions of heavenly bodies.

Lisa Downing also re-evaluates the relationship between Locke and Newton; she does so by way of Maupertuis's analysis of the nature of attraction. Her paper helps explain both how Locke and Newton came to be seen as fellow travelers, and how philosophers drew on Lockean resources to defend Newtonian natural philosophy. In particular, she shows how Maupertuis transforms s'Gravesande's claim that laws as regularities are the ultimate aim of Newtonian knowledge into a claim in which experience is in principle capable of settling the existence of attraction as an inherent quality of body.

Finally, two papers investigate how the nature of philosophy was reconfigured through responses to Newton. Michael Friedman emphasizes the importance of metaphysical and theological issues – about God, his creation of the material world in space, and the consequences that different views of such creation have for the metaphysical foundations of physics. Friedman argues that Kant's differences with Newton over these issues constitute an essential part of Kant's radical transformation of the very meaning of metaphysics as practiced by his predecessors. Friedman shows that since Newtonian absolute space is viewed as a regulative idea of reason, there is also an associated reconfiguration, for the critical Kant, of the relationships among space, the interactions of matter, and the idea of God. For the idea of God, too, is a regulative idea of reason. Indeed, there is an important sense in which it is the ultimate such regulative idea. For the critical Kant the only possible meaning that the idea of divine omnipresence (and divine providence) can now have is a purely *practical* one, in so far as we unconditionally obey the command of morality to strive to realize the realm of ends here on Earth, and, accordingly, we take the whole of that material nature of which we are a part to be in principle *capable* of such a realization (or, more precisely, its successive approximation). On Friedman's account, Kant thereby brings the characteristic mode of metaphysical investigation into the relationships among space, God, and matter practiced by his predecessors to a close, and transforms it into transcendental philosophy.

Eric Schliesser explores how the most able eighteenth-century Scottish Newtonian, Colin MacLaurin, uses the authority of Newton to attack Spinoza on empirical and moral grounds. MacLaurin argues from the empirical success of Newtonian natural philosophy to the *rejection* of alternative positions, methodologies, and foundations within philosophy. At the same time, MacLaurin argues for a certain form of self-limitation: aiming for completeness is likely to get us into trouble. Schliesser argues that in MacLaurin's hands Newtonian science recommends a lowering of expectations – it favors piecemeal progress over the demands of systematicity. MacLaurin thereby subordinates application of Newton's science to his religious and moral outlook. Schliesser shows that MacLaurin constructed a tradition in which Descartes, Spinoza, and Leibniz are linked as a threesome not in opposition to empiricism, but in opposition to a tradition of mathematical-empirical research stretching back to Galileo. Thus Schliesser's analysis echoes Daniel Garber's.

PART I

Newton and his contemporaries

1

Newton's law-constitutive approach to bodies

a response to Descartes

KATHERINE BRADING

1.1 Introduction

In his *Principia* Newton offers us a science of bodies in motion. Such a science has bodies as its subject-matter, but what are these bodies? If Newton's three laws of motion are to say anything, then there must be bodies for them to refer to. I shall label this the 'problem of bodies'. In this chapter I outline the 'problem of bodies' as Newton finds it in Descartes's *Principles of Philosophy*. I claim that while there is no obvious solution *explicit* in Descartes's writings, an implicit solution is strongly suggested. I argue that Newton was acutely aware of the problem, and addressed it explicitly by adopting the strategy implicit in Descartes. My claim is that Newton offers a *law-constitutive* solution to the problem of bodies, according to which the definition of bodies is *incomplete* prior to the specification of the laws of nature, and *completed by* those laws of nature.

1.2 Descartes and the problem of bodies

Taken together, Descartes's laws of nature concern the behaviour of 'bodies'.[1] Here are the laws as he stated them in his *Principles of Philosophy* (Part II, paragraphs 37, 39, and 40):[2]

> The first law of nature: that each thing, as far as is in its power, always remains in the same state; and that consequently, when it is once moved, it always continues to move.
>
> The second law of nature: that all movement is, of itself, along straight lines; and consequently, bodies which are moving in a circle always tend to move away from the center of the circle which they are describing.

1 The first law has a more general scope, concerning 'things' in general. I am grateful to Eric Schliesser for drawing my attention to this.

2 Quotations are from Descartes (1991), the Miller and Miller translation of the *Principles of Philosophy*. The *Principles* was first published in Latin in 1644.

> The third law: that a body, on coming in contact with a stronger one, loses none of its motion; but that, upon coming in contact with a weaker one, it loses as much as it transfers to that weaker body.

The 'problem of bodies' is this: what are the 'bodies' to which these laws apply? For Descartes, the answer is 'parts of matter'. Famously, however, this answer masks a difficulty that Descartes never satisfactorily resolved, and which arises as follows.

1.2.1 Descartes on matter and its parts

In Part I of the *Principles*, Descartes argues that extension is the sole principal attribute of material substance, and Part II opens with an argument that leads from extended matter (mass noun) to bodies plural (count noun), repeating essentially the same argument as in the *Meditations*.[3] He writes (Descartes, *Principles of Philosophy*, II.1) that we

> clearly and distinctly perceive, a certain matter which is extended in length, breadth and depth; the diverse parts of which are endowed with various shapes and subject to various movements, and which also cause us to have sensations of color, odor, pain, etc.

Then, since God is not a deceiver, we are entitled to conclude that extended matter is indeed divided into parts of various shapes and movements, affecting our senses in this way. However, while Descartes has argued earlier for the claim that matter is extended, he offers no argument in support of the claim that we clearly and distinctly perceive *parts* of matter that are endowed with various shapes and subject to various movements, and also affect our senses. This knowledge of bodies plural is dependent on what comes to us through our senses, and the faculty of the imagination; whenever Descartes considers the nature of the bodies that affect our senses, he takes us back to the sole principal attribute of body, pure extension, and to a conception of body in general[4] that contains nothing corresponding to a division of extended matter into parts.[5] There is therefore an apparent gap between what is known via the

3 Descartes (1985), Meditation VI. The *Meditations* were first published in 1641.

4 The term 'body in general' should be understood to refer to the nature of any body (that it is extended), and need not refer to the entirety of Descartes's indefinite extension (see Kaufman, 2008; Schmaltz, 2008b). However, the conception of 'body in general' understood as referring to a part of matter (a body, any body), presupposes that Descartes's extended matter is divided into parts (bodies). The slide from extended matter to bodies plural (a body or any body) via 'the nature of body in general' is vividly seen in the *Principles* II.1&4. Just how the division into parts is achieved is the issue we are interested in here.

5 The claim that our knowledge of bodies plural is located only in the imagination and not the intellect might seem in conflict with the wax example of Meditation II. However, in the wax example bodies plural are presupposed as given, and the issue is our knowledge of these bodies.

intellect (the nature of body in general, as extended), and what is known via the senses (that extended matter is divided into parts, having various shapes and motions, and that these parts correlate with our experience of bodies (plural) and their various sensory properties). To close this gap, Descartes must explain in virtue of what extended matter is divided into parts *such that* we can clearly and distinctly perceive that it is indeed so divided. If Descartes is to solve the 'problem of bodies', he must provide within his metaphysical system the resources for this division.[6]

After several passages discussing matter as extension, Descartes returns to the topic of the *parts* of matter in motion in paragraph 23, where he writes:

> all the properties which we clearly perceive in it [that is, in extended matter] are reducible to the sole fact that it is divisible and its parts movable . . . all the variation of matter, or all the diversity of its forms, depends on motion.

There are two possibilities here: either motion is the principle by which matter is divided into parts, or matter is divided into parts by some other principle and then the motions of the parts account for the 'diversity of its forms'.

This paragraph leads naturally into a discussion of motion, in which Descartes famously distinguishes between 'What movement is in the ordinary sense' (paragraph 24) and 'What movement properly speaking is' (paragraph 25). Newton's vehement criticisms of Descartes's 'proper definition of motion' in 'De Gravitatione'[7] are now legendary, and I will have reason to return to them later. The definition offered in the paragraph reads as follows (emphasis in the original):

> it is *the transference of one part of matter or of one body, from the vicinity of those bodies immediately contiguous to it and considered as at rest, into the vicinity of some others.*

There are two points here that are important for my argument. First, the motion of a given body is defined with respect to other (special) bodies. For a body to move *is* for it to move with respect to these other special bodies. Second, Descartes clearly equates 'one part of matter' with 'one body'. He immediately goes on to say more about this second point:

> By *one body*, or *one part of matter*, I here understand everything which is simultaneously transported; even though this may be composed of many parts which have other movements among themselves.[8]

6 Descartes's God is so powerful that he could divide matter into parts in ways incomprehensible to us, presumably, but that won't do here because Descartes requires that we clearly and distinctly perceive that matter is so divided. Therefore, on Descartes's own terms, God must be dividing matter into parts in a way that is intelligible to us and can be accounted for within Descartes's metaphysical system.

7 Newton (2004).

8 Note that Descartes also emphasizes at this point that motion is a mode of a body, just as is shape.

Here, Descartes offers an account of the division of indefinite extension into parts or bodies *through* motion: one body, or one part of matter, is everything that is 'simultaneously transported'. Thus, it seems that motion is the principle by which matter is divided into parts. The resulting view is that, in short, motion is defined in terms of bodies, but the division of indefinite extension into bodies is achieved through their relative motions. This is, at best, a rather tight circle.[9]

1.2.2 Strategies for solving the 'problem of bodies'

The question we are trying to address is this: Given Descartes's laws of nature, what are the 'things' and 'bodies' to which these laws apply?

Here is one way to understand Descartes's general strategy: first, provide a metaphysical account of bodies, and then provide the laws of nature governing the behaviour of these bodies, consistent with

(a) various principles, including God not being a deceiver and God acting constantly in the world, and
(b) our experience of change.

Construed this way, one understanding of his proposed solution is that there is a substance, body, and this is divided into individual bodies (the term 'individual body' is used in paragraph 31) by its *modes* (especially shape and motion). In other words, on this interpretation Descartes seeks to provide a solution to the problems of individuation and identity of bodies, following which he sets out the laws of nature applying to those bodies.

What exactly are we looking for, when we ask Descartes for his solution to the problems of individuation and identity of bodies?[10] The following

9 The ensuing paragraphs (26–35) elaborate on this definition of motion, emphasizing that true motion is reciprocal (paragraph 29) – hence why, when we are asking about the motion of a single body, we must consider the reciprocal bodies to be at rest – and that there is only one true motion associated with each body. Thus, paragraph 31 begins 'Each individual body has only one movement which is peculiar to it . . .' Notice the term 'Each individual body'. The remainder of Part II of the *Principles* continues to make reference to bodies, and no further explicit information is given concerning how the division of indefinite extension into parts is achieved. Paragraph 36 turns our attention to the causes of motion, the primary cause being God, and the secondary cause being the laws of nature.

10 In his introduction to the collection *Individuation and Identity in Early Modern Philosophy* (Barber and Gracia 1994), Barber notes that in Cartesian philosophy the problem of individuation loses the prominence it had enjoyed in much medieval philosophy, but writes that (p. 2) 'since philosophers in the early modern period were for the most part systematic, presenting ontologies rivalling their medieval counterparts in comprehensiveness if not in detail, one can ask how within their systems the problem of individuation could or should have been resolved even where explicit discussion of the issue is minimal'.

distinctions will be helpful in thinking about what might constitute such a solution. First, there is the metaphysical problem of individuation, and here we should distinguish between necessary and sufficient conditions of individuality, and a principle of individuation. For example, one might subscribe to the view that the necessary and sufficient condition of individuality is being distinct from all other individuals, and that the principle of individuation that grounds this is being distinct *in virtue of* being a unique bundle of properties. (Other conditions that might be thought to be necessary and/or sufficient include indivisibility, noninstantiability, and ontological independence. Other principles of individuation include haecceities, or essence, or Aristotelian prime matter, and so forth.) The problem of individuation is to be distinguished from the problem of identity over time, where we ask *in virtue of what* is this individual the very same individual at another time. These metaphysical problems have epistemological counterparts, where we ask about our access to the individuating features of these metaphysical individuals (how we distinguish them from one another), and our warrant for according them the status of individuality.[11]

What does Descartes offer us, as regards physical bodies? Let's begin with the problem of individuation. I think that we can offer the following interpretation. The necessary and sufficient condition of individuality for physical bodies is being a part of matter that is divided from the rest of matter. In virtue of what is a region of matter so divided? Answer: in virtue of being in motion from

11 Barber (1994, p. 5) discusses the relationship between the metaphysical and epistemological problems, noting that they are often in tension:

> the epistemologist complains about the cavalier attitude of his ontologically inclined brethren who generate entities and distinctions in an unconscionable manner, while the ontologist in turn dismisses the epistemologist as one blinded to the richness of the universe through a neurotic fixation on a few favorite sense organs.

He distinguishes between a 'strong model', whereby 'epistemological considerations serve as criteria for the adequacy of an ontological system: putative candidates for inclusion in the catalogue of existents must first pass a test for knowability and, once included, their classification in terms of categorical features must again meet the same rigorous standard', and a 'weak model', which distinguishes between the ontological question of 'what it is *in objects* that *individuates* those objects' versus the epistemological questions of how we can differentiate among objects through our experience of them, but requires only that these should be compatible – neither has a veto over the other. Barber goes on to say that, 'broadly speaking, the weak model is dominant in medieval philosophy', but that 'By 1641, however [the year Descartes published his *Meditations*], the strong model has replaced its weaker medieval counterpart.' He writes (p. 6): 'the epistemological turn is significant for its effect on the *content* of those discussions [i.e. of individuation and identity]. What could possibly count as solutions to those problems is restricted by the imposition of new criteria; solutions formerly held to be uncontroversial are rendered puzzling, incomprehensible, or in conflict with newly discovered "truths" about the world.'

the vicinity of immediately contiguous bodies considered to be at rest into the vicinity of some others.

This account faces immediate challenges as to its adequacy, both as a proposed solution to the problem of individuation, and with respect to providing bodies that are suitable as the subject of Descartes's laws of nature. First, it is not clear that Descartes's definition of motion, without parts of matter that are prior to motion, is coherent. Second, even if it is, we have as Garber (1992, pp. 178–179) puts it 'a rather unwelcome consequence' that rules out the possibility of two bodies being at rest with respect to one another,[12] and therefore of any body ever being at rest. This leads directly to problems when we attempt to apply Descartes's laws of nature, most obviously in his rules of collision that supplement his third law. As Garber goes on to discuss (1992, pp. 179–180), the rules rely crucially on the distinction between bodies at rest and in motion.[13]

Turning now to the problem of identity over time, the only remaining resource seemingly available is shape. The shape of a part of matter would help in giving identity over time, but isn't enough, unless every part of matter has a different shape from every other part.

It seems to me that the strategy of *first* providing a metaphysical account of bodies (i.e. a solution to the problems of individuation and identity of bodies), and *then* providing the laws of nature governing the behaviour of these bodies, does not succeed given the resources that Descartes provides.

Suppose we agree that in the first half of Part II of the *Principles* Descartes does not solve the problem of individuation for the bodies that are the subject-matter of his physics. Nevertheless, the second half of Part II proceeds as if the problem has been solved – it assumes that there are individual bodies that satisfy the laws of nature. But if there *are* no bodies in Descartes's system, then there is nothing for his physics to be about, which to me at least casts something of a shadow over the entire exercise.[14]

Fortunately, there is a very different way to read what happens in Part II of the *Principles*. I am not advocating it as an exegesis of what Descartes took himself to be doing, but I do think the strategy I outline is implicit in the text, I think it is broadly successful, and I will argue below that it is a strategy that

12 Thus, Descartes's discussion (paragraph 55) of the cohesion of the parts of solid bodies in terms of their being at rest relative to one another is, strictly speaking, nonsense: solid bodies cannot have any parts.

13 Garber further notes that Descartes was aware of this difficulty, and yet failed to recognize how problematic it is, merely saying that a body at rest is a 'part' of a larger body.

14 Garber (1992, p. 181) concludes his discussion of motion and individuation and Descartes writing, 'I shall continue to talk as if Descartes is dealing with a world of individual bodies, colliding with one another, at motion and at rest with respect to one another. But, in the end, I suspect that this is something that he is not entitled to, and this is something that, if true, would seriously undermine his whole program.'

Newton explicitly takes up. The strongest, and most straightforward, version of the solution is this:

> The necessary and sufficient condition for the individuality and identity of physical bodies is that they satisfy the laws of nature.

So: instead of *first* solving the problem of generating bodies, and *then* applying the laws of nature to those bodies, physical bodies *are* whatever satisfy the laws.[15] We expand that rather tight circle where motion and body are inter-defined, and thereby hope to create a virtuous circle.[16]

A weaker version of the solution would drop the claim to sufficiency, as follows:

> A necessary condition for the individuality and identity of physical bodies is that they satisfy the laws of nature.

Even on this weaker version, one consequence is that the account of bodies has no wider applicability than that of the laws. That is, an account of bodies is available to us at best only in those circumstances where the laws are applicable; if there are circumstances for which it is inappropriate to apply the laws, then we will also lack an account of bodies in those circumstances. What we have is *a law-constitutive solution to the problem of bodies.*

Notice also that we have limited our goal to giving an account of *physical bodies*, rather than *bodies considered in general.* This is consistent with Garber's point (1992, pp. 176–177) that Descartes's definition of 'one body' in paragraph 25 has a restricted application. He writes (p. 176): 'it is important to note, first of all, that this definition should be understood as limited to a special kind of individuality, that which pertains to body as such, what we might call physical individuality, to distinguish it from a broader notion of individuality'. He emphasizes that this notion is not appropriate for other fields of interest (such as 'morality, property law, medicine, animal husbandry, agriculture, etc.', p. 177), and states (p. 177): 'The notion of an individual body he is concerned to define there is concerned with the notion of a physical individual, the sort of thing that can enter into the basic laws of nature.'

15 Notice the change that this makes to the problem of individuation. Traditionally, the challenge is to specify one constituent of an individual that is not present in any other individual. In this way, the world can be created one individual at a time. But on the approach I have outlined here, the challenge is to carve the given undifferentiated world up into individuals 'all at once', and the resulting account of individuality does not include the resources for creating the world one individual at a time. But this is not to say that it is not a coherent strategy for creating individuals (*pace* Leibniz).

16 I will discuss the principle of individuation below.

I want to push this further, by arguing that the definition of physical bodies is *incomplete* prior to the specification of the laws of nature, and *completed by* those laws of nature.

In my opinion, this way of proceeding is strongly suggested by the text of the *Principles*, because of the failure (arguably) of Paragraphs 1–35 to provide a complete solution to the problem of bodies, combined with the fact that Descartes's next move is to introduce his laws of nature. Paragraphs 36 onwards present the laws of nature for bodies and concern their elaboration and consequences: these contain additional resources for individuating bodies, including the laws themselves and refinements of the concept of motion (introducing 'determination', for example). If we accept that Paragraphs 1–35 are insufficient by themselves, and we are seeking a solution to the problem of bodies using the resources Descartes offers, then a natural move is to make use of the laws in attempting to complete the solution. This is a law-constitutive approach to the problem of bodies.

I have talked about necessary and sufficient conditions for a region of matter to be a physical body, and I have talked about identity over time, but I have said nothing about the principle of individuation. On this account, a principle of individuation would tell us *in virtue of what* a body satisfies the laws of nature. It seems that either there is no further question here (and principles of individuation are dispensed with), or the only possible further response is 'God'. It is consistent with Descartes's philosophy that the principle of individuation is, indeed, God. But the other option is also available: the above approach to solving the 'problem of bodies' makes philosophically viable the abandonment of principles of individuation for physical bodies.

1.3 Newton and the 'problem of bodies'

We know from Newton's early writings that he also asked about the division of uniform matter into parts (McGuire and Tamny, 1983, p. 339): 'Suppose the first matter one uniform mass without parts; how should that body be divided into parts, as we see it now is, without admission of a vacuum?' Of course, the central topic here is the discussion of atoms and the void. But rather than following this line of Newton's thinking, my interest is in how the problem of bodies shows up in other writings, specifically in the manuscript generally referred to as 'De Gravitatione' and in the *Principia*.

In his *Principia* Newton, like Descartes, offers us a science of bodies in motion, with laws that apply to those bodies. What are these bodies? To answer this question, I will begin by looking at 'De Gravitatione', and I will argue that in this text Newton criticizes Descartes's account of body *as a solution to* the 'problem of bodies', that he offers his own solution to the problem, and that this solution is explicitly *law-constitutive* (in the sense explained above). I will then argue that we should understand this solution as being present in the *Principia*, and I will do this by looking at the text, but also primarily at some draft revisions. My overall message is that in Newton's work we can find a powerful philosophical solution to the 'problem of bodies'.[17]

On the account offered so far, a necessary condition for the individuality and identity of physical bodies is that they satisfy the laws of nature. This is the weak version of the law-constitutive solution. The strong version asserts that *the* necessary *and sufficient* conditions are satisfaction of the laws of nature. My claim is that the weak version of the solution is explicit in Newton, and that, in stating what the physical bodies are that are the subject-matter of his laws, criteria additional to satisfaction of the laws are to be given. I am also willing to argue that the strong version is implicit (although I will say little to support this here), and that it offers important insights into the notion of body at work in physics.

1.3.1 Newton's criticisms of Descartes's account of bodies, in 'De Gravitatione'

The Newton manuscript 'De Gravitatione' contains explicit criticisms of Descartes's account of bodies in motion, as he understood it from reading Descartes's *Principles of Philosophy*. Within the current philosophy of physics literature, a great deal of attention has been paid to Newton's criticisms of Descartes's definition of motion.[18] However, the paragraph that introduces these criticisms makes clear that the target is also *the account of body* along with the definition of motion. Newton writes (2004, p. 14, my emphasis):

> when I suppose in these definitions that *space is distinct from body*, and when I determine that *motion is with respect to the parts of that space, and not with respect to the position of neighboring bodies*, lest this should be taken as gratuitously contrary to the Cartesians, I shall venture to dispose of his fictions.

The two things (the account of body, and the definition of motion) are intimately tied together. What I want to place centre stage is this: Newton's diagnosis of the reason why 'Cartesian motion is not motion' (2004, p. 20) is that

17 When discussing Newton on body, the main focus of interest has been on Newton's matter theory, but my interest is different from this. As with Descartes, Newton offers us laws that apply to bodies, and our question is: what are the bodies to which these laws apply? Supplying a theory of matter could indeed answer this question, but – as I have argued – that is not the type of solution that Descartes ended up offering, and nor – as I shall now argue – is it the solution that Newton offers in his *Principia*. Nevertheless, it is a genuine solution, and one which is (in an important way) *complete* even in the absence of a theory of matter.

18 See Slowik (2002, chapter 1) for example.

Descartes has offered an inadequate account of *body*, where 'inadequate' means 'inadequate for the purposes of a science of bodies in motion'.

There's a lot bundled up there, and I want to unpack it. Before discussing Newton's criticisms of Descartes's account of body, let me first review the familiar criticisms Newton makes of Descartes's definition of motion. The standard philosophy of physics story about 'De Gravtitatione' focuses on Descartes's relational definition of motion. Newton offers several criticisms of this definition, many of which are united by a central theme that, according to Newton, we should be looking for a systematic connection between the presence of forces and changes in states of motion, and that Descartes's account fails to offer this.[19] However, the argument that has received the most attention is one that doesn't rely on appeal to the presence or absence of forces, and the conclusion Newton draws is much stronger. He argues *not* that Descartes has given a definition of motion that fails when we try to apply it, but that *he has failed to give a definition of motion at all* (Newton 2004, p. 20):

> Now since it is impossible to pick out the place in which a motion began – that is, the beginning of the space traversed – for this place no longer exists after the motion is completed, that the traversed space, having no beginning, can have no length; and since velocity depends upon the length of the space passed over in a given time, it follows that the moving body can have no velocity, just as I wished to show at first. Moreover, what was said regarding the beginning of the space passed over should be understood concerning all the intermediate places; and thus, as the space has no beginning nor intermediate parts, it follows there was no space passed over and thus no determinate motion, which was my second point. It follows indubitably that Cartesian motion is not motion, for it has no velocity, no determination, and there is no space or distance traversed by it.

I don't want to dwell on this argument against the Cartesian definition of motion. Instead, I want to shift attention to the very next sentence, which is this (Newton 2004, pp. 20–21):

> So it is necessary that the definition of places, and hence of local motion, be referred to some motionless being such as extension alone or space in so far as it is seen to be truly distinct from bodies.

So, there is a criticism of Descartes's account of body here, and the thrust of it is this: Descartes's account of body is inadequate *in the sense that* it is inadequate to the purposes of a science of bodies in motion. Why? Because if body is

19 For example, Newton points out apparent problems with reconciling in a consistent manner when a body has a 'tendency to recede' and when it is in motion or at rest (2004, p. 15), and that we can have changes in motion of a body even when there are no forces acting on that body, and vice versa (2004, p. 18).

identified with extension, then we can't give an adequate account of what it is for a body to move. Thus, the 'problem of bodies' – of specifying a concept of body that is adequate to the purposes of a science of bodies in motion – is explicitly at stake in Newton's criticisms of Descartes on body and motion.

There is also strong evidence that this is exactly one of the problems that Newton is trying to solve in 'De Gravitatione'. Prior to the attack on Descartes's definition of motion, Newton states four definitions of his own (Newton 2004, p. 13), the second of which is a definition of body. It reads: 'Body is that which fills place.' Newton further elaborates on this as follows (my emphasis added): 'Note. I said that body fills place, that is, so completely fills it that it wholly excludes other things of the same kind or other bodies, *as if* it were an impenetrable being.' He then goes on to state the purpose that this notion of body is intended to fulfil, writing that 'body is here proposed for investigation not in so far as it is a physical substance endowed with sensible qualities, but only in so far as it is extended, mobile, and impenetrable'. That is, the notion of body is intended to be 'adequate to' the task Newton has in mind: he writes that he has 'postulated only the properties required for local motion' (Newton 2004, p. 13).

In conclusion, the main content of this first part of 'De Gravitatione' is Newton's detailed arguments as to why space and body must be distinct from one another: body cannot be merely extension because then we cannot give a satisfactory account of what it would be for bodies to move. Descartes's account of body is inappropriate for the purposes of a theory of bodies in motion.

1.3.2 Newton's solution to the 'problem of bodies' in 'De Gravitatione'

In the second part of 'De Gravitatione' (beginning towards the end of p. 21 of Newton, 2004), Newton offers his positive account of space and body, the most familiar aspect of which is Newton's insistence that space and body are distinct, having a very different ontological status from one another. In addition to a rich account of space,[20] these passages are where we find evidence that Newton's solution to the 'problem of bodies' is a law-constitutive solution of exactly the kind found implicitly in Descartes. Newton does not *first* give a general account of bodies, and *then* show that it is satisfactory for the purposes of a science of bodies in motion (among other things). Rather, a necessary condition for something to be a body is that it satisfy certain laws. The textual evidence for this claim is as follows.

Two properties that Newton attributes to bodies are mobility and impenetrability (see Newton 2004, p. 27). A region of space that is impenetrable will be 'impervious to bodies', and 'by hypothesis' the implication of this is that it will

20 For discussion see DiSalle (2006), McGuire (1978), Stein (2002).

'assume all the properties of a corporeal particle, except that it will be regarded as motionless' (Newton 2004, p. 28) Crucially, this includes being sensible, or 'tangible' (Newton 2004, p. 28). Newton then goes on to introduce what he means by mobility (Newton 2004, p. 28, my emphasis):

> If we should suppose that that impenetrability is not always maintained in the same part of space but can be transferred here and there *according to certain laws*, yet so that the quantity and shape of that impenetrable space are not changed, there will be no property of body which it does not possess.

Newton sums up his position (p. 28) by saying that 'these beings will either be bodies, or very similar to bodies', and if they are bodies then we can define them as 'determined quantities of extension' that are (1) mobile, (2) impenetrable, such that they reflect off one another '*in accord with certain laws*', (3) sensible, and movable by us. The appeal to laws is emphasized by Janiak (2006), where he notes that 'in a clever and crucial twist, Newton adds that the region's mobility would be lawlike'. Newton is explicit that a necessary condition for something to be a body is that it move in accordance with the laws.[21]

In sum, I have shown that in 'De Gravitatione' Newton's criticism of Descartes's concept of body claims that it is inadequate to the purposes of a science of bodies in motion, and I have argued that he offers an explicitly law-constitutive solution to this problem (the 'problem of bodies'). In the following section I will argue that this same solution is also at work in the *Principia*. Before doing so, however, I will address a criticism of Newton's account of body, as offered in 'De Gravitatione', made by Bennett and Remnant (1978).

In their paper 'How matter might at first be made', Bennett and Remnant (1978) argue that the account of body offered by Newton in 'De Gravitatione' is a failure. They focus on the criterion of impenetrability, and object that Newton does not have the resources to say what he means by impenetrability.[22]

21 There is a subtlety here. In a clear jibe at Descartes, Newton is cautious about saying 'positively what the nature of bodies is' since he has 'no clear and distinct perception of this matter': he leaves open the possiblity that God could create bodies that appeared to us in every way as Newton has described them and yet differ in nature from those Newton describes (see Newton, 2004, p. 27). This could be read as casting doubt on the law-constitutive approach as providing necessary conditions, but I think that this isn't right. In 'De Gravitatione', space and body differ in their epistemic status. The exclusively a posteriori character of our enquiries into the nature of body render the results less certain, and this includes the possibility that God has created bodies with a nature different from that described by Newton. Nevertheless, Newton ends the relevant paragraph by concluding that his description of body will be such that 'we can hardly say that it is not body'. It seems to me that the necessary conditions should be understood as inheriting this modest epistemic status.

22 I am grateful to Eric Schliesser for suggesting that I revisit Bennett and Remnant's position in the light of my reading of 'De Gravitatione'.

There are many issues raised by their discussion, and I will pick up only one thread that relates directly to the law-constitutive interpretation I have been advocating here. According to Bennett and Remnant, impenetrability can do the job of ensuring that two shapes, once delineated and distinguished from space (such that they can move around with respect to space) never overlap. What impenetrability cannot do, they say, is delineate a shape (so that it is distinct from space such that it has the possibility of being mobile with respect to space) in the first place. They claim that impenetrability is the only resource that Newton has to delineate shapes, and that his account is therefore a failure.

This criticism rests on the requirement that Newton's account of bodies as regions of impenetrable space underwrites *in virtue of what* bodies are mobile. However, on the view I propose, it is a condition on being a body that – in addition to being impenetrable and sensible – it be mobile. Mobility is itself one of the criteria that Newton stipulates, independently of impenetrability, so impenetrability was never intended to confer mobility. The condition of mobility is itself a stipulation, and Newton is not attempting to explain in virtue of what a shape is delineated in space such that it has the possibility of being mobile. As Newton himself says, body is that which *fills* place; a portion of matter, or a body, is not identified as an impenetrable region of space. It is, rather, a perceptible shape that in fact moves around according to certain laws. So we can grant Bennett and Remnant this much: Newton has not given an account of how matter might at first be made in the sense that he has not given an account of what makes possible the mobility of an impenetrable and sensible region. Nevertheless, Newton has given a clear set of conditions that, if satisfied in the making of matter, would deliver a world such that 'if all of this world were constituted out of these beings, it would hardly seem to be inhabited differently'.[23] And this is all that the law-constitutive approach requires.

1.3.3 The 'problem of bodies' in Principia

Definition 1 of Newton's *Principia* is famous:[24]

> Definition 1: Quantity of matter is a measure of matter that arises from its density and volume jointly.

In elaborating on the definition, Newton says: 'I mean this quantity whenever I use the term "body" or "mass".' So 'body' is a 'quantity of matter', and the first part of the definition tells us that this quantity is a measure of matter. The first part of the definition introduces a new quantity into physics, and the second part relates this newly introduced concept to the pre-existing concepts

23 Newton, 'De Gravitatione', p. 28.

24 Quotations are from Newton (1999), the Cohen and Whitman translation of the *Principia*.

of density and volume. But still, this isn't hugely informative. What else does Newton give us? Definition 3 attributes a property to bodies, inertia,[25] and in the Scholium to Definition 8, Newton repeats an assertion familiar from 'De Gravitatione': 'Place is the part of space that a body occupies.' And, of course, we have Newton's laws of motion:

Axioms, or the Laws of Motion

Law 1: Every body perserveres in its state of being at rest or of moving uniformly straight forward, except insofar as it is compelled to change its state by forces impressed.

Law 2: A change in motion is proportional to the motive force impressed and takes place along the straight line in which that force is impressed.

Law 3: To any action there is always an opposite and equal reaction; in other words, the actions of two bodies upon each other are always equal and always opposite in direction.

What Newton is doing in these opening sections of the *Principia* is specifying the notion of body that is needed for his project, his science of bodies in motion, to get off the ground. But the question I am interested in is the same as the one I discussed with respect to Descartes. Does Newton intend to offer an account of body that is independent of the laws, or is the account of body *incomplete* prior to the specification of the laws, and *completed by* those laws? I think the latter. The material from 'De Gravitatione' discussed above, and which pre-dates the *Principia*, points in this direction, and material from after the *Principia* also points the same way, or so I will now argue.

The relevant later materials are drafts published by McGuire in 1966.[26] McGuire dates the drafts at 'some time towards the end of 1716', but in any case, they were done in preparation for the third edition of the *Principia*, which came out in 1726. According to McGuire (p. 115) the intended positioning in the third edition is just after the Rules of Reasoning, and indeed much of what is at stake in the drafts for Newton concerns his claim that he is 'arguing from the phenomena'. But for our purposes I want to highlight the following aspects:

From Draft 1 (McGuire, p. 113):[27]

> thus body and vacuum are here defined [*not in order that we deny that other bodies exist but in order that we may show in what sense these words are to be*

25 Definition 3: 'Inherent force of matter is the power of resisting by which every body, so far as it is able, perseveres in its state either of resting or of moving uniformly straight forward.'

26 Page references are to the reprinted version, McGuire (1995).

27 Square brackets were used by Newton to indicate passages that were to be omitted when the document was copied. Italics indicate passages Newton crossed out. I have used underlining to add my own emphasis.

understood in what follows. The propositions which follow are understood of bodies of this kind. About other bodies let authors in other sciences dispute.]

From Draft 2 (McGuire, p. 114):

> **Definition I** Body I call everything tangible in which there is resistance to tangible things, and whose action *resistance*, if it is great enough, can be perceived.
>
> It is indeed in this sense that the common people always accept the word body. And of this sort are The Earth, Planets, Comets, metals, stones, sand, clay . . . These emit and reflect light and are weighed down by their constituent parts *and are numbered among the phenomena and in their motions observe the laws of bodies*. [Mathematical solids are not perceived by touching nor cause a resistance nor are they usually called bodies.] Vapours and exhalations on account of Their rarity lose almost all perceptible resistance, and in the common acceptance often lose even the name of bodies and are called spirits. And yet they can be called bodies in so far as *they are the effluvia of bodies* and have a resistance proportional to density. [But if the effluvia of bodies were to change thus in respect of their forms so that they are to lose *all* power of resisting, *and cease to be numbered among the phenomena*, these I would no longer call bodies: for I speak with the common people.]

From Draft 3 (McGuire, p. 115):

> **Definition II** Body I call everything which can be moved and touched, in which there is resistance to tangible things, and its resistance, if it is great enough, can be perceived.
>
> It is indeed in this sense that the common people always accept the word body. And of this sort are The Earth, Planets, Comets, metals, stones, sand, clay . . . I add the heavenly bodies. These emit and reflect light . . . and in their motions observe the laws of bodies. Mathematical solids are not *perceived* do not move by touching nor cause a resistance, nor are they usually called bodies . . . . At the beginning of the first book I have defined the quantity of matter so that it may be treated in mathematical terms; here I have defined body composed of such matter in order that it may be treated in physical terms.

I think there is clear evidence in these drafts of two things:

(1) The definition of body is intended for the specific purposes of Newton's project.
(2) The definition includes the requirement that for something to *be* a body of this kind it move according to the laws.

This is a law-constitutive solution to the 'problem of bodies'. I think that the solution is deliberate on Newton's part, and that it is already present in the first edition of the *Principia*.[28]

The conclusion we should draw is this: it is explicit in Newton's writings, from 'De Gravitatione' through the *Principia*, that a necessary condition for the individuality and identity of physical bodies is that they satisfy the laws of nature: in answering the question 'What are the bodies that are the subject-matter of Newton's laws?', we must make reference to those laws.

1.3.4 An interpretative consequence: transmutation

I believe that this approach to bodies is explicit in Newton's writings, and I am proposing it as an interpretation of one (very small) aspect of what Newton took himself to be doing. If we accept this then, even though the topic itself is narrow, there will be wider implications through its connections to other areas of Newton's work, including his views on atoms and the void, and the divisibility of matter, and on universal and essential properties and the distinction between them. Another example is his views on transformation and transmutation, and how we should understand Rule 3 of his Rules of Reasoning (in the second edition, 1713) and the abandoned Hypothesis 3 (of the first edition). Hypothesis 3 reads as follows:[29]

> Hypothesis 3. Every body can be transformed into a body of any other kind and successively take on all the intermediate degrees of qualities.

The apparent problem is that Hypothesis 3 allows the degree of a quality to vary, which entails, in the extreme, that the quality might disappear altogether.

28 One might object that – being later – these drafts are irrelevant to the claim that the solution is already present in the first edition, leaving open the possibility that this view is a post-hoc rationalization. I have already urged that 'De Gravitatione' provides evidence that the position I am advocating on body was already in place prior to publication of the *Principia*. I also suggest that this law-constitutive view of bodies is what Newton relies on when, after the publication of the first edition of the *Principia*, the controversy over the reality of the void gets going, but I will not argue for this here. It is also interesting to note that Newton's definition of motion as being with respect to space appears to long pre-date his offering a definition of bodies, and indeed Newton doesn't include any definition of bodies in his manuscripts until quite close to the time of writing the *Principia*. The first time it appears seems to be in 'On the Motion of Bodies in uniformly yielding media', which Herivel dates to the 1680s. Here, Newton defines absolute and relative time and space and then states (Herivel 1965a, pp. 309–310): 'Definition 5 By common consent bodies are movable things unable to penetrate each other.' The next definition of body to appear is one that is a clear pre-cursor to Definition 1 of the *Principia*. For those interested in the dating of 'De Gravitatione', the presence of a definition of body in this manuscript might add credence to the 1680s dating.

29 See Newton (1999, p. 198) for this translation of Hypothesis 3.

And this seems to allow that we might transform a body into something that is not a body. McMullin (1978, p. 7) puts the point as follows:

> Newton gradually came to believe that he would have to limit his original transformation hypothesis in order that mechanical properties remain invariant. After all, if solidity could be 'remitted' (decreased) at all, it was conceivable that it could be 'taken away' entirely, yet this must clearly be excluded, since it would entail that a body could cease to be subject to mechanics, that is, could cease to be a body.

I agree with the consequences McMullin states here: if solidity could be taken away entirely, then the body would cease to be subject to the laws of motion, and this implies that it would cease to be a body. But I don't agree that Newton gradually came to believe that he would have to limit his transformation hypothesis. Rather, I think that what happens is that Newton makes precise and explicit a view he was already committed to, in particular that things have to have certain features in order to count as bodies. In his revisions, he adds a new hypothesis about qualities which cannot be intended or remitted, but for a while he retains the old Hypothesis 3 alongside this new one. I agree with McGuire (1967) that this is because Newton saw no conflict between the two.[30]

1.4 Solving the 'problem of bodies'

I have argued that in the work of Newton we find a solution to the 'problem of bodies' according to which a necessary condition for the individuation and identity of physical bodies is that they satisfy the laws. This is the weak version of the 'law-constitutive' solution that I have been advocating, and it allows that the sufficient conditions can be completed from resources outside the laws themselves. The strong version states that the necessary *and sufficient* condition for the individuation and identity of physical bodies is that they satisfy the laws. Both the weak and strong versions are limited in the same way: they provide a solution to the 'problem of bodies' for *physical bodies* rather than *bodies in general*; that is, in each case the solution picks out bodies of a certain kind, while leaving open the possibility that there may be other kinds of bodies that are of relevance for other interests.

This solution to the philosophical problem leaves us with a research programme: to fill out the details of the laws, and of any additional conditions, and

30 I am also sympathetic to McGuire's position that the reason Newton eventually abandoned Hypothesis 3 during his revisions of the *Principia* was because he didn't want to have to explain the compatibility and the details of his atomism and transformation thesis, partly because it would have been a distraction from his main point, which was to argue for the universality of gravitation as a quality of bodies.

to demonstrate that the resulting package is indeed a complete and coherent account of the physical bodies that are the subject-matter of the laws.

One way to interpret the *Principia* would be that the definition of mass and the three laws of motion complete the task of filling out these details. We might read Stein (2002) as endorsing the weak version of the law-constitutive solution, and as viewing Newton as having completed the filling out of the details, when he writes (p. 275),

> we have a perfectly clear conception of these attributes of bodies that the mechanical, corpuscular, philosophy has conceived as fundamental, including laws governing the interactions of those bodies: the laws of impact. *That* means, in Newton's view, that we have a sufficiently clear conception of *what bodies are* if the mechanical philosophy is true.

Be that as it may regarding the interpretation of Newton's own position, implicit in the *Principia* is a strategy for filling out the details that is very different, and which shifts us from the weak to the strong solution (although I am definitely not advocating this as an interpretation of Newton's own position). According to this approach, Newton's three laws of motion begin the project but do not complete it. The left-hand side of Newton's second law is a place-holder for force-functions associated with whatever forces there happen to be in the world. Completing our account of physical bodies requires the specification of all these force-functions. Newton's law of universal gravitation provides one such force-function, and thereby moves us closer to an account of physical bodies. Newton believed that there were more force-functions to be found (associated with electrical phenomena, for example), but he did not know what they were or how many more remained to be found. Filling in the details of our solution to the 'problem of bodies' will be complete only when all the force-functions have been found. The strong version of the law-constitutive approach maintains that the laws are both necessary and sufficient, and – with this as a guiding heuristic – the research programme it engenders is the search for the specific forms of the laws that provide the details of this solution to the 'problem of bodies'.[31]

There are no guarantees that a research programme guided by the strong version of the law-constitutive approach will succeed: perhaps we will always be left with some additional features of bodies that need to be specified antecedently to the laws, in order for the laws to have bodies that can serve as their subject-matter. Furthermore, there is no guarantee that this strategy will generate one

31 This approach appears to ride roughshod over Newton's distinction between universal and essential qualities; I think that the distinction can be maintained even while pursuing the strong programme (McMullin 1978, for example, argues that the role of *universal* qualities is to ensure that the bodies that are the subject-matter of Newton's mechanics *remain* bodies (by remaining solid, etc.)), but in any case I am not advocating the strong programme as an interpretation of Newton's own thinking.

unified kind of physical body: perhaps the bodies that serve as the subject-matter of the laws when gravitation is included will turn out not to be identical to those that serve as the subject-matter of the laws when electrical phenomena are at issue. Thirdly, there is no guarantee that the law-constitutive approach to physical bodies will deliver *individuals*. While I have formulated the law-constitutive approach offered in this chapter in terms of necessary and sufficient conditions for the individuation and identity of physical bodies, a more general formulation of the law-constitutive approach (in its strong version) says that the necessary and sufficient condition for some region of a world to be a physical body is that it satisfy the laws of that physical theory. This formulation is neutral as to whether the bodies to which the laws of a given theory apply will turn out to be individuals. Finally, there is no guarantee that the 'bodies' that we end up with are sufficiently close to our pre-theoretic account of bodies that we will be willing to call them bodies: a generalization to the 'entities' that are the subject-matter of a given theory is therefore natural. All these become matters that can be decided only by including the details of a particular physics, and not in advance.

When we see this, we realize just how radical is this solution (in its weak or strong version) to the 'problem of bodies'. Metaphysics and physics become entangled: not all metaphysical questions about what bodies are can be settled prior to doing physics, and that doesn't mean 'physics in general', it means that some metaphysical questions are not independent of a specific and *specified* physical theory (and which questions are independent and which are not depends on the specific theory in question).[32]

My claim is that the law-constitutive approach (weak or strong) is successful as a philosophical solution to the problem of bodies. I have argued that there is no guarantee of success when we work out the details with the specific laws we find in this, the actual, world. But that is a different matter from its philosophical viability as a candidate generic solution. Thus, independently of whether we accept my account as an interpretation of what Newton took himself to be doing, one of the implications of his work for philosophy is that it offers a solution to a problem found in Cartesian philosophy.

The philosophical consequences of this solution should be taken seriously. One motivation for contemporary structural realism stems from the fact that quantum mechanics fails to determine whether its particles are individuals or non-individuals. According to French and Ladyman, if quantum mechanics is interpreted as being about objects (rather than about structure), then it fails to adequately specify the entities that are its subject-matter. This is because, according to French and Ladyman, objects must be determinately individuals or

32 In Brading (2011) I further explore the metaphysical ramifications of the law-constitutive approach, treating composite systems, their unity, and the actual/potential parts debate.

non-individuals.[33] Clearly, this includes in the necessary conditions of quantum object-hood a requirement that goes beyond being an entity that satisfies the laws of quantum mechanics. Thus, how we respond to their challenge will depend in part on the extent to which we are willing to say: to be an entity that serves as the subject-matter of a theory *is* to satisfy the laws of that theory; no less, and also no more.

1.5 Conclusions

I have drawn conclusions of increasing strength during the course of this paper, and there are various points at which one might want to get off the boat. However, there are some key claims that I would press for, as follows.

Descartes fails to offer an adequate account of the bodies that are the subject-matter of his laws. This is by his own criteria: there are not the resources within his metaphysics to underwrite his claim that we have a clear and distinct idea of bodies (plural) as opposed to body (Cartesian indefinite extension). It is also by Newton's criteria: Descartes fails to specify a concept of body that allows him to go on and provide an account of what it is for a body to move. In these ways, Descartes fails to solve the 'problem of bodies'. Newton's solution to the problem involves distinguishing body from space, and stating that a necessary condition for something to be a physical body is that it move according to certain laws. It is this latter claim that I have focused on, arguing that while this law-constitutive account of body can be read implicitly in Descartes, it is explicit in Newton.

I have also gone on to claim that this is a powerful and effective solution to the 'problem of bodies', one which challenges the need for a principle of individuation distinct from necessary and sufficient conditions, and which has significance for discussions of the entities that are the subject-matter of contemporary physics. About these latter claims, there are surely more arguments to be made.

Acknowledgements

This paper developed from work that Dana Jalobeanu and I began several years ago. I am also grateful to the editors of this volume and for the comments and suggestions of Filip Buyse, Daniel Garber, those present at the 'Newton and/as Philosophy' workshop in Leiden in 2007, and all those who have offered feedback when I have presented aspects of this work at various places over past years. Thank you.

33 See Ladyman (1998) for the first statement of this position. See French and Ladyman (2003) for further elaboration. Brading and Skiles (forthcoming) appeals to the law-constitutive approach in responding to French and Ladyman's argument.

2

Leibniz, Newton and force

DANIEL GARBER

Leibniz developed a conception of the physical world that was grounded in the notion of force. So did Newton. In this essay I would like to explore the different ways that these two contemporaries conceived of this notion, and the different ways in which the notion of force functions in their thought. My main focus in the essay will be Leibniz. Leibniz's view of the physical world is much less well known (and much less familiar to readers of this volume) than is Newton's, and his metaphysical conception of force much more central to his thought than it was to Newton's. Consequently, I will begin with a rather extensive development of Leibniz's conception of force, showing where it comes from and what it is supposed to do for him. After we have a clearer idea about Leibniz's conception of force, I would like to turn to Newton and compare the way the two treat the common notion that is so important to both.

2.1 Leibniz: body and force

Before beginning the project, I would like to make a brief comment on Leibniz's thought about the natural world. I'm not going to talk about monads in this essay. Leibniz had well-developed views about the physical world, about motion and its laws, about body, the nature of body, and the structure of body, and about force. This is the level at which I would like to work here. At a metaphysical level below that of body, Leibniz also had views about the ultimate nature of reality, what we might call the level of fundamental metaphysics. It is at this level that one finds his discussion of monads. While the level of fundamental metaphysics is connected with Leibniz's thought about the physical world, the two are not as closely connected as one might think. We can talk about body, motion, and force without directly engaging the metaphysical subbasement. And that's what we will do in this context.

I would like to thank the audience at the Leiden Newton Conference for a very helpful discussion that considerably clarified the issues. I would like especially to thank Eric Schliesser for his comments on earlier versions of this paper, even if he may disagree with how I have responded to them in the final version. Much of the material in this paper is a summary of material that I discuss at greater length in Garber (2009).

In order to understand Leibniz's notion of force, we must begin with his early speculations about the notion of body. And in order to understand Leibniz's early speculations about body, we must see them in the context of his early – and continuing – infatuation with the mechanical philosophy. In one of his earliest writings, the "Confesio naturae contra atheistas" ("The Confession of Nature against Atheism"), published in 1669 when Leibniz was only 23 years old, he writes:

> For through the admirable improvement of mathematics and the approaches which chemistry and anatomy have opened into the nature of things, it has become apparent that mechanical explanations – reasons from the figure and motion of bodies, as it were – can be given for most of the things which the ancients referred only to the Creator or to some kind (I know not what) of incorporeal forms.[1]

The version of the mechanical philosophy that most attracted him at this time was that of Thomas Hobbes, as given in the *De corpore* of 1655.[2]

For Hobbes, as for Descartes, the notion of extension is basic to the idea of body. It is almost definitive of the Cartesian school of philosophy that bodies are the objects of geometry made real, and contain nothing but extension and its modes. Hobbes goes even farther and actually identifies fundamental physics with geometry itself.[3] One feature of Hobbes' physics that attracted the young Leibniz was its treatment of force in the physical world. Descartes recognizes force in the physical world, grounded in the continual activity of God in sustaining the world from moment to moment.[4] But God has no substantive role to play in Hobbes' conception of the physical world. And as a consequence, there is no place for a concept of force or resistance outside of motion itself in Hobbes' physics: "Rest does nothing at all, nor is of any efficacy; and... nothing but Motion gives Motion to such things as be at Rest, and takes it from things moved."[5]

1 A6.1.489 (L 109–110) (see the list of abbreviations at the end of the chapter).

2 Beeley (1995), e.g., emphasizes the connections between Hobbes' thought about the physical world and that of the young Leibniz.

3 On Hobbes' conception of body, see *De corpore*, chapter 8.1. On the interconnection of physics and geometry, see *De corpore*, chapter 6.6. Geometrical and physical considerations are freely intermixed in parts II and III of *De corpore*. However, one should note that he proceeds differently in part IV, where he treats sense, light, heat, cold, etc., starting with experience and inferring causes.

4 On Descartes's derivation of the laws of motion from God and the grounding of force, see Garber (1992, chapters 7–9). Though I still hold to the view as presented in these chapters, in the last few years there have emerged a variety of alternative views on the notions of force and activity in Descartes. For a recent discussion that sets out some of these recent alternative positions, see Schmaltz (2008a).

5 *De corpore*, 15.3.

When, in the late 1660s and early 1670s Leibniz becomes interested in the laws of motion and impact, it is Hobbes' conceptions of body and force that are at the root of his project.[6] The works in question are the *Theoria motus abstracti* and the *Hypothesis physica nova*, both published in 1671. The heart of the early theory of motion in these writings is an account of the collision of two bodies; for Leibniz, as for Hobbes, collision is the only way in which the motion of a body can be changed naturally. Because bodies exert no force except through motion, the outcomes of collisions are determined by simply combining the instantaneous motions (conatus) of the two bodies at the moment of collision; body as such offers no resistance to motion and so the mass or size of the bodies in question plays *no role whatsoever* in the outcome of a collision. As Leibniz put it in the *Hypothesis physica nova*, "all power in bodies depends on the speed."[7] If two bodies with unequal speed collide, then, Leibniz argues, the two will move together after the collision with a speed which is the difference between the two, and in the direction of the faster. One interesting special case is when one of the bodies is at rest. In that case, a moving body, no matter how small, can set a resting body, no matter how large, into motion without losing any of its own motion. As Leibniz noted in a document written during this period (1675), "We have assumed by a kind of prejudice that a greater body is harder to move, as if matter itself resisted motion."[8] Another interesting special case is when two bodies with the same speed collide directly. In this case, both come to a halt and all motion is lost.[9]

These laws of motion, reasonable as they might be in the abstract, fit very poorly with the world we see around us, as Leibniz knew; in particular, the bodies of our world do seem to offer resistance to being set into motion. Leibniz reconciled these abstract laws with experience through an hypothesis about the makeup of the world; this is the new hypothesis of the *Hypothesis physica nova*. Using this hypothesis that the bodies of everyday experience are made up of tiny parts, corpuscles, Leibniz is able to convince himself that he can explain why it is that larger bodies will appear to resist new motion in collision more than smaller bodies will.[10] And so, the laws bodies appear to obey in our world are the result of abstract and geometrical laws, very different from what we experience in day-to-day life, operating in a complex world that

6 It should be noted that when Leibniz became interested in the laws of motion in 1669, it was initially the Huygens/Wren laws of impact that caught his attention. On this see Garber (2009, pp. 14ff). But it seems to have been to Hobbes that the young Leibniz turned to clarify the question. Descartes and his laws didn't really enter the picture until somewhat later.

7 A6.2.228. 8 A6.3.466 (RA 31) 9 See A6.2.269 § 12.

10 See on this A6.2.164 (§ 33); *Hypothesis physica nova* §§ 22–23, A6.2.228–232. For a more detailed discussion of this, see Hannequin (1908, pp. 103–107); Duchesneau (1994, p. 63f). This may well have been inspired by Hobbes, *De corpore* 15.8. (I would like to thank Kathryn Morris for this observation and for the reference.)

God created for his ends. Leibniz was later to see the deep problems with this account, but in 1670 and 1671, the young Leibniz was very pleased with his first serious excursion into physics.

This early world begins to crumble in the mid 1670s, when Leibniz is about 30 years old. Attractive as his earlier picture of the physical world is, he begins to notice some uncomfortable features. Though at the macroscopic level Leibniz is able to save the phenomena and account for the fact that it seems to take more effort to move a larger body than a smaller one, at the level of fundamental physics, this isn't true. And as a consequence, at the fundamental level, no basic conservation principles would seem to hold. But in 1676, Leibniz comes to accept a metaphysical principle that will shape his physics for the rest of his life: the principle of the equality of cause and effect.[11] On this principle there must be exactly the same ability to do work in the total cause as there is in the full effect. He quickly realizes that this seriously undermines the strategy of his early physics. Immediately after giving an extensive formulation of the principle, he writes:

> It has been established through experience that the cause why a larger body is moved with difficulty even on a horizontal plane is not [always] heaviness, but massiveness.[12] Unless body were to resist, perpetual motion would follow, since a body resists in proportion to its bulk [*moles*], since there is no other factor that would limit it [*nulla alia ratio determinandi*]. That is to say, since there is no other factor [*ratio*] which would hinder it from rebounding to less than its [original] height, since in itself, without an extrinsic impediment through the impulse of [another] body, it would give [the other body] its whole motion, and retain it as well.[13]

With these considerations, we can no longer separate out the basic abstract laws of physics from the ones that are obeyed by the bodies of everyday experience: unless the principle of the equality of cause and effect holds at *every* level, we are subject to the possibility of perpetual motion. And as a consequence of that, we must recognize that there *must* be genuine resistance to motion in body, something by virtue of which bodies can resist the acquisition of new motion.

At the same time that Leibniz was worrying about resistance, he was also beginning to worry about another feature of his earlier physics, the coherence

11 The key document here is a remarkable essay, "De Arcanis Motus et Mechanica ad puram Geometriam reducenda," "On the secrets of motion and the reduction of mechanics to pure geometry," probably written in the summer of 1676. It can be found in Hess (1978, pp. 202–205). It is discussed, and its significance underscored in Fichant (1978).

12 The word here is "soliditas." According to Robert Estienne's *Dictionarium latinogallicum* (Paris, 3rd edn., 1552), "soliditas" means "massivité, solidité, fermeté." Of these, I think that the first is closest to what Leibniz has in mind here.

13 Hess (1978, p. 205).

of the notion of motion and the distinction between motion and rest. When two bodies are in motion with respect to one another, what does it mean to say that one is *really* in motion and the other is *really* at rest? The solution he comes to is that the body that is really in motion is the one that is the real cause of the change. This idea is set out in a series of fragments that the Akademie editors date from Summer 1678 to Winter 1680/1. In these fragments, Leibniz explains and defends the following view:

> A body in motion is one which is the proximate efficient cause for why each of its parts changes position with respect to other bodies; otherwise it is said to be at rest.[14]

In another fragment from the series, he gives the cause in question a general name:

> And so we attribute motion to that which has a force of acting [*vis agendi*]. Whence it is also obvious that those who have said that what is real and positive in motion is equally found in both contiguous bodies receding from one another have spoken falsely. For the force of acting can only be in one of them, and therefore it is also the cause [of the change of position].[15]

The 'vis agendi,' the force of acting, then, is what will enable us to say that there is a distinction between motion and rest.

Leibniz's discovery of the 'vis agendi,' the active force by which motion and rest are distinguished from one another, together with his realization that bodies must have a passive force by virtue of which they resist the acquisition of new motion, led Leibniz to a new view about the nature of body. Hobbesian (or Cartesian) bodies, the objects of geometry made real, have nothing in them from which such forces could arise. So, Leibniz concludes, *there must be something to bodies over and above their geometrical properties.* But what is it? Leibniz's answer is simple: to ground these forces in bodies we must revive the forms that the schoolmen had posited, and that the mechanists had rejected. Here is how Leibniz puts it in a document written shortly after his decision in 1679 to revive substantial forms:

> [W]hen I considered how, in general, we could explain what we experience everywhere, that speed is diminished through an increase in bulk [*moles*] as, for example, when the same boat carried downstream goes more slowly the more it is loaded down, I stopped, and all my attempts having been in vain, I discovered that this, so to speak, inertia of bodies cannot be deduced from the initially assumed notion of matter and motion, where matter is understood as that which is extended or fills space, and motion is understood as change of space or place. But rather, over and above that which is deduced from extension and its variation or modification alone,

14 A6.4.2011. 15 A6.4.2019.

> we must add and recognize in bodies certain notions or forms that are immaterial, so to speak, or independent of extension, which you can call powers [*potentia*], by means of which speed is adjusted to magnitude. These powers consist not in motion, indeed, not in conatus or the beginning of motion, but in the cause or in that intrinsic reason for motion, which is the law required for continuing. And investigators have erred insofar as they considered motion, but not motive power or the reason for motion, which even if derived from God, author and governor of things, must not be understood as being in God himself, but must be understood as having been produced and conserved by him in things. From this we shall also show that it is not the same quantity of motion (which misleads many), but the same powers that are conserved in the world.[16]

Leibniz's idea seems to be this. Resistance and the ability to do work are kinds of activity in bodies, and therefore cannot be derived from bare matter, which is inert. And therefore, to inert extended matter we must add something that can be the source of these kinds of activity, both resistance and the positive ability to do work. That is, to inert matter we must add form. The view that he is trying to articulate seems to be that for there to be activity – both resistance and the positive activity by which one body acts on another – there must be form in bodies. Form is the source of this activity, both resistance and positive activity.

Leibniz's idea seems to have been that both the active force that bodies have to cause changes in other bodies and which grounds motion, and the passive forces of resistance that bodies exert in opposition to the active forces that act on them derive from the form, which is added to inert Cartesian (or Hobbesian) matter. But this passes quickly over into a subtly different view. While continuing to locate the active forces in the soul or form, Leibniz moved the passive forces into matter. In a passage from the essay "On the Method of Distinguishing Real from Imaginary Phenomena," now dated as between Summer 1683 and Winter 1685/6, Leibniz wrote:

> Concerning bodies I can demonstrate that not merely light, heat, color and similar qualities are apparent but also motion, figure, and extension. And that if anything is real, it is solely the force of acting and suffering [*vim agendi et patiendi*], and hence that the substance of a body consists in this (as if in matter and form). Those bodies, however, which have no substantial form are merely phenomena or at least only aggregates of the true ones.[17]

16 A6.4.1980 (AG 249). Cf. also the following passage from 1680, where Leibniz goes as far as to identify activity and active force with the essence of body: "[I found] that there are substantial forms, and that the nature of body consists not in extension, but in an action which is related to extension, since I hold that a body cannot be without effort . . . " (Leibniz to de La Chaise (April/May 1680), A2.1^2.798).

17 A6.4.1504 (L 365).

Leibniz is somewhat more expansive on this theme in another essay that probably dates from the same period, just before the composition of the *Discourse on Metaphysics*, an essay that the Akademie editors have entitled "De mundo praesenti":

> Corporeal substances have parts and species. The parts are matter and form. Matter is the principle of being acted on [*principium passionis*] that is, the primitive force of resisting, which is commonly called bulk or antitypy, from which flows the impenetrability of body. The substantial form is the principle of action or the primitive force of acting. Furthermore, there is in every substantial form a certain knowledge [*cognitio*] that is an expression or representation of external things in a certain individual thing, in accordance with which a body is *per se* one, namely in the substantial form itself. This representation is joined with a reaction or conatus or appetite which follows this thought of acting. This substantial form must be found in all corporeal substances which are *per se* one.[18]

With this we have a new and coherent view of body. Bodies are no longer the objects of geometry made real, as they are for Descartes and Hobbes. They are Aristotelian unities of substantial form and primary matter. But even more interesting than that, the newly revived form and matter of the Aristotelians is identified with *force*: form with the active force associated with motion, and matter with the passive force associated with resistance.

This view gets its fullest statement in the essay, *Specimen dynamicum* which Leibniz published in the *Acta eruditorum* in 1695, a kind of summary treatment of elements of the science of force (which he entitled 'dynamics') that he had developed in the intervening years. Though it was published after the publication of Newton's *Principia* in 1686/7, one can argue that the metaphysics of force that Leibniz presents there grows largely out of the view that he first articulated in the writings beginning in the late 1670s. In the *Specimen dynamicum* and related writings Leibniz presents a conception of force that involves two important distinctions, the distinction between primitive and derivative forces, and the distinction between active and passive forces. So in all, there are four principal varieties of force, primitive active and passive force, and derivative active and passive force. Leibniz writes:

> *Active force* (which might not inappropriately be called *power* [*virtus*], as some do) is twofold, that is, either *primitive*, which is inherent in every corporeal substance *per se*... or *derivative*, which, resulting from a limitation of primitive force through the collision of bodies with one another, for example, is found in different degrees. Indeed, primitive force (which is nothing but the first entelechy) corresponds to the *soul or substantial*

18 A6.4.1507–1508 (RA 285–287). It should be noted that Leibniz is rather unclear about what exactly the species are.

> *form*... Similarly, passive force is also twofold, either primitive or derivative. And indeed, the *primitive force of being acted upon* [*vis primitiva patiendi*] or of *resisting* constitutes that which is called *primary matter* in the schools, if correctly interpreted. This force is that by virtue of which it happens that a body cannot be penetrated by another body, but presents an obstacle to it, and at the same time is endowed with a certain laziness, so to speak, that is, an opposition to motion, nor, further, does it allow itself to be put into motion without somewhat diminishing the force of the body acting on it. As a result, the *derivative force of being acted upon* later shows itself to different degrees in *secondary matter*.[19]

The active and passive forces in question are the ones that we discussed earlier: the forces connected with motion (active), and the forces connected with resistance (passive). One might add here the distinction Leibniz makes between two kinds of active force: living and dead. Living force is associated with actual motion. Dead force, on the other hand, is associated with what Leibniz calls the "solicitation to motion," the acceleration involved in a spring or a bow.[20] But also interesting is the distinction Leibniz draws here between primitive and derivative forces. Leibniz characterizes the primitive active force as corresponding to "the soul or substantial form"; the primitive passive force, on the other hand, is characterized as constituting "that which is called *primary matter* in the schools, if correctly interpreted."[21] In this way, primitive forces are conceived of as *things*, or at least as *constituents* of things, the way we might talk about the Church as a *force* in society. Derivative forces, in contrast, are the momentary and quantifiable modes of the primitive forces, both active and passive. It is in terms of these forces that the laws of nature are framed. As Leibniz wrote to Johann Bernoulli in 1698:

> If we conceive of soul or form as the primary activity from whose modification secondary [i.e. derivative] forces arise as shapes arise from the modification of extension, then, I think, we take sufficient account of the intellect. Indeed there can be no active modifications of that which is merely passive in its essence, because modifications limit rather than increase or add.[22]

In this way form and matter, primitive active and passive force constitute the grounds for the particular modifications that are the particular magnitudes of active force and resistance that we can observe in bodies. In this way, Leibniz's conception of force is connected in one direction with his doctrine of the nature

19 GM VI 236–237 (AG 119–120). A very similar account is given in "On Body and Force, May 1702"; see G IV 395 (AG 252).
20 See GM VI 238 (AG 121).
21 GM VI 236–237 (AG 119–120).
22 GM III 552 (AG 169).

of body, while at the same time it is connected with our understanding of the behavior of body and the laws of motion.

2.2 Newton vs. Leibniz on Force

Leibniz's doctrine of force and body is directed solidly against the mechanist doctrines of Descartes, Hobbes and their followers. There is every reason to think that it was largely in place by 1686 when, seemingly out of nowhere, Newton published his *Philosophiae naturalis principia mathematica*, what came to be known simply as the *Principia*, and articulated a conception of force that was very different from what Leibniz had developed. The ultimate success of the Newtonian program has all but driven Leibniz's conception of force off of the playing field. But I would like to reflect a bit about the places in which the two conceptions of force come together, and the ways in which they differ.

Now, there are at least two sets of notes Leibniz wrote on Newton's *Principia* that survive, in addition to the numerous comments that he made in letters and essays on his great contemporary and rival.[23] Interestingly enough, none of them are addressed directly at the relation between their accounts of force. Neither is the issue addressed in the Leibniz–Clarke exchange, which many (though not all) consider a near-direct exchange between Leibniz and Newton. My account here is thus less a report on how each saw the other in relation to his own views, than it is the remarks of a modern commentator trying to make historical and philosophical sense of the relations between them on this issue.

The various notes we have that precede the final composition of the *Principia* show that Newton struggled with the articulation of the notion of force. Though it would be interesting to trace the development of Newton's conception and compare it with the considerations that drove Leibniz,[24] I am going to limit myself to a consideration of the final account as it appears in the *Principia*.

The star of the *Principia* is Newton's notion of impressed force. Newton defines as follows in the *Principia*:

> *Impressed force is the action exerted on a body to change its state either of resting or of moving uniformly straight forward.*
>
> This force consists solely in the action and does not remain in a body after the action has ceased. For a body perseveres in any new state solely by the

23 Leibniz (1973) was thought to be the only such set of notes for quite a while. Then Bertoloni Meli (1993) published another set of notes in his appendix 1, and in chapter 5 argued very convincingly that Leibniz had read Newton's *Principia* before writing his "Tentamen de motuum caelestium causis."

24 For a recent account that gives an elegant account of the last stages of Newton's development of the notion of force, see Bertoloni Meli (2006a), § 2.

> force of inertia. Moreover, there are various sources of impressed force, such as percussion, pressure, or centripetal force.[25]

(The reference to the force of inertia is significant here, and we will address that shortly.) It is this kind of force that is at issue in Newton's second law: "A change in motion is proportional to the motive force impressed and takes place along the straight line in which that force is impressed."[26] The project of book I of the *Principia* is to give us the general tools that we need to infer the existence of various impressed forces from the phenomena. As Newton writes in the preface to the first edition, "the basic problem of philosophy seems to be to discover the forces of nature from the phenomena of motions and then to demonstrate the other phenomena from these forces."[27] In book III Newton works out one important example, that of gravitational force. But there is an explicit assumption from the beginning that this is not the only kind of force in nature. Again, in the preface to the first edition Newton writes:

> For many things lead me to have a suspicion that all phenomena may depend on certain forces by which the particles of bodies, by causes not yet known, either are impelled toward one another and cohere in regular figures, or are repelled from one another and recede. Since these forces are unknown, philosophers have hitherto made trial of nature in vain.[28]

In the *Principia*, Newton gives us the methods we need to find these other forces, perhaps chemical forces, electrical forces, magnetic forces, etc. All of these other impressed forces can be handled by the same mathematical tools, etc., but they are in their nature rather different. What they have in common is simply that they are causes of the change in velocity of bodies. Force answers the question not of what bodies are, but how they behave.

This conception of force seems importantly different from Leibniz's conception. For Newton, the focus is on the impressed force that changes the motion of a body. His interest in the *Principia* is in the common cause of the impressed forces that explain the trajectories of the heavenly bodies, as well as the fall of bodies on Earth. That cause is what he calls gravity. Newton is thus interested in causes in the *Principia*, but only in determining that there is a common cause for a variety of phenomena: otherwise the cause in question remains unknown in its nature, though one can, perhaps, read him as entertaining the possibility that gravitation (i.e., the underlying cause of the impressed forces that are examined in the *Principia*) is essential to body as such. At the same time, Newton recognizes other forces in nature (chemical, electrical, magnetic, etc.) which will have their own distinctive (but as of yet unknown) causes, causes that might be present in some bodies but not in others. These other causes will impress other forces on bodies in other circumstances.

25 Newton (1999, p. 405). 26 *Ibid.*, p. 416. 27 *Ibid.*, p. 382.
28 *Ibid.*, pp. 382–383. Cf. also *Ibid.*, p. 588.

Now, it is not impossible that all of the Newtonian impressed forces can be given a mechanical explanation, as Descartes or Hobbes or Leibniz would want to do. That is, it is not impossible that all of the causes of the impressed forces that explain the different phenomena of interest to Newton are, at root, mechanical causes. It is also possible for Newton that gravitation could be an essential property of body as such. While one can find many apparently tentative speculations about this question in his notes and papers,[29] this wasn't Newton's interest, at least not in the *Principia*: his first problem was to establish the existence of the forces of different kinds. It was only after that that one should investigate the underlying causes, and that only to the extent that experience and experiment reveal them to us. To the best of our knowledge, Newton never got to that part of his investigation. The framework of the *Principia* doesn't commit him to mechanism or to anything else: it is an open framework that allows one to entertain a larger vision.

Leibniz, unlike Newton, is doing what we might call fundamental physics, trying to characterize the physical world at its physically most basic level. And in this connection, Leibniz's interest in the notion of force is linked to a strict and rigorous version of the mechanist program: at root there is one kind of material stuff, all of which obeys the same laws. Leibniz is working within a rigidly mechanist framework. And like Descartes and Hobbes before him, his problem is grounding the physical world in an appropriate conception of that material stuff. This is where the notion of force enters into his project. Force is not just a tool to explain the empirically observed behavior of bodies: the conception of force that interests Leibniz operates at the physically most fundamental level and reveals the underlying nature of body as such. In Newton's world one can contemplate different *kinds* of forces, with different underlying causes. Leibniz begins with the commitment that in the world, there is just body of one sort. And in this world, there is force of one sort, at root. Again, it might turn out for Newton that all the different forces can be reduced to one kind of force, the mechanist's force that acts through collision. But while Newton is not committed to that position for a priori reasons, Leibniz is. It is this one, unitary conception of force that he identifies with body as such. For Newton, the force treated in the second law and as the focus of attention in the rest of the *Principia* is a generic notion, as it were, a general kind of thing, characterized by its mathematical structure and connection with other notions such as mass and acceleration; it can thus come in different varieties, gravitational, chemical, electrical, magnetic, and perhaps others. For Leibniz, on the other hand, force is something very specific and very concrete: it is what is ultimately grounded in the fundamental make-up of a mechanist world. It is force with a capital 'F', the ultimate stuff that grounds the physical world.

Indeed, in a very strict sense, there *are* no external, impressed forces for Leibniz. It is well known that for Leibniz, there is no genuine causal

29 See the introduction to part III in Newton (1962).

communication between substances: all of the activity of a corporeal substance derives from its own internal states. Thus Leibniz writes in Part II of the *Specimen dynamicum*: "*every passion of a body is of its own accord*, that is, *arises from an internal force, even if it is on the occasion of something external*."[30] In this way, from a metaphysical point of view, at least, the central notion of force, that of an impressed force, is strictly speaking unintelligible from the point of view of a Leibnizian conception of force.

But even if Newton's impressed force doesn't have much in common with Leibniz's conception of force, there is another Newtonian notion that seems much closer. In the *Principia* Newton writes:

> *Inherent force of matter is the power of resisting by which every body, so far as it is able, perseveres in its state either of resting or of moving uniformly straight forward.*
>
> This force is always proportional to the body and does not differ in any way from the inertia of the mass except in the manner in which it is conceived. Because of the inertia of matter, every body is only with difficulty put out of its state either of resting or of moving. Consequently, inherent force may also be called by the very significant name of force of inertia . . .[31]

The inherent force of matter is very closely identified with matter itself: it is something that pertains to body as such, by virtue of which body is capable of resisting an impressed force imposed upon it. In this respect it is rather different than the more central notion of impressed force in Newton, as commentators have noted with some puzzlement. In his recently published commentary on the *Principia*, I. B. Cohen remarks that this definition is "in many ways the most puzzling of all the definitions in the *Principia*." He continues:

> Today's reader will . . . be struck by the fact that Newton uses the word "force" in relation to "inertia" ("vis inertiae"), although – as Newton is at pains to explain – this is an internal force and not the kind of force which (according to the second law) acts externally to change a body's state of rest or of motion. Unless we follow Newton's instructions and make a sharp cleavage between such an internal "force" and external forces, we shall fail to grasp the Newtonian formulation of the science of dynamics.[32]

The fact that Cohen consistently puts this use of force into scare-quotes suggests that he doesn't think that it is really force, properly speaking. If what we mean by force is a physical magnitude that satisfies Newton's second law, then the *vis insita* certainly isn't a force.

30 GM VI 251 (AG 134–135). 31 Newton (1999, p. 404).
32 Cohen's "Guide" in Newton (1999, p. 96).

But Newton's inherent force of matter is very similar to Leibniz's primitive passive force; both are expressions of the basic nature of body as such.[33] And both are connected with the resistance to change in a body's state of motion or rest. Here is how Leibniz characterizes primitive passive force in the *Specimen dynamicum*, as quoted above:

> And indeed, the *primitive force of being acted upon* [*vis primitiva patiendi*] or of *resisting* constitutes that which is called *primary matter* in the schools, if correctly interpreted. This force is that by virtue of which it happens that a body cannot be penetrated by another body, but presents an obstacle to it, and at the same time is endowed with a certain laziness, so to speak, that is, an opposition to motion, nor, further, does it allow itself to be put into motion without somewhat diminishing the force of the body acting on it. As a result, the *derivative force of being acted upon* later shows itself to different degrees in *secondary matter*.[34]

And here is Newton, from Definition 3 of the *Principia*:

> Moreover, a body exerts this force only during a change of its state, caused by another force impressed upon it, and this exercise of force is, depending on the viewpoint, both resistance and impetus: resistance insofar as the body, in order to maintain its state, strives against the impressed force, and impetus insofar as the same body, yielding only with difficulty to the force of a resisting obstacle, endeavors to change the state of that obstacle. Resistance is commonly attributed to resting bodies and impetus to moving bodies, but motion and rest, in the popular sense of the terms, are distinguished from each other only by point of view, and bodies commonly regarded as being at rest are not always truly at rest.[35]

Let me begin by emphasizing some striking similarities between these two conceptions. First of all, Leibniz's passive force seems to be responsible for doing pretty much what Newton's *vis insita* is supposed to do: it is the force that resists change both in bodies at rest and bodies in motion. In this way it is the force that breaks the speed of a colliding body. Unlike Leibniz, Newton does not emphasize the difference between primitive and derivative forces. But one can say that something very like Leibniz's distinction is inherent in Newton's definition insofar as he distinguishes between the "inherent force" which is "always proportional to the body" and "inherent" in it, and the *exercise* of this force on the occasion of an impact.

But despite the similarities, there are some profound differences as well. One profound difference concerns the relation between resistance and impetus, the force exerted by a body resisting change in its motion, and the force exerted by

33 On the relation between Newton's *vis insita* and Leibniz's notion of inertia, see Bernstein (1981).

34 GM VI 236–237 (AG 119–120).

35 Newton (1999, 404–405).

a body in motion on another body. For Newton, the two are the same force, and differ only in our point of view:

> this exercise of force is, depending on the viewpoint, both resistance and impetus: resistance insofar as the body, in order to maintain its state, strives against the impressed force, and impetus insofar as the same body, yielding only with difficulty to the force of a resisting obstacle, endeavors to change the state of that obstacle.

Though Newton, of course, recognizes a real distinction between absolute motion and absolute rest, this distinction does not enter into his account of the *vis insita*: it is both resistance and impetus. But for Leibniz, the arch-relativist in other respects, there is a real distinction between resistance and impetus, between (primitive) passive force and (primitive) active force. Indeed, this distinction is identical for Leibniz to a fundamental distinction in his metaphysics between matter and form, the two distinct principles that make up corporeal substance. Though a relativist about space and time, Leibniz is *not* a relativist about motion and rest, and thus not a relativist about impetus and resistance. For him they are radically distinct, and ground a fundamental dichotomy in his metaphysical account of body.

There is another, perhaps more subtle difference between Newton and Leibniz on this issue worth noting. Newton's is a *vis insita*, a force inherent in body. This is important to Newton, no doubt. But Newton is less interested in the contribution this makes to the understanding of the *nature* of body, and more interested in the consequences that it has for our understanding of the *behavior* of body. What is important about it being an *inherent* force is simply that it is always available to us in understanding the behavior of bodies. Impressed forces come and go as bodies find themselves in different external circumstances. But you can *always* count on a body to resist the change in its state. For Leibniz, though, the inherence of passive forces, important as it is for understanding the behavior of bodies, is also central for his account of their nature. The primitive passive force of a body is a central constituent of its underlying nature.

And with this we reach what is a very deep difference between Leibniz and Newton and their conception of force. In the *Principia*, Newton is interested in demonstrating certain theorems about force and motion and in demonstrating from them and from certain empirical phenomena the existence of a universal law that explains planetary motion and terrestrial gravitation and the connection between the two. While in various places he speculates about the nature of matter, his thoughts about matter are just that, speculations, and in the *Principia*, at least, he does not want to advance a solid doctrine of body and its makeup. But Leibniz's project is quite different. Leibniz, too, is interested in motion and the behavior of bodies in motion. But his interest in force is broader than that. What Leibniz seeks is the big picture: the nature of body as a grounding for an account of motion and its laws. Force plays a role in Leibniz's account of the behavior of bodies. But he is just as interested, if not more, in the way the notion of force can illuminate the *nature* of body. In that way, force plays a central role in his proposed replacement for the Cartesian/Hobbesian account of the nature of body.

Let me put this difference in broader terms still. Earlier in the century, there is a tension between two great traditions in thinking about the natural world. Descartes is working in a broadly Aristotelian tradition of natural philosophy. His aim is ultimately to give a view of the world that includes an account of the behavior of bodies as such, but grounded in an understanding of the true first causes: the nature of bodies, the causes of their motion, the way in which the laws that govern their behavior are grounded in the first cause, i.e. God. A different strand was the Galilean project. Galileo's project was within the domain of mixed mathematics, as it was called, a quantitative account of the world that favored mathematical description over an account of the ultimate first causes. I would claim that Leibniz is a heritor of the natural philosophical tradition of Descartes, and Newton is a heritor of the mathematical tradition that Galileo followed. The very different ways in which Leibniz and Newton treat the notion of force are, I would claim, reflections of that fundamental difference.

Abbreviations

A	Leibniz, Gottfried Wilhelm (1923–). *Sämtliche Schriften und Briefe.* Deutsche Akademie der Wissenschaften zu Berlin (eds.) Berlin: Akademie-Verlag. References include series, volume, and page. So 'A6.4.1394' is series 6, volume 4, p. 1394. Note that 'A2.1^2.123' refers to the second edition of series 2, volume 1, p. 123.
AG	Leibniz, Gottfried Wilhelm (1989). *Philosophical Essays.* Roger Ariew and Daniel Garber (eds. and trans.) Indianapolis: Hackett Pub. Co.
G	Leibniz, Gottfried Wilhelm (1875–90). *Die philosophischen Schriften.* C. I. Gerhardt (ed.) 7 vols. Berlin: Weidmann. References include volume and page. So 'G VII 80' is volume 7, p. 80.
GM	Leibniz, Gottfried Wilhelm (1849–63). *Leibnizens mathematische schriften.* C.I. Gerhardt (ed.). 7 vols. Berlin: A. Asher. References include volume and page. So 'GM VII 80' is volume 7, p. 80.
L	Leibniz, Gottfried Wilhelm (1976). *Philosophical Papers and Letters.* Leroy E. Loemker (ed. and trans.) 2nd edn. Dordrecht: Reidel.
RA	Leibniz, Gottfried Wilhelm (2001). *The Labyrinth of the Continuum : Writings on the Continuum Problem. 1672–1686*, Richard Arthur (ed. and trans.). New Haven: Yale University Press.

3

Locke's qualified embrace of Newton's *Principia*

MARY DOMSKI

3.1 Introduction

> The Commonwealth of Learning, is not at this time without Master-Builders, whose mighty Designs, in advancing the Sciences, will leave lasting Monuments to the Admiration of Posterity; But every one must not hope to be a Boyle, or a Sydenham; and in an Age that produces such Masters, as the Great – Huygenius, and the incomparable Mr. Newton, with some other of that Strain; 'tis Ambition enough to be employed as an Under-Labourer in clearing Ground a little, and removing some of the Rubbish, that lies in the way to Knowledge.
>
> (*Essay*, pp. 9–10)[1]

The above, now famous passage is included in the Epistle to the Reader that begins the first edition of John Locke's *An Essay Concerning Human Understanding* (1689). For interpreters of the *Essay* who have granted attention to the impact of Locke's natural philosophical context on the development of his empiricist stance towards knowledge,[2] Locke's self-appointed role as an "Under-Labourer" has offered something of a mystery. For based on these

Portions of the work completed for this paper have been presented at several venues over the last few years. I especially thank audience members at the 2006 History of Science Society Annual Meeting, the Sixth Biennial Congress of the International Society for the History of the Philosophy of Science (HOPOS), the UC, San Diego History of Philosophy Roundtable, the University of Houston, and SPAWN 2009 for their comments. Special thanks are owed to Margaret Atherton, Nico Bertoloni Meli, Daniel Garber, Helen Hattab, Sam Rickless, Don Rutherford, and George Smith for their helpful and constructive feedback. Finally, my thanks to Eric Schliesser for his insightful comments on the penultimate version of this paper and to Ken Winkler for sharing his paper with me prior to its publication. I hope in what follows I have done his arguments justice.

1 All citations to the *Essay* refer to Locke (1975), edited with an introduction by Peter H. Nidditch. Aside from references to the Epistle to the Reader, citations refer to book, chapter, and section.

2 Recently, there have been several such commentators. See, for instance, the references cited in Winkler (2008, note 6) as well as works by G. A. Rogers (1978a), (1978b), (1979a), (1979b), (1982).

brief remarks, it is not altogether clear what lessons we should draw about Locke's understanding of the "Master-Builders" who were "advancing the Sciences" in the late seventeenth century. Specifically, and as will be my focus in this chapter, it is not altogether clear what Newton's inclusion on this list of "Master-Builders" indicates about Locke's interpretation of Newton's natural philosophical achievements in the *Principia mathematica* (1687), which was published just two years prior to the first edition of the *Essay*.[3]

One popular reading of this passage, and of Locke's general attitude towards Newton's natural philosophical achievements, points to Newton's experimentalism as a primary influence on Locke. Given Newton's association with naturalists such as Boyle and Sydenham in the Under-Labourer passage,[4] Locke's strong ties to the natural history tradition (primarily developed through his close association with Boyle), and Locke's emphasis on the use of experiment and observation in natural philosophy in Book IV of the *Essay*, it seems that, for Locke, Newton's great achievement lay in grounding his natural philosophy on an observational and experimental method.[5] Though prima facie a reasonable reading, such a view leaves us questions regarding Locke's explicit advocacy for Newton's *mathematical* method of natural philosophy, which he voices in his correspondence with Stillingfleet as well as in other texts written in the 1690s.

Taking these passages regarding Newton's mathematical approach to nature seriously, Kenneth Winkler (2008) has recently offered an alternative reading of Locke's embrace of Newtonian natural philosophy. Drawing on evidence from Locke's writings during the 1690s as well as revisions to the fourth (1700) edition of the *Essay*, Winkler aims to show that Locke did in fact appreciate the distinctively mathematical character of Newton's *Principia* achievement and, moreover, made revisions to the *Essay* that were intended to defend Newton's "mathematical physics." Thus, according to Winkler's account, neither Newton's association with natural historians such as Boyle and Sydenham in the Under-Labourer passage, nor Locke's emphasis on natural history in the *Essay*, should determine how we read Locke's interpretation of Newton's achievement in advancing the sciences. As he puts it, "What I want to show in this paper is that in spite of this natural-historical emphasis [in Book IV of the *Essay*], Locke provides a way of interpreting – and defending – a mathematical

3 Though there is the reference to Newton in the "Under-Labourer" passage of 1689, Rogers reveals that the body of the *Essay* was in near finished form when Locke first read Newton's masterpiece and thus discounts claims that Newton's *Principia* had a substantial influence on the stated doctrines of the first edition *Essay* (Rogers 1978b). As I discuss below, there are signs that Newton's work influenced the later fourth edition of the *Essay*.

4 In what follows, I follow Feingold (2000) and use the term "naturalist" in a very narrow sense to refer to those working in the natural history tradition.

5 Commentators such as John Yolton and Roger Woolhouse propose just such a reading of Locke's attitude toward Newton's *Principia* achievement. See especially Yolton (1969, 1970) and Woolhouse (1994).

physics, and that there is a reason to believe that in doing so, he was responding directly to Newton's achievement in the *Principia*" (Winkler 2008, p. 232). Moreover, on Winkler's score, Locke's revisions and writings from the 1690s reveal that he "voices an optimism about mathematical natural philosophy that goes beyond the as-good-as-we'll-get sort of praise that he confers on natural history" (Winkler 2008, p. 233).

While Winkler is right to bring Locke's writings from the 1690s to bear on our interpretation of Locke's attitude toward Newton's natural philosophical achievements, his strategy is not without problems. On the one hand, it is not clear why Locke would maintain his emphasis on natural historical methods in all editions of the *Essay* if, as Winkler claims, Locke had in his later career come to embrace Newton's mathematical methods as a viable alternative to the naturalist investigation of bodies. The very fact that there are no extensive revisions to the discussion of natural philosophy in the *Essay* already raises questions about just how far Locke's acceptance of Newton's method extended.[6] On the other hand, and most importantly, the texts to which Winkler appeals do not definitively show Locke trying to defend Newton's *physics*. As I will discuss in greater detail in Section 3.2, Locke lays emphasis on Newton's success as a mathematician (not a mathematical natural philosopher) in the fourth edition of the *Essay*, and in other writings from the 1690s, Newton is praised for his application of mathematics to a very specific domain of nature, namely, the motions of planetary bodies. In light of these qualifications, and given Locke's career-long advocacy of natural historical methods for natural philosophy in the *Essay*, I suggest in what follows that we temper our assessment of Locke's acceptance of Newton's *Principia* methods, and specifically, read Locke's acceptance of Newton's mathematical natural philosophy in light of his naturalist commitments. On such a reading, we find that, *pace* Winkler, Locke took Newton's work to be emblematic, not of a general physics, but of a sub-discipline of natural philosophy dealing only with the forces and motions of heavenly bodies.[7]

To make my case, I draw on Robert Boyle's attempt to reconcile the naturalist mode of investigation with the use of mathematics in natural philosophy, and

6 My thanks to George Smith for emphasizing this point to me at SPAWN 2009.

7 There is, of course, the related question of whether Locke's reading of Newton's *Principia* achievement agrees with how Newton himself understood the significance of his work. While an important issue, I lay focus on how we ought to read Locke's remarks regarding Newton's *Principia* method in light of the account of human knowledge and natural philosophy that we find in the *Essay*. Whether Newton would agree with this assessment is a question that goes beyond the limited scope of this chapter, though at the end of Section 3.4 I offer some suggestions about how to understand the difference between their readings. For an interpretation of how we might contrast the methods of natural philosophy proposed by Newton in the *Principia* and by Locke in the *Essay*, see Section 10.1 of De Pierris (this volume).

specifically, his distinction between the tasks and methods of practical and speculative naturalists. The practical (or experimental) naturalist investigating terrestrial bodies, which we can sense as well as manipulate, is assigned a method of historical observation and experiment as she attempts to collect a store of facts that will inform the construction of principles and theories. In contrast, the speculative naturalist investigating celestial bodies is granted a theoretical method, and specifically, a mathematical method to deepen our understanding of an area of nature where experiment cannot carry us forward. To put this differently, practical naturalists take a "bottom-up" approach to nature and use sense data to fashion a more complete idea of body by reference to the qualities and properties knowable through observation and experiment. Speculative naturalists instead take a "top-down" approach to nature and begin with laws and principles, which rest on observational data collected by the practical naturalist, and then apply these laws and principles to other areas of nature. On my reading, Locke invokes this Boylean framework for natural philosophy, where the proper methods for investigation are determined based on our experimental access to the objects investigated, to reconcile Newton's success with the natural historical methods promoted in the *Essay*.

3.2 Locke on Newton's mathematical natural philosophy

In his effort to show that Locke was both familiar with and an advocate for the "mathematical physics" presented in Newton's *Principia*, Winkler marshals textual evidence from Locke's correspondence with Stillingfleet in the 1690s and the revised fourth (1700) edition of the *Essay*.[8] In the Stillingfleet correspondence especially, we find Locke highlighting the distinctiveness of Newton's mathematical-demonstrative method and promoting it as a method that can enlarge our knowledge of nature. For instance, when Stillingfleet voices worry about the use of demonstrations in natural philosophy, Locke offers a defense of the demonstrative method.[9] Though he agrees with Stillingfleet that *Descartes* was ultimately unsuccessful in establishing the truth of his natural philosophical system, Locke thinks it wrong for Stillingfleet to place blame on Descartes's chosen method. Descartes failed, according to Locke, not because he used demonstrations, but because he lacked demonstrations that adequately captured planetary motions (1823, IV, p. 427).[10] And to support his claim, Locke refers

8 The second and third editions of the *Essay* appeared in 1694 and 1695, respectively.

9 Winkler points out that there are important differences between Stillingfleet's stated position in his letters to Locke and Locke's construal of Stillingfleet's position in his replies (Winkler 2008, pp. 239–241). In my discussion I focus on Locke's sometimes misleading interpretation of Stillingfleet's claims, for it is in the context of his replies to what he interprets as Stillingfleet's objections that we gain insight into Locke's reading of Newton.

10 Excluding references to the *Essay*, citations to Locke's writings refer to the volume and page numbers of the 1823 edition of *The Works of John Locke*.

Stillingfleet to Newton's demonstration, from Book II of the *Principia*, that Descartes's vortex theory of planetary motion is incorrect.[11]

Locke also appeals to Newton's mathematical approach to nature to support his critical evaluation of Scholastic natural philosophy. In response to Stillingfleet's alleged Scholastic worries about using mathematics in natural philosophy, Locke accuses the Scholastics of subscribing to a program of natural philosophy that is no help in combating skepticism and suggests, in a rather strong tone, that they resist a mathematical approach to nature because it is difficult to master.[12] He claims moreover that Newton's success has authorized the use of mathematics in our investigations of material things and finds it "a great pity that Aristotle had not understood mathematics as well as Mr. Newton, and made use of it in natural philosophy with as good success" (1823, IV, p. 427).

Aside from appealing to Newton's *Principia* achievement to challenge Cartesian and Scholastic natural philosophy, Locke also admits to Stillingfleet that, in light of the results presented in "Mr. Newton's incomparable book," he must revise the claim made in earlier editions of the *Essay* that bodies act upon one another only by impulse (1823, IV, pp. 467–468).[13] Locke also revises and expands other sections of the *Essay*, which, according to Winkler, reveal the imprint of Newton's mathematical natural philosophy on its fourth edition. Specifically, Winkler identifies changes to the fourth edition that Locke allegedly made "precisely in order to defend mathematical physics (particularly Newtonian physics) against (what he took to be) Stillingfleet's criticisms" (Winkler 2008, p. 243). Winkler admits that

> This is a conjecture I cannot possibly establish, but the elements of a defense of Newton are undeniably present in the fourth edition, and absent from earlier editions. The controversy with Stillingfleet, which reached its

11 Based on Locke's remarks, it is not altogether clear what Locke thinks Newton actually demonstrated. As Winkler points out, it could be that Newton establishes a conditional claim ("If my assumptions about nature are correct, then Descartes's theory is wrong.") or an affirmative denial of Descartes's vortex theory. Given the strictures of Locke's account of knowledge, Winkler claims that Locke can only sustain the demonstration of the conditional claim (cf. Winkler 2008, p. 237).

12 Locke remarks to Stillingfleet, "Mathematics in gross, it is plain, are a grievance in natural philosophy, and with reason: for mathematical proofs, like diamonds, are hard as well as clear, and will be touched by nothing but strict reasoning. Mathematical proofs are out of the reach of topical arguments, and are not to be attacked by the equivocal use of words or declamations, that make so great a part of other discourses; nay, even of controversies. How well you have proved my way of ideas guilty of any tendency to scepticism, the reader will see; but this I will crave to leave say, that the secluding mathematical reasoning from philosophy, and instead thereof reducing it to Aristotelian rules and sayings, will not be thought to be much in favour of knowledge against skepticism" (1823, IV, p. 428; cited in Winkler 2008, p. 238).

13 I will return to the passage sent to Stillingfleet in Section 3.4 below.

> climax as Locke was contemplating changes in the fourth edition, might well explain their appearance there.
>
> (Winkler 2008, p. 243)

While I agree there are good reasons to connect the changes to the *Essay* that Winkler cites to Newton's *Principia* achievement, upon closer examination, it is not altogether clear that these changes reveal Locke's attempted defense of Newton's *physics* as Winkler maintains.

For instance, Winkler turns our attention to the four pages added to Locke's chapter on maxims. As in his correspondence, Locke pits Aristotle against Newton and presents himself as a proponent of Newton's method. However, he does not reiterate the point made to Stillingfleet that Newton has authorized the use of mathematics in *natural philosophy*. Rather, in the passage from the fourth edition *Essay* that Winkler cites, Newton is presented as a *mathematician* who successfully used a demonstrative method to establish true and certain propositions. Locke writes,

> Maxims are not of use to help Men forwards in the Advancement of Sciences, or new Discoveries of yet unknown Truths. Mr. *Newton*, in his never enough to be admired Book, has demonstrated several Propositions, which are so many new Truths, before unknown to the World, and are farther Advances in Mathematical Knowledge: But for the Discovery of these, it was not . . . general *Maxims* . . . that help'd him. These were not the Clues that lead him into the Discovery of the Truth and Certainty of those Propositions. Nor was it by them that he got the Knowledge of those Demonstrations; but by finding out intermediate *Ideas*, that shew'd the Agreement or Disagreement of the *Ideas*, as expressed in the Propositions he demonstrated.
>
> (IV.vii.11; cited in Winkler 2008, p. 239).

Winkler does not expand on Locke's appeal to "intermediate ideas" here, but we can get a better handle on the method Locke is attributing to Newton by considering Locke's example of demonstrating that the three angles of a triangle are equivalent to two right angles (IV.ii.2). When we attempt to connect these ideas, namely, the idea of "three angles of a triangle" and the idea of "two right angles," the mind "cannot by an immediate view" discover their "Agreement or Disagreement," and thus, in this case, "the Mind has no immediate, no intuitive Knowledge." The mind is therefore "fain to find out some other angles, to which the three Angles of a Triangle have an Equality; and finding those equal to two right ones, comes to know their Equality to two right ones" (IV.ii.2). Though Locke is not explicit, the "intermediate idea" we must "discover" to complete the demonstration is the idea of "angles that lie along a straight line." For by appealing to the properties of parallel lines, we can show that the angles of a triangle are equivalent to the angles lying along a straight line, and since two right angles are supplementary, these angles also are equivalent to angles lying along a straight line. Given the transitivity of the equality relation, we can thus establish that the angles of a triangle are equivalent in measure to two right angles.

In line with the above example as well as Locke's claim that Newton *demonstrated* "new Truths" that are "farther Advances in *Mathematical* Knowledge," what Locke is claiming of Newton is that he used a method of discovering "intermediate ideas" to establish mathematical propositions, not natural philosophical ones. Importantly, this assessment of Locke's remarks is consistent with his pronouncement that demonstrative knowledge is only possible in mathematics and morality (IV.iii.1–7), and not in natural philosophy (IV.xii.10). Admittedly, in the discussion dedicated to improving our knowledge, Locke heralds the method of discovering intermediate ideas as a means of enlarging our knowledge of nature. As Locke puts it in this context, rather than "relying on Maxims, and drawing Consequences from some general propositions" (IV.xii.15) in our natural philosophical endeavors, we should instead "practice the Art of finding out those Intermediate Ideas, which may shew us the Agreement, or Repugnancy of other Ideas, which cannot be immediately compared" (IV.xii.14).[14] On my reading of these remarks, it is not Newton's natural philosophical achievements that Locke is trying to defend. Rather, Locke is praising Newton the mathematician and claiming that Newton's mathematical achievements can and should serve as a model for natural philosophers to emulate, even though truth and certainty are out of reach in this domain.[15]

Further questions arise for Winkler's reading when we consider other texts from the 1690s in which Locke offers his explicit endorsement of Newton's *Principia* achievement. For instance, in Section 43 ("Fundamental Verities") of his posthumously published *Of the Conduct of the Understanding* (1697) Locke reiterates his acceptance of universal gravitation and deems it as "the basis of natural philosophy." He writes,

> There are fundamental truths that lie at the bottom, the basis upon which a great many others rest, and in which they have their consistency. These are teeming truths, rich in store, with which they furnish the mind, and, like the lights of heaven, are not only beautiful and entertaining in themselves, but give light and evidence to other things, that without they could not be seen or known. Such is the admirable discovery of Mr. Newton, that

14 While no specific example of a natural philosophical "demonstration" is offered by Locke, we can assume that at least some of the "intermediate ideas" needed in this context would be ideas drawn from our sensory experience. We would not, that is, simply rely on ideas generated by our study of mathematics to establish natural laws. My thanks to Eric Schliesser for urging me to consider this point more carefully.

15 On this score, I think Winkler has to say more to strengthen his argument against Phemister, who claims that Locke "refuses to countenance demonstration in natural science" (Phemister 1993, p. 243). For Winkler's discussion of Phemister's position, see Winkler (2008, pp. 233 and 237).

> all bodies gravitate to one another, which may be counted as the basis of natural philosophy; which, of what use it is to the understanding of the great frame of our solar system, he has to the astonishment of the learned world shown; and how much farther it would guide us in other things, if rightly pursued, is not yet known. Our Saviour's great rule, that "we should love our neighbors as ourselves," is such a fundamental truth for regulating human society, that, I think, by that alone, one might without difficulty determine all the cases and doubts in social society.
>
> (1823, IV, p. 282; cited in Winkler 2008, p. 243)

According to Locke, universal gravitation can serve as a starting point for understanding nature; it is, in fact, the foundational and "fundamental truth" on which Newton's program rests. But notice as well that, for Locke, Newton's great success was to apply this principle to a specific province of nature, namely, he used universal gravitation to improve our "understanding of the great frame of our solar system." According to Winkler, this claim points to Locke's acceptance of Newton's general physics – his account of both celestial and terrestrial bodies.[16] However, there is strong evidence from *Some Thoughts concerning Education* (1693) that Locke is intentionally distinguishing Newton's work from the investigation of terrestrial bodies.

As Locke presents it in this earlier work, it was via the application of mathematics to "our planetary world and the most considerable phenomena in it" that Newton was able to offer a "good and clear" account of this particular province "of the incomprehensible universe." Locke also explicitly contrasts Newton's *Principia* achievement with "the systems of physics" that he has encountered thus far, claiming that Newton has done what physicists have not: used mathematics to deepen our understanding of "the motions, properties, and operations of the great masses of matter in this our solar system." He writes in section 194,

> Though the systems of *physics*, that I have met with, afford little encouragement to look for certainty, or science, in any treatise, which shall pretend to give us a body of natural philosophy from the first principles of bodies in general; yet the incomparable Mr. Newton has shown, how far mathematics, applied to some parts of nature, may, upon principles that matter of fact justify, carry us in our knowledge of some, as I may so call them, *particular*

16 Though Winkler does not state this explicitly, his reading of this passage and his general characterization of "mathematical physics" point to astronomy as a branch of physics. He claims, in particular, that "By a 'mathematical physics' I mean a physics that makes essential use of mathematics in formulating its fundamental claims, and in deriving other claims – explanations or predictions of particular facts, for example – from them" (Winkler 2008, p. 241). While in general, and from our contemporary standpoint, it is natural to say that the physics of Newton's *Principia* encompasses study of planetary forces and motions, my point above is that, for Locke, physics and astronomy were importantly and essentially different sub-disciplines of natural philosophy in general.

> *provinces* of the incomprehensible universe. And if others could give us so good and clear an account of other parts of nature, as he has of this our planetary world, and the most considerable phaenomena observable in it, in his admirable book "Philosophiae naturalis principia mathematica," we might in time hope to be furnished with more true and certain knowledge in several parts of this stupendous machine, than hitherto we could have expected. And though there are very few that have mathematics enough to understand his demonstrations; yet the most accurate mathematicians, who have examined them, allowing them to be such, his book will deserve to be read, and give no small light and pleasure to those, who, willing to understand the motions, properties, and operations of the great masses of matter in this our solar system, will but carefully mind his conclusions, which may be depended on as propositions well proved.
>
> (1823, IX, pp. 186–187; emphasis added)[17]

On my reading of Locke's remarks, it is the first line of this passage that is most telling, for here, Locke makes clear the contrast between physics and Newton's study of nature. Specifically, what Locke indicates in the passage above is that *physics*, as it is currently practiced, offers little hope for establishing "a body of natural philosophy from the first principles of bodies in general," whereas Newton's mathematical examination of planetary motions gives us reason to believe that, in time, we can "be furnished with more true and certain knowledge in several parts of this stupendous machine, than hitherto we could have expected." In other words, it is Newton's study of the heavens that engenders a hope that the areas of nature studied by the "physicist" can be brought to the level of certainty Newton has already achieved. With these sentiments, Locke is thus pointing not only to Newton's *Principia* as a work of astronomy. He is also pointing to an important, and, I claim, essential difference between the methods and objects of astronomy and physics, which I discuss in further detail below.

3.3 Dividing the labor of natural philosophers

At this point, careful readers of the *Essay* may have suspicions about the reading of Locke's attitude toward Newton's program that I am forwarding, since the division of natural philosophy into physics and astronomy that I claim Locke accepts in the 1690s is not present in the *Essay*. Rather, in all editions Locke promotes an experimental, natural historical program for improving our knowledge of nature. In Book IV in particular, Locke heralds the use of

17 Winkler appeals to this passage in support of his claim that Locke was trying to defend Newton's mathematical physics (Winkler 2008, p. 244), though importantly, he omits the first sentence of the passage where we find Locke contrasting physics with Newton's mathematical program of natural philosophy.

experience, experiment, and natural history as the means by which we can better catalogue the qualities of bodies and in turn, better understand the causal operations between bodies and the causal processes that give rise to the qualities we observe. He says, for instance,

> In the Knowledge of Bodies, we must be content to glean, what we can, from particular Experiments: since we cannot from a Discovery of their real Essences, grasp at a time whole Sheaves; and in bundles, comprehend Nature and Properties of whole Species together. Where our Enquiry is concerning Co-existence, or Repugnancy to co-exist, which by Contemplation of our *Ideas* we cannot discover; there Experience, Observation, and natural History, must give us by our Senses, and by retail, an insight into corporeal Substances. The Knowledge of Bodies we must get by our Senses, warily employed in taking notice of their Qualities, and Operations on one another.
>
> (IV.xii.12)

In a similar vein, Locke remarks that crafting "rational and regular experiments" and amassing "historical observations" will allow us greater (albeit not perfect) insight into the properties of those bodies we are investigating. Though these methods ultimately prohibit us from establishing a perfectly scientific natural philosophy characterized by demonstrative knowledge and certainty, they are the best we have at our disposal:

> I deny not, but a Man accustomed to rational and regular experiments shall be able to see farther into the Nature of Bodies, and guess righter at their yet unknown Properties, than one, that is a Stranger to them; But yet, as I have said, this is but a Judgment and Opinion, not Knowledge and Certainty. This *way* of getting, and *improving our Knowledge in Substances only by Experience* and History, which is all that the weakness of our Faculties in this State of *Mediocrity*, which we are in in this World, can attain to, makes me suspect, that natural Philosophy is not capable of being made a Science. We are able, I imagine, to reach very little general Knowledge concerning the Species of Bodies and their several Properties. Experiments and Historical Observations we may have, from which we may draw Advantages of Ease and Health, and thereby increase our stock of Conveniences for this Life: but beyond this, I fear our Talents reach not, nor are our Faculties, as I guess, able to advance.
>
> (IV.xii.10)

According to Anstey (2002, 2003), the above pronouncements in Book IV of the *Essay*, as well as evidence from Locke's own investigations into the properties of air and blood, reveal that Locke accepted the construction of natural histories as *constitutive* of natural philosophy (Anstey 2002, p. 68; Anstey 2003, p. 27). That is, he promoted and employed a method of historical observation that, following Bacon and Boyle, demanded the compilation of "vast collections of facts about particular objects or qualities" (Anstey 2002, p. 71). These collections of facts in turn supply the foundation both for our knowledge of particular things and for the construction of "true axioms" that express causal explanations for the qualities of and interactions between bodies, which we witness in nature.[18]

Though I agree with Anstey that claims made in the *Essay* support Locke's embrace of natural historical methods, Locke's claims from the 1690s reveal that natural history is constitutive of one arena of natural philosophy. Namely, natural historical methods ought to be used when investigating those objects on which we can conduct experiments, that is, when doing physics. When, in contrast, we investigate bodies on which we cannot conduct experiments, as Newton does when investigating the forces and motions of planetary bodies, Locke opens the door for the natural philosopher to adopt different methods of inquiry.

To be clear, I am not suggesting that Locke took a step away from his naturalist commitments. Rather, I am suggesting that Locke came to embrace a general framework for natural philosophy set out by Robert Boyle in the early 1660s, a time when, Milton (1994) notes, "Locke seems to have been reading everything that Boyle was having published" (Milton 1994, p. 37).[19] During this period, as Boyle attempted to account for the use of mathematics in natural philosophy, he developed an important distinction between the methods and objects of the "practical" and "speculative" naturalists. As I argue in Section 3.4, it is this distinction that we see at play in Locke's later writings as he was trying to meet the challenge Newton's *Principia* posed to his earlier account of natural philosophy.

18 My account of natural history is drawn from Anstey (2002), whose more detailed account relies on the "Preparative toward a Natural and Experimental History" that Bacon appended to the *Novum Organon* (cf. Anstey 2002, pp. 71–72). Anstey's overall aim in Anstey (2002) is to establish the influence of Bacon's account of natural history, as filtered primarily through the writings and work of Boyle, on Locke's interpretation of how to properly practice natural philosophy. While I do not dispute the importance of Bacon's work for Locke, I lay emphasis below on the impact of a distinctively Boylean feature of natural history on Locke's understanding of natural philosophy. As a note, it is to Anstey's work that Winkler appeals when he admits that the primary image of natural philosophy presented in the *Essay* is natural-historical rather than mathematical (Winkler 2008, p. 232). For a reading of Locke's method for natural philosophy in the context of his embrace of a deductive ideal for scientific knowledge, see Section 10.1 of De Pierris (this volume).

19 Boyle was the natural historian who arguably exerted the greatest direct influence on Locke's account of natural philosophy: Locke shared a close professional association with Boyle, and Boyle was the most well represented author in Locke's library. Recent commentators who discuss the impact of Locke's connection to Boyle on his philosophical outlook include (but are certainly not limited to) Alexander (1985), Anstey (2002), Rogers (1966), Stewart (1981), and Woolhouse (1971).

Now, according to Boyle, the practical (or experimental) naturalist and the speculative (or theoretical) naturalist play complementary roles in natural philosophy: the practical naturalist conducts experiments and gathers empirical evidence, while the speculative naturalist posits hypotheses concerning the causal interactions that give rise to what we observe.[20] As Francis Bacon before him, Boyle emphasizes that the order of events is crucial to the success of natural philosophy: the key is to put the practical and speculative naturalist in proper conversation with each other.[21] And as his naturalist predecessor, Boyle grants the practical naturalist the lead role so that the speculative naturalist can formulate better, more accurate causal hypotheses. Thus, in the Preface to "A Defense of the Doctrine Touching the Spring and Weight of the Air" (1660), Boyle explains his own role as a practical naturalist:

> For first, as I elsewhere declare, it was not my chief design to establish theories and principles, but to devise experiments, and to enrich the history of nature with observations faithfully made and delivered; that by these and the like contributions made by others, men may in time be furnished with a sufficient stock of experiments to ground hypotheses and theories on.
>
> (Boyle 1772, I, p. 121)[22]

The same sentiment is repeated in his "Proëmial Essay" (1661), where Boyle admits his "disabilities" as a speculative naturalist and appoints himself as a practical "under-builder" who will establish natural histories that can serve as the foundation for those more adept at positing causal hypotheses.[23]

20 See Anstey (2005) for a nice overview of the different ways the distinction between "practical" (or "experimental") and "speculative" natural philosophers was used during the second half of the seventeenth century.

21 My reading of the ties between Bacon and Boyle is indebted to Rose-Mary Sargent's very important work on this topic (see Sargent 1994, 1995). Sargent draws three important parallels between the naturalism of Bacon and Boyle: both inverted the order of proof and discovery, both accepted the link between knowledge and power, and where Bacon promoted the marriage of "rationalism" and "empiricism," Boyle promoted the marriage of the "speculative" and "practical" parts of learning (Sargent 1994, pp. 58–59).

22 Citations to Boyle's writings refer to the volume and page numbers included in *The Works of Robert* Boyle (1772).

23 In a modest tone, he writes, "I have often found such difficulties in searching into the cause and manner of things, and I am so sensible of my own disability to surmount those difficulties, that I dare speak confidently and positively of very few things, except matters of fact. And when I venture to deliver any thing, by way of opinion, I should, if it were not for mere shame, speak more diffidently than I have been wont to do . . . I am content, provided experimental learning be really promoted, to contribute even in the least plausible way to the advancement of it; and had rather not only be an under-builder, but even dig in the quarries for materials towards so useful a structure, as a solid body of natural philosophy, than not do something towards the erection of it" (Boyle 1772, I, p. 307).

The division of labor that Boyle establishes for the practical and speculative naturalists informs his attitude toward the integration of mathematics with natural philosophy.[24] Boyle places clear limits on how we should integrate mathematics with our study of nature and resists an approach to nature that takes a mathematical view of natural bodies as its starting point. For instance, he cautions that

> we must not expect from mathematicians the same accurateness, when they deliver observations concerning such things, wherein it is not only quantity and figure, but matter and its other affections, that must be considered.
>
> (Boyle 1772, I, p. 347)

He elaborates further that

> the phenomena which the mathematician concurs to exhibit, do really belong to the cognizance of the naturalist. For when matter comes once to be endowed with qualities, the consideration how it comes by them, is a question rather about the agent or efficient, than the nature of the body itself.
>
> (Boyle 1772, III, p. 427)[25]

Nonetheless, Boyle recognizes the usefulness of mathematics for natural philosophy as detailed in his aptly titled "Of the Usefulness of Mathematicks for Natural Philosophy" (1663; Boyle 1772, III, pp. 425–434). At the very outset of the tract, he notes that intense study of mathematics can "much improve reason, by accustoming the mind to deduce successive consequences" (Boyle

24 Boyle's attitude toward the integration of mathematics with natural philosophy is also informed by his nominalist commitments. For more on this issue, see McGuire (1972), who suggests that in subscribing to an ontology of particular entities and rejecting the reification of concepts, Boyle "dissociates himself from the Paracelsians, the Helmontians, the Cambridge Platonists, and by implication from Newton himself who was very much in the tradition of the Platonists" (McGuire 1972, p. 528). As Eric Schliesser has stressed to me, given Boyle's distaste for the mathematical natural philosophy of the Cartesians, Descartes should also be included on McGuire's list.

25 Following Shapin (1988), the hesitancy Boyle voices towards mathematical natural philosophy in these passages is intertwined with Boyle's embrace of a "particularist" ontology according to which "[w]hat had real physical existence in nature were particulars: particular things, particular bodies, particular events" (Shapin 1988, p. 39). In other words, because of Boyle's commitment to an ontology of particulars, "the legitimate search for natural regularities implied . . . the use of observational and experimental methods and the exclusion or limitation of rationalist practices" (*ibid.*, p. 39). In this respect, "mathematical representations of reality pointed to an improper ontology" (*ibid.*, p. 33), because there are, for Boyle, no natural entities that correspond to the abstract a priori objects employed by the mathematicians. Nonetheless, as we will see immediately below, there are cases where Boyle does in fact promote the use of mathematics in natural philosophy, namely, in our investigations of celestial bodies.

1772, III, p. 426) – a skill, of course, required by the *speculative* naturalist in his attempts to establish causal laws that hold between natural bodies. However, the *practical* naturalist can also reap rewards from mathematics by appealing to the diverse properties of figures and the proportions that hold between them as she endeavors to construct experiments. In general, Boyle urges that the doctrine of proportions assume a central place in the practical naturalist's examination of observational data, for "as it is the soul of the mathematics themselves," it not only "helps the naturalist . . . to understand diverse phenomena of nature, . . . it may enable him to perform diverse things, which he could not perform without it" (Boyle 1772, III, p. 432).

Boyle's strongest support for the use of mathematics is voiced in his discussion of astronomy, where he endorses the use of a mathematical method to chart the motions and positions of heavenly bodies. On his score, we will only be able to decide whether the Ptolemaic or the Copernican system is the true system of the world by appealing to the doctrine of the sphere and applying a mathematical framework to the heavens (Boyle 1772, III, p. 429).[26] And as his remarks indicate, mathematics can be of greater help in this arena, because using mathematics, we are able to to catalog the positions and motions of distant celestial bodies without relying on hypotheses about the true nature of these remote bodies or the causes of their motions. In other words, mathematics enables the astronomer to generate a map of the heavens that can serve as a reliable, metaphysics-free basis upon which causal hypotheses can be established. He writes,

> That then the knowledge of celestial bodies is not well to be attained, nor consequently the theories proposed of them, to be intelligently judged of, without arithmetick and geometry (those wings on which the astronomer soars as high as heaven) he must be very little acquainted with astronomy, and particularly with the various and too often intricate theories of planets, that can doubt. And truly, when I consider the astonishing distance and immensity of the celestial bodies, and those almost numberless fixed stars (each of them perhaps much vaster than the whole earth) which in a clear night I take pleasure to gaze at through the better sort of telescopes, both

26 He writes, "indeed what satisfactory account can be given of the varying lengths and vicissitudes of days and nights, and eclipses of the sun and moon, the stations and retrogradations observed in planets, and other familiar celestial phenomena, without supposing these great mundane bodies to have such situations in respect to one another, and to move in such lines, or at least to be made to appear to move in them by the motion of the earth in such a position, and in such lines? Nay, how without the knowledge of the doctrine of the sphere will the naturalist be able to make any sober and well grounded judgment in that grand and noble problem, which is the true system of the world? Which is endeavored to be solved after such differing manners by the Ptolomaens and Peripateticks, by the Tychonians and by the Copernicans, both less and more modern" (Boyle 1772, III, p. 429).

> in the milky way, and in other parts of the sky, that seem not so much as whitish to our eyes; I cannot but highly prize a science that acquaints us, that what we know of so much of the universe as the globe we inhabit and call the world, is but a point to it, taking up a little more room in it, than physical center in the sphere.
>
> (Boyle 1772, III, p. 429)

As indicated by these remarks, the proper methods of astronomy are determined by the types of objects under investigation. Given the "astonishing distance and immensity of the celestial bodies," and specifically, their distance from our senses, we should not speculate about the metaphysical constitution of these bodies in our astronomical pursuits (as, for instance, Scholastic natural philosophers do when they propose the fifth element of ether to explain planetary motion). Rather, we should rely on what we can observe of these bodies, namely, their shapes and positions, and use these observations as the foundation of our investigations. And, as Boyle has it, these observations will best serve the speculative naturalist, who is attempting to establish the "true system of the world," if they are rendered in mathematical form. For just as geography offers a helpful mathematical framework for navigation (Boyle 1772, III, pp. 429–430), so too does astronomy offer a mathematical, metaphysically neutral framework that allows us to understand our terrestrial place among the stars.

According to Boyle's discussion of the practical and speculative naturalists, the type of object under investigation helps determine the extent to which mathematics can be fruitfully employed in natural philosophy. When we investigate terrestrial objects with which we have direct sensory access and upon which we can conduct experiments, the practical naturalist ought to remain cautious about reducing these bodies to their quantitative features. However, when we investigate celestial objects that are more removed from our senses and upon which we cannot conduct experiments, the practical naturalist can catalog the observed positions and motions of these bodies more effectively when these positions and motions are placed in a mathematical framework. For without immediate or direct sensory contact with these objects, geometry and arithmetic – "those wings on which the astronomer soars as high as heaven" – allow the astronomer to map the positions and motions of these heavenly spheres without also having to speculate about the metaphysical constitution of the objects under investigation.[27]

27 In this respect, Boyle appears to share Bacon's general stance toward the relationship between observation, mathematics, and astronomy. According to Bacon – and consistent with his pronouncement in the *Novum Organon* that "inquiries into nature have the best result when they begin with physics and end in mathematics" (II, p. 8) –, the natural historian ought to take a lead role even in our investigations of the heavenly bodies. Though Bacon admits that the current success of astronomy "owes little to observation

3.4 Locke's qualified embrace of Newton's mathematical natural philosophy

Though Locke does not explicitly employ Boyle's distinctions between the practical and speculative naturalists or appeal to the discussion of mathematics in the 1663 "Of the Usefulness," Boyle's remarks can help us make better sense of Locke's simultaneous advocacy for the natural history model of natural philosophy and Newton's mathematical model of natural philosophy. For texts from the *Essay* and Locke's later writings suggest that the natural historical method is the method for the practical naturalist investigating terrestrial bodies, whereas the mathematical-demonstrative method is the method for the speculative naturalist investigating celestial bodies.

Notice, for instance, that when Locke endorses natural historical methods in Book IV of the *Essay*, the natural investigations he describes take our sensory access to nature as their starting-point, and the methods Locke recommends – observation, experiment, and natural history – are methods that enable us to improve and better organize the data that our senses report. These data, in turn, enable the naturalist natural philosophy to "see farther into the Nature of Bodies, and guess righter at their yet unknown Properties" (IV.xii.10), that is, they enable the naturalist to craft a more complete idea of body and thereby enlarge our knowledge of nature (IV.xii.14). The insight we gain about the nature of corporeal substances by applying the naturalist method is thus not grounded on conjectures or principles;[28] it is rather focused on cataloging the properties of those bodies we can observe and on which we can conduct experiments. In other words, it is a "bottom-up" project assigned to the practical naturalist who aims to gather and organize empirical evidence that will inform the construction of principles and theories.

Importantly, Newton's methods and aims are not couched in the same terms. First and foremost, as we saw above, Newton's work is characterized by Locke as a work of astronomy – as an investigation of bodies which are more distant from our senses. Certainly, we can observe these bodies and carefully chart their positions, but a method of "rational and regular *experiment*" will be of no help in this domain. And thus we find Locke situating Newton among the

and axioms of nature," he suggests that astronomy, like all other sciences, can be improved by assimilating it with the observations and axioms of the naturalist (I, 85). For more on Bacon's attitude towards astronomy, see Rees (1986).

28 Locke does make room for the use of hypotheses when describing naturalist natural philosophy in the *Essay* (IV.xii.13), but following Anstey (2003), Locke is highly cautious and limits the use of hypotheses. They are, for Locke, aids for our memory and can, when used correctly, direct us to new discoveries (cf. Anstey 2003, pp. 31–33). As such, my main point above remains: the naturalist natural philosopher is urged not to begin her investigations with hypotheses and must take care not to ground her knowledge claims on anything other than the evidence gained via sensory interaction with natural bodies.

speculative naturalists that Boyle describes, that is, as a natural philosopher who takes a theoretical, "top-down" approach to celestial bodies and begins his investigation of planetary motions with laws and principles in hand.

For instance, in his *Elements of Natural Philosophy* (1698ff.), Locke remarks that "It appears, as far as human observation reaches, to be a settled *law of nature*, that all bodies have a tendency, attraction, gravitation towards one another"; this "fact is made evident to us by experience" and, Locke claims, can safely be taken as "a *principle* in natural philosophy" (1823, III, pp. 304–305; emphasis added). He claims as well in *Of the Conduct of the Understanding* (1697) that "all bodies gravitate toward one another" is included among the "fundamental truths that lie at the bottom, the basis upon which a great many others rest." Locke's descriptions of gravitation as a "fundamental truth" and "principle" of natural philosophy are important, because they help separate Newton from the practical naturalist. While the practical naturalist aims to increase the store of "facts" regarding the particular bodies and qualities in nature, and should do so using the natural historical method outlined in Book IV of the *Essay*, Newton instead establishes a law of nature – the law of universal gravitation – based on empirical evidence, namely, based on his empirically derived laws of motion, and then uses this law to explain other natural phenomena. That is, on Locke's reading, Newton establishes and accepts gravitation as a "fundamental truth," and this truth, or principle, then guides his further investigations of nature.[29]

This reading of Newton's method also helps us make sense of why Locke suggests to Stillingfleet that Newton's proposal of universal gravitation is not grounded on the same sort of empirical evidence that we use to improve our idea of body. Whereas the practical naturalist is focused on the properties of bodies that we learn via sense experience, Newton proposes a law that helps make sense of how bodies interact, and also grants us a broader view of how God may govern bodies than our senses and understanding may indicate. Thus, in explaining why he is willing to revise his understanding of how bodies act upon one another, Locke writes:

> It is true, I say [at *Essay* II.viii.8], "that bodies operate by impulse, and nothing else." And so I thought when I writ it, and can yet conceive no other way of their operation. But I am since convinced by Mr. Newton's incomparable book, that it is too bold a presumption to limit God's power, in this point, by my narrow conceptions. The gravitation of matter towards matter, by ways inconceivable to me, is not only a demonstration that God can, if he pleases, put into bodies powers and ways of operation, above

29 The proposal of attraction-at-a-distance was, of course, historically contentious. Even Newton himself appears to back away from the proposal, or at least become more cautious about action-at-a-distance, in the letters written to Bentley during 1692 to 1693. For more on this issue, see Schliesser (2010b).

> what can be derived from our idea of body, or can be explained by what we know of matter, but also an unquestionable, and every where visible instance, that he has done so. And therefore in the next edition of my book, I shall take care to have that passage rectified.
>
> (1823, IV, pp. 467–468; cited in Winkler 2008, 243)

What is striking about Locke's remarks is that he is contrasting the notion of attraction proposed by Newton with what we learn about bodies from our sensory interaction with matter. As he suggests, to claim that bodies share a gravitational attraction is to claim at the same time that "God can, if he pleases, put into bodies powers and ways of operation, *above what can be derived from our idea of body, or can be explained by what we know of matter*" (emphasis added). In line with the interpretation I forward above, Locke is suggesting that Newton starts with a principle of nature – the law of attraction – that is not grounded simply on what we learn about the properties and qualities of matter from our sensory experience.[30] While there is empirical evidence that supports Newton's assumption of gravitational attraction, Newton has taken a step beyond what careful observation reveals about those bodies before our senses. He has, in particular, proposed a principle of nature that places these bodies into a certain kind of causal relation and thereby grants us insight into how God may govern these very bodies.[31]

30 Here I depart from Winkler, who takes the passage to Stillingfleet as evidence that Locke credits Newton with improving our idea of body by discovering that bodies obey an inverse-square law (cf. Winker 2008, p. 242). My reading of this passage does agree with that presented by Downing (2007, p. 375).

31 We also notice in these remarks Locke's willingness to extend gravitation to terrestrial bodies as he claims that there are "every where visible" instances of attraction in nature. However, the emphasis of my treatment remains: on Locke's reading, Newton adopted a "speculative" methodology in his astronomy, one that commences with the proposal of a "fundamental truth" of nature. In this respect, the contrast between Lockean "physics" and Lockean "astronomy" remains intact. However, to be clear, I am not attributing to Locke the claim that Newton has successfully transformed natural philosophy into a demonstrative science. Even though, on Locke's reading, Newton has established universal gravitation as a "principle" and "fundamental truth" of natural philosophy, his results will still retain the status of "judgment and opinion," because he is applying his claims to imperfectly known natural bodies. Reiterating the point originally made in the *Essay* (IV.xii.10), Locke states in Section 190 of *Some Thoughts Concerning Education* (1693/1989), entitled "Natural Philosophy," "Natural philosophy, as a speculative *science*, I imagine, we have none; and perhaps I may think I have reason to say, we never shall be able to make a science of it. The works of nature are contrived by a wisdom, and operate by ways, too far surpassing our faculties to discover, or capacities to conceive, for us ever to be able to reduce them into a science. Natural philosophy being the knowledge of the principles, properties, and operations of things, as they are in themselves, I imagine there are two parts of it, one comprehending spirits, with their nature and qualities [i.e., metaphysics]; and the other bodies" (1823, VII, p. 182; emphasis added).

Appreciating the qualifications Locke uses to characterize Newton's *Principia* achievement, we are in a better position to see how Locke can maintain his simultaneous commitment to a natural historical model of natural philosophy and Newton's mathematical-demonstrative natural philosophy. Since on my reading Locke was defending Newton's mathematical *astronomy*, not his mathematical physics, Newton could use a speculative, mathematical method for investigating the motions of heavenly bodies that does not violate the naturalist commitments presented in the *Essay*.[32] While the *practical* naturalist should not invoke hypotheses and theories as she observes the particular bodies and qualities in the *terrestrial* world, the speculative naturalist is assigned the very task of proposing hypotheses and theories in order to establish laws of nature based on the evidence amassed by the practical naturalist. And in line with Boyle's remarks, the speculative *astronomer*, who examines objects on which experiments cannot be conducted, should invoke a mathematical system in order to better chart their positions and motions. For in this arena, where experiment cannot help natural philosophy progress, mathematics offers a way of reaching a better understanding of the solar system that does not force the astronomer to make unnecessary metaphysical assumptions. On my reading, this is precisely what Locke's Newton does: he first establishes the law of universal gravitation as a "fact . . . made evident to us by experience" and then, by accepting this law as a "fundamental truth," applies mathematics to the heavenly realm of nature and thereby provides us a "good and clear account . . . of this our planetary world, and the most considerable phaenomena observable in it."[33]

32 On this point I depart from Stein (1990), who claims that Locke's acceptance of the truth of universal gravitation, as evidenced in the Stillingfleet correspondence, "suggests that [Locke] is less firmly committed to his 'official' epistemology (and metaphysics)" as presented in the *Essay* (Stein 1990, p. 33). My claim is that by appeal to Boyle's general framework for natural philosophy, Locke can consistently accept universal gravitation as a "fundamental truth" of natural philosophy, as qualified above, and maintain the stated doctrines of the *Essay*.

33 While a detailed treatment of the *accuracy* of Locke's reading goes well beyond the scope of this chapter, some brief remarks are certainly in order. To some readers, the reading I attribute to Locke may seem extremely misguided if not categorically wrong, since now, some 300 years after the first edition *Principia* appeared, it is standard to look to Newton's masterpiece as a work that collapsed the divide between the celestial and terrestrial. In other words, from our contemporary standpoint, Newton was not doing astronomy but a general physics and mechanics that encompassed all the motions and forces in nature. However, what we have to bear in mind is that Locke had access only to the first edition *Principia*, and immediately after its publication, the full significance of Newton's achievement was not yet appreciated. For instance, Locke's qualified embrace of Newton's use of gravitation receives some support from the position taken by Christiaan Huygens, who was not convinced that, in the first edition *Principia*, Newton had succeeded in confirming the existence of *universal* gravitation. Namely, Huygens claimed, on empirical grounds,

3.5 Conclusion: revisiting the mystery of the "Under-Labourer" Passage

What we have then from Boyle and, I claim, also from Locke, is a general framework for natural philosophy where the proper methods for investigation are determined based on our sensory access to the objects investigated. The practical naturalist investigating terrestrial bodies, which we can sense as well as manipulate, is assigned a method of historical observation and experiment as she attempts to collect a store of facts that will allow us to improve our ideas of bodies. In other words, their "historical observations" help ground our systems of physics on the "first principles of bodies in general" – on the qualities and properties that all bodies seem to share. As such, the practical naturalist takes a "bottom-up" approach to nature and uses sense data to fashion a more complete idea of body by reference to the qualities and properties gathered from observation and experiment. In contrast, the speculative naturalist, on the other hand, takes a "top-down" approach to nature and begins with laws and principles, which rest on observational data collected by the practical naturalist, and then applies these laws and principles to other areas of nature. And as we see in both Boyle's comments about the speculative naturalist investigating celestial bodies and Locke's characterization of Newton's success in natural philosophy, the speculative naturalist investigating *celestial* bodies on which experiments cannot be conducted is granted a theoretical and mathematical method to deepen our understanding of an area of nature where experiment cannot carry us forward.

Adopting this general framework for natural philosophy allows Locke to embrace the mathematical-demonstrative method Newton employed in the *Principia* while still holding firm to the natural historical commitments voiced in the *Essay*. It also allows us to address the apparent mystery posed by the "Under-Labourer" passage with which I began the paper. On the face of it, the association of Newton with Boyle – a mathematical natural philosopher with a naturalist natural philosopher – seems odd, if not grounded on some gross misunderstanding of Newton's work in the *Principia*. However, as I have argued, the key to unlocking this puzzle is to resist trying to locate a common feature of the methods which Newton and Boyle employed. The key is rather to focus on Locke's claim that they are "Master-Builders" who are "advancing the

that Newton's first edition succeeded in establishing only inverse-square *celestial* gravity, not terrestrial gravity. (See, for instance, on this issue, Schliesser and Smith (forthcoming) and Section 2 of Maglo (2003).) When later editions emerge (the second in 1713 and the final third edition in 1726), the evidence for terrestrial gravity is strengthened, as detailed in Smith (2002a) and (2002b); however, these are editions to which Locke did not have access (he died in 1704).

Sciences" – in the plural. On my reading, Locke's Newton and Locke's Boyle are advancing different sciences – Newton astronomy and Boyle physics. To place them together as Master-Builders is thus not to attribute to them a common method or even a common domain of inquiry. It is rather to underscore that each has mastered the methods appropriate to their chosen sub-discipline of natural philosophy.

4

What geometry postulates

Newton and Barrow on the relationship of mathematics to nature

KATHERINE DUNLOP

An outstanding challenge facing Newton scholars is to understand his Preface to the *Principia* as a unified introduction to the work. The first half of the Preface explains how geometry is related to mechanics: it is a "part of universal mechanics." In the second half of the Preface, Newton asserts the relevance of mechanics to natural philosophy. He outlines how he will solve "the whole difficulty of natural philosophy," which is "to discover the forces of nature from the phenomena of motion." He will first (in Books I and II) demonstrate propositions mathematically, then (in Book III) use them to derive actual gravitational forces from celestial phenomena. He can be taken to hint that forces are already "discovered from phenomena" in Books I and II when he says of Book III only that it "illustrates their results" as it explains [*explicare*] the "system of the world" (p. 382). To find unity in the Preface, we must relate Newton's discussion of the relationship between geometry and mechanics to his approach to forces in Nature.

What emerges most clearly from these remarks is Newton's opposition to Descartes. Descartes famously failed to derive mathematical laws of motion from the "mechanical" interactions (characterized more precisely in Section 4.3.2 below) to which his science limited itself. Newton promises to discover fundamental interactions from quantitative relations of motions and forces, deduced in the first two Books. Newton's opposition to Descartes's classification of curves, which features in important recent commentary, provides one link between the Preface's halves. In (1637) Descartes claims that "mechanical" curves – those not generated by sufficiently determinate motions – are *not* "geometrical," meaning that their magnitude cannot be known exactly. The distinction is inimical to Newton's project because it curtails mathematical treatment of the phenomena represented by curves.

I wish to acknowledge very helpful comments by Helen Hattab, Andrew Janiak, Charles Larmore, George E. Smith, and especially Douglas Marshall and Eric Schliesser.

Without taking away from the importance of Descartes for Newton's remarks, I wish to consider them in relation to another historical figure, Isaac Barrow. Newton claims that geometry is specifically "that part of universal mechanics which reduces the art of measuring to exact propositions and demonstrations" and thus has its "foundation" in "mechanical practice." Before him, Barrow contended that the physical import of mathematical sciences is guaranteed by their postulates. My thesis is that in emphasizing the practice that grounds the postulates, Newton pursues Barrow's strategy for demonstrating mathematics' relevance to causality. He can thus make good on his bold claims for the "mathematical" treatment of force contained in the *Principia*'s first two Books.[1]

4.1 Synopsis of the chapter

Barrow contends that mathematics qualifies as science by (broadly) Aristotelian standards. He is thus obliged to show that the properties it invokes in demonstration function as causes of its conclusions. He argues (as I explain in Section 4.2.1) that the (so-called auxiliary) constructions used in geometrical proof are just the processes that give rise to its objects, and so cause the objects to have the properties shown of them. To make clear that these processes are presupposed as a foundation, he emphasizes that they occur in geometry's definitions and postulates. But definitions and postulates as Barrow understands them pertain to a wide range of activities, not limited to the generation of objects. Barrow introduces a notion of "formal" causality (which would not be recognized by an Aristotelian), on which all of these qualify as causes. He maintains that this notion is preferable to that of efficient causation for purposes of scientific explanation. Barrow further argues, although his purpose does not require it, that objects answering to geometrical conceptions can be found *in nature* through the activities enjoined by postulates. In Section 4.2.2, I present his case and suggest that his liberal understanding of postulates leads to difficulty. Where the activities licensed by postulates do not first create objects, it is not clear how they ensure the objects' conformity to geometrical conceptions.

Section 4.3.1 presents evidence of the relevance of Barrow's views for Newton's Preface. Beginning in Section 4.3.2, I explain how Barrow's views are useful for Newton's purposes. I first outline a difficulty faced by Newton. As Andrew Janiak has made especially clear, Newton's readers would expect him either to explain gravity in terms of interactions at the surface of bodies, or to retract his claim to have discovered the cause of motions. To escape the dilemma, Newton

1 To be sure, Newton later distinguishes the content of Books I and II from "physical" argument, and claims that only the latter can establish the "physical varieties and relationships" and "physical causes and seats" of forces (Def. VII, p. 407) or attribute them "in a true and physical sense" to loci (p. 408). But he does not pronounce on the question whether mathematical argument is capable of relating forces in any way to effects.

claims to pursue "mathematical" investigation of force – specifically to *measure* things, to geometry's standards of precision – while deferring "physical" investigation of the conditions under which they are possible. An important difference between his view and Barrow's is that Barrow aims to supplant the prevailing conception of causality, while Newton leaves open that the conception assumed by his readers will apply at a later stage of investigation. Still (as I argue in Section 4.3.3), Newton follows the precedent set by the division of labor between geometry's first principles, as both he and Barrow understand them. What Newton thus derives from Barrow is at least a way to distinguish his approach from those that then dominated natural philosophy. This demarcation of research topics has also been singled out as the element that gives Newton's method its power.[2]

Next, I consider how Newton can justify his claim to capture causal relationships. Because Newton frankly acknowledges that mathematical results can fail to correspond to Nature,[3] he is obliged to find conditions under which a result can be regarded as physically significant. I draw on George E. Smith and William Harper's accounts of how, according to Newton, phenomena can be taken as measures of gravitational force. Smith, in particular, finds Newton to be deeply concerned with the conditions on scientific measurement. My aim is to embed Newton's procedure still more deeply in a philosophical context: in, namely, reflection on the way geometry's first principles secure physical significance for its conclusions. In Section 4.3.4, I show that the approach attributed to Newton by these scholars exemplifies the use of postulates in geometry, as conceived by Barrow. In Section 4.3.5, I argue that on Newton's more precise understanding of postulates, they justify treating results as measurements. I hope thereby to make clear how the beginning of the Preface, with its focus on postulates as the link between geometry and mechanics, lays the ground for the results that follow.

4.2.1 Barrow on the definition of geometrical objects

In his first year as Lucasian Professor at Cambridge, Barrow gave a series of lectures "aimed at reviving interest in mathematics" (Mahoney 1990, p. 181). The content of the lectures is what would now be recognized as foundations of mathematics. They primed the audience not to do mathematics, but to consider

2 By I. Bernard Cohen, to take one prominent example. See for instance his (1980, p. 75). Similar views are expressed by Charles Larmore (1987, p. 94) and Patrick Suppes (1962, p. 118).

3 Most explicitly in a draft of Book II of the *Principia*, published in 1728 as *A Treatise on the System of the World*. Newton contrasts "mathematical" and "natural" points of view, claiming that the "distinction between" attracted and attracting bodies belongs "more" to the former than the latter. (Thanks to Eric Schliesser for calling my attention to this passage.)

it "as a body of learning and as a mode of reasoning," thus "fixing its place and role in the catalogue of the sciences" (Mahoney 1990, p. 183).

Barrow's overarching goal is to show that metaphysics qualifies as a science by traditional standards. These criteria, which derive from the *Posterior Analytics*, concern the manner in which conclusions are drawn. They require inferences to be explanatory in at least two ways. The principles of the science must be more general than its conclusions, and whatever is deductively prior must also be "prior in the order of nature," meaning that it functions as the (formal, material, final, or efficient) cause of its consequence. In an Aristotelian framework, inferences can be rigorously expressed only as syllogisms. Thus, to demonstrate a conclusion with scientific rigor was to link it to the major premise by a "middle term" that functions as its proximate cause. Typically, the middle term involves properties that belong to the subject (of which the conclusion is predicated) in virtue of what it is. Middle terms thus function as answers to the question of what a thing is, i.e. (real) definitions; they articulate essence, in the sense of an inner principle that explains all properties of the object that are relevant for the science. (However, they cannot be assumed to pertain to natures, i.e. the inner principles *of change* that comprise the subject-matter of *natural* science.)

At the basis of Barrow's defense of the scientific status of mathematics is his conception of definition. He points out that "Mathematicians do sometimes define [magnitudes] and deduce their properties" by "their Generations." In these cases, names are affixed to the objects that result from certain processes. For example, a circle is defined as the figure "described by the carrying about of a Right Line one of whose extremes is fixed" (p. 61) as follows: "*The plane Figure which is produced from the Rotation of a Right Line may be called a* Circle: Or a Circle *is a plane Figure which is produced by the Circumduction of a Right Line*" (p. 129).

Barrow's point is that the constructions that occur in definitions of this kind, and are therefore essential to the objects, are also the steps by which geometrical proof proceeds. In the preceding century, Jesuit writers familiar to Barrow had argued that mathematical argument is not fully rigorous in form, in particular, that the middle terms of mathematical syllogisms do not relate the conclusions to the objects' essences.[4] They took as an example the proof that the internal

4 The debate over the scientific status of mathematics was shaped by the arguments of Alessandro Piccolomini. In his 1547 *Commentarium de certitudine mathematicarum disciplinarum*, Piccolomini challenges the view, common to Averroës and Latin commentators on Aristotle, that the reasoning used in mathematics is so rigorous as to place it first among the sciences in certainty. His attack was carried further by Benedictus Peyrera in *De communibus omnium rerum naturalium principiis et affectionibus* (1576) and answered by Christopher Clavius in his commentary on Euclid's *Elements* (1591) and Josephus Blancanus (Giuseppe Biancani) in *De Mathematicarum Natura Dissertatio* (1615), all fellow Jesuits. On this debate, which became known as the *Quaestio de Certitudine Mathematicarum*, see Mancosu (1996, chapter 1) and Feldhay (1998). Barrow was exposed to it during

angles of a triangle sum to 180° (proposition 32, Book I of Euclid's *Elements*). In the proof, the base of the triangle is extended to create an "external" angle, this angle is divided by constructing a parallel to the triangle's opposite side, and the sum of the internal angles is shown to equal the sum of the angles thus formed. The objection is that the extension of the base and the construction of the parallel are not essential to the triangle, since it could perfectly well exist without them.[5] Barrow replies:

> [B]ecause a Triangle is constituted of Right Lines, the Properties of a Right Line do so far pertain to it. But it is the Property of a Right Line that it may be produced; therefore this Production is not altogether accidental or extrinsecal to a Triangle. In like manner it is also the Property of a Right Line, as was before demonstrated, that another Right Line may be drawn parallel to it through any Point without itself. Therefore this also agrees essentially with a Triangle, as far as Right Lines (*i.e.* its Sides) do enter its Constitution.
>
> (p. 98)

But as Barrow was well aware, in geometry objects are often defined without mention of the processes by which they are generated. (Indeed, Euclid defines a circle as "a plane figure contained by one line such that all the straight lines falling upon it from one point," the center, "along those lying within the figure are equal to one another" (p. 153)). Barrow maintains that every definition is nonetheless "a Proposition wherein a Name is imposed, or ascribed from some possible Supposition of a Thing clearly resulting; which Supposition, being expressed in the Proposition, determines and circumscribes that Name" (p. 129). So definitions, in general, determine that names are to apply to whatever results from "suppositions," of which generative procedures are a special case. It matters only that the "supposed" possibilities "agree with their Subject both *necessarily* and *solely*, *i.e.*... do so reciprocate with their Subject, that if they be supposed, it is also supposed of Necessity." For instance, it agrees with a circle "*that if any Point be assumed in the Diameter, and a Right Line be erected perpendicular to the said Diameter meeting the Circumference, the Square of the intercepted Line is equal to the [Product] of the Segments of the*

travels in Italy, and in the *Mathematical Lectures* he refers to Peyrera and Blancanus by name.

5 Thus, Piccolomini endorses the objection of ancient geometers that that from which the angle-sum property is deduced is not essential to the triangle: "For even though there be no exterior angle, the interior angles are equal to two right angles; for it is a triangle even if its side is not extended" (in Proclus's commentary on the *Elements*, quoted in Cozzoli (2007, p. 166)). Peyrera charges, similarly, that the demonstrated property "will belong" to the triangle "whether the side is produced and the external angle is formed or not, or rather even if we imagine that the production of the one side and the bringing about of the external angle is impossible." Quoted in Mancosu (1996, p. 15).

Diameter", and "*that every two Right Lines that can be drawn from the Extremities of the Diameter to any Point in its Circumference will make a right Angle*", and conversely "every Figure" that agrees with either of these is a circle (p. 85). In order to show that mathematical proof is causal, Barrow must explain how these defining properties can be causes.

The alternative definitions of a circle are equivalent in the important sense that any one of them can be deduced from the others. Barrow claims that whenever an "Affection be taken at pleasure before others for the Definition of its Subject," so that the others "necessarily follow and become known" through it, it "so far supplies the place of a Cause" (p. 86). He contends that the "most close and intimate Connection" or "mutual *Causality* and Dependence" of "the Terms of a *Mathematical Demonstration*... may be called a *formal Causality*, because the remaining Affections do result from that one Property, which is first assumed, as from a Form" (88). In supposing that the defining property determines the others as a formal cause, Barrow sides with the Aristotelian tradition (Mancosu 1996, p. 14) rejected by the Jesuits.

Yet an orthodox Aristotelian conception does not have room (at any one level of description) for alternative forms, each of which can be arbitrarily taken as the explanatory basis for other properties. And Barrow shows little interest in reconciling his view with tradition. Instead, he argues that nothing more can be demanded of causal argument: it is not possible to link an effect with a unique cause. Barrow claims, in particular, that "there can be no such Connection of an *external, ex. gr. efficient Cause* with its *Effect*" through which either an effect or a "determinate" cause is "necessarily supposed" by supposing the other. His attack on efficient causation relies on a strongly voluntarist conception of connections in nature, according to which "the *Free-Will* and Power of *Almighty God*" to "hinder the Influx and Efficacy of any *Cause*"[6] or produce an effect by any means "at his Pleasure" makes "every Action of an *efficient Cause*, as well as its consequent *Effect*, depend" on Him (p. 88).[7] While Barrow knew other ways to understand God's power of choice, which would reconcile it with the necessity of finite causes, he does not acknowledge them here.[8] On his

6 For instance, it "does not follow that the Moon undergoes an Eclipse" in "that most celebrated and trite Example of a Demonstration from the *Efficient Cause*... of the Earth's Interposition between the Sun and Moon." For "if God please, the solar Rays may pass through the Body of the Earth, or reach the Moon by an indirect Passage without touching the Earth; or otherwise the Moon may be enlighted some other Way" (p. 90).

7 As Antoni Malet makes clear, it is by giving God a "decisive role in the day-to-day workings of his creation" that Barrow can hold that His "'free-will and power' preclude us from attributing necessity to any would-be efficient causal connection" (1997, p. 268).

8 For instance, prominent Scholastics held that God's power can be understood in two ways, "absolutely" (*potentia absoluta*) or as "ordained" (*potentia ordinata*) (see Osler 1994). Taken absolutely or "in itself," God's power extends to all that is not logically contradictory. The power to block the efficacy of causes is attributed to God in this sense. To understand God's

view, the connections between final and material causes and their effects are likewise without the necessity required for demonstrative argument. Barrow concludes that since "there is [no] other Causality in the Nature of Things, wherein a necessary Connection can be founded" (p. 88), we have "nothing to wonder at, that . . . no Demonstrations in Geometry" pertain to any other kind of cause (p. 90). The thrust of his remarks is that if a framework for scientific reasoning cannot accommodate mathematical argument, so much the worse for that framework.

In dismissing efficient causes and reinterpreting formal causality, Barrow seems to clear the way for an alternative conception of scientific explanation. It has been suggested that by putting God "in direct control of the operations of nature" and emphasizing His immutability, Barrow "provides a greater guarantee of the regularity of nature than when nature had been governed by immanent powers and countless intermediaries".[9] Mathematical inference, with its characteristic necessity, would then be an appropriate model for the workings of Nature. Barrow's rather cavalier treatment of the traditional framework can thus be taken to reflect his confidence that mathematical reasoning pertains to the natures of objects.[10] In the context of the dispute over mathematics' status as a science, the question is whether mathematical inference as he conceives it can legitimately take the place of more traditional notions of causality.

As we have seen, the properties (or "affections") stated in definitions can be causal in the more familiar sense of bringing the defined object into being. Barrow claims that definitions which "shew [the] possible Existence, and evidently discover the Method of" constructing a magnitude are of all "the most lawful and the best" (p. 223; cf. p. 87), and accordingly focuses on them in his attempt to exhibit the relevance of mathematical reasoning to nature. But the generations of mathematical objects need not be understood as processes in the natural world.[11] To understand Barrow's position, we must know *what* is

power in the second way is to relate it to what He has actually chosen in creating the natural order. It is not necessarily a defect of His power, thus understood, to be constrained by finite causal relations. In his sermons, Barrow acknowledges a distinction between the "special interposition of [God's] hand" and "the natural power" or "ordinary course of inferior causes" (Hughes 1831, p. 378). However, his considered view appears to be that "God performs miracles with the same actions that He takes care of the world," as Malet puts it (1997, p. 271).

9 Harrison (2002, p. 16). See also Malet (1997, p. 274).

10 As Mancosu puts it, Barrow "begins with the basic presupposition that mathematics is the science par excellence. Nothing can be more remote from his perspective than the subtle scholastic distinctions that had characterized the Renaissance contribution to the *Quaestio*" (1996, p. 23).

11 This is shown by the example of Blancanus, who holds that the properties studied in mathematics are essential to *its* objects, but denies that they pertain to objects in the domain of natural science. See Mancosu (1996, p. 180).

supposed, as a possibility, in a mathematical definition. It takes care to specify the kind of possibility at issue, for Barrow's text points in two directions.[12]

Barrow's account includes two conditions for the possibility of the things supposed in mathematical definitions. As a positive condition on possibility, Barrow maintains that a thing can exist only in virtue of its efficient cause. With his voluntarist conception of efficient causation, he is in position to assert that all demonstrative knowledge depends on God, "not only on the part of the knowing . . . Faculty, but also on the part of the knowable Object."[13] He also states a negative condition: that "nothing hinders, but there may be such" (p. 111), or more formally, that what is supposed contains "nothing impossible or inconsistent" (p. 108). Like the positive condition, the negative one is given content by "the infinite and incomprehensible" power of God. Barrow interprets it to mean that something is possible as long as God can create a world in which it is the case. This is clear from the way he defends the possibility of Galileo's supposition "*that heavy Things are naturally carried towards the Center of the Earth with a Motion uniformly accelerated*" and thus his right to call his work "a new Science."

> But if it be false (as I think it not always true concerning many Causes) that there is such a Motion in the present OEconomy of Nature; yet because such a Motion may exist at the Pleasure of God, as implying nothing in it contrary to Possibility, therefore the Conclusions, which result by a lawful Inference from such a Supposition, ought to be accounted for lawful Demonstrations.[14]
>
> (p. 110)

Similarly, the demonstrations that astronomers base upon the Suppositions "*That the Motion of the Stars is in perfect Circles or Ellipses*, and *That they are every Way regular and equable*, also *That they keep the same Periods of the*

12 Cf. Mancosu (1996, p. 141).

13 Because "all Possibility intrinsically denotes a Respect to the Cause or Power by which the Things do exist, which are called or conceived to be possible; therefore a Demonstration supposes the Power which effecteth all Things that are conceived or supposed under the Notion of Possibles, i.e. the infinite and incomprehensible Power of God, which can produce whatever Effects we are able to conceive as possible, and innumerable other beyond our Comprehension" (p. 110).

14 Barrow takes Galileo to correctly distinguish two kinds of investigation, but to locate his own work on the wrong side of the contrast. In the Third Day of the *Discorsi*, Galileo endorses the way of "pretending" by which some "have laudably demonstrated" the "essentials" of "spiral and conchoidal lines" derived "from "certain motions," paths of which "nature makes no use." Galileo claims, however, that he is not merely theorizing about "some kind of motion invented at pleasure," because his own definition of accelerated motion "agrees with the essence of naturally accelerated motion" (that is, with nature's use of "a certain kind of acceleration for descending heavy things") (Galileo 1974, p. 153).

Times, and the same Orbits, with a perpetual Constancy" are "most true, and their Astronomy true, not indeed of this World, but of" one which "God may create . . . , where the Stars will exactly agree with such Motions" (p. 111). Moving from the mixed sciences to geometry, Barrow asserts that "though no such Motions be ever found in the Nature of Things, as Geometricians suppose to be described by *Spiral Lines, Quadratrices, Conchoids, Cissoids, &c.*," whatever follows from these suppositions is "rightly demonstrated." For "God has given us the Power of creating innumerable imaginary Worlds in our Thoughts, which himself, if he please, can cause to be real" (p. 111). The objects of mathematical demonstration thus seem to exist only in the intellect, as ideas of worlds that God can create. So the causality Barrow attributes to mathematical reasoning seems to fall short of relevance to Nature.[15]

4.2.2 *Barrow on the justification of postulates*

Yet Barrow adamantly denies that mathematical objects should "have no other Existence in the Nature of Things than in the Mind alone."[16] He holds "what is most opposite to it," "*viz.* That all imaginable Geometrical Figures are really inherent in every Particle of Matter, I say really inherent in Fact and to the utmost Perfection" (p. 76). Barrow is particularly concerned to deny that mathematics is about "things intelligible" as opposed to "things sensible," for in the debates of the preceding centuries, the involvement of the senses in reasoning was at least a sign that it pertains to objects and powers in Nature. Those who denied mathematics' relevance to nature typically held that it "abstracts" from the "sensible matter" characteristic of objects in the natural world.[17] But Barrow maintains that mathematical abstraction is in fact "such as agrees with

15 Accordingly, David Sepkoski takes Barrow to regard the objects of mathematics as "fictional entities" in the mathematician's mind which need not exist in the domain of "material substance" (2005, p. 50). Thus mathematical reasoning proves of principles "only that they might" apply "to this world" (p. 51). Sepkoski does not consider Barrow's claim that manual practice "sensibly prov[es] the Reality and Possibility" of mathematical suppositions (*ML* 188, quoted at length in Section 4.2.2 below). The affinity between the "constructivism" attributed by Sepkoski to Barrow and the "Newtonian style" as Cohen describes it (see note 42, below) has been noted by Guicciardini (2003, p. 418).

16 He objects specifically to the view held by Blancanus. See note 11 above.

17 In the *Posterior Analytics*, Aristotle claims that because the mixed sciences consider objects as possessed of sensible matter, they follow a "perceptual" reasoning distinct from that of mathematics (79a2–6). Abstraction from sensible matter is taken as the distinguishing feature of mathematical reasoning by Blancanus (among other Thomists). It does not have the same significance for the Averroists Piccolomini and Peyrera. Yet they hold that in mathematics, the intellect does not deduce the essential traits of particulars perceived through the senses (with their accidents), as it does in natural science (see Cozzoli 2007, p. 164). Rather, it derives universal principles from quantities that "are formed in the imagination, the occasion being afforded by quantities found in sensible matter"

all other Sciences," namely "a distinct Consideration of certain things more universal, others less universal being omitted and as it were neglected" (p. 14). As a "Doctrine of Generals," there is "no reason [why] it should be separated from the Consideration of Particulars"; why one science should "treat of an intelligible Sphere, and another of a sensible one," which are "altogether the same" as to "the Verity of the Thing." All of mathematics' objects are, "in reality," "at the same time both intelligible . . . as the Mind apprehends and contemplates their universal Ideas, and sensible as they agree with several particular Subjects occurring to the Sense" (p. 19).

Because Barrow holds that nothing can be "attributed to the intelligible Sphere (i.e. one understood universally) which does not perfectly agree with the sensible (i.e. with every particular one)" (19), he faces the problem of how figures conceived with geometrical precision can "perfectly agree" with the irregular surfaces apprehended by sense. He explicitly poses the question "Who ever did see or perceive by Sense an *exact Right Line* or a *perfect Circle*?". In response, he concedes that "the Occasion of contemplating" mathematical objects is "taken from the Senses," but the things themselves are not "immediately and directly" presented (p. 75). But we can make precise shapes apparent to the senses by taking away the material stuff that overlies them. According to Barrow, the figures "really inherent in matter" are hidden from "the Sense" in the same way that "the Effigies of Caesar lies hid under the unhewn Marble,"

> and is no new Thing made by the [sculptor], but only is discovered and brought to Sight by his Workmanship, i.e. by removing the Parts of Matter which involve and overshadow it. . . . So if the Hand of an Angel (at least the Power of God) should think fit to polish any Particle of Matter without Vacuity, a Spherical Superfice of a Figure exactly Round would appear to the Eyes; not as created anew, but as unveiled and laid open from the Disguises and Covers of its circumjacent Matter.
>
> (pp. 76–77)

But this appeal to the practice of arts and crafts will appear ad hoc unless these activities have a place in mathematical science. To resolve the difficulty,[18]

(Piccolomini 1565, p. 95). Thus the objects of mathematical reasoning never come before the senses. Cf. Feldhay (1998, pp. 83–84).

18 Although I cannot make the case here, I suspect Barrow intends to improve on Galileo's view. Galileo famously claims in the *Dialogue* that the "geometrical philosopher" must "deduct the material hindrances" in order "to recognize in the concrete the effects which he has proved in the abstract," in the same way as "the computer who wants his calculations to deal with sugar, silk, and wool must discount the boxes, bales, and other packings" (Galileo 1967, p. 207). The question this raises is why the activities of tradesmen and artisans should be part of mathematical science. In his appeal to postulates, Barrow has an answer.

Barrow argues that the activity by which material things are brought into accord with mathematical description is enjoined by geometry's first principles, specifically its postulates.

In Euclid's *Elements*, proofs begin with principles of three kinds, namely axioms, postulates, and definitions. Because Euclid does no more to characterize them than to list them under these headings,[19] their taxonomy was a matter of philosophical dispute. Barrow shows little compunction in departing from classical views, or even the surface structure of the *Elements*. For instance, he demotes axioms from the ranks of first principles, claiming instead that they are proved from definitions and postulates (pp. 77–80). His view of postulates is also revisionary. Barrow defines a postulate as a proposition "assuming or affirming some evidently possible Mode, Action, or Motion of a Thing" (p. 128). This understanding of postulates does not fit the last two principles (namely, the assertion that all right angles are equal and the infamous Postulate of Parallels)[20] of the five Euclid lists under this heading. Barrow accordingly narrows the list to three in his edition of the *Elements*: "From any given point to any other given point to draw a right-line; To produce a finite right-line, strait forth continually; Upon any center, and at any distance, to describe a circle."[21]

We can understand what kind of possibility is supposed in mathematical definitions by considering how, on Barrow's view, first principles are justified. First principles are traditionally held not to require proof because they are self-evident. Barrow does not think it wrong to say that demonstration "reaches . . . to some thing simply indemonstrable, confirmed by its own Force, and evident from its own Light" (p. 104). But he cautions that the principles need not "appear necessarily true in themselves, or immediately evident to every Capacity; but only to him who comes ready and prepared for learning that particular Science, to which the said Principles are subservient, i.e. to the studious and teachable Mind" (p. 105). This reflects the traditional conception of ("synthetic") argument from first principles as both maximally rigorous

19 They can be distinguished, but only roughly, by their content and form. The definitions and axioms differ in that the definitions concern geometrical objects (points, lines, figures, and planes), while the axioms hold of mathematical objects in general, including quantities treated in arithmetic and algebra. The postulates also concern specifically geometrical objects, and are further distinguished by their grammatical form. They are infinitive constructions, preceded by a passive imperative (Ἠιτήσθω) which can be translated "Let it be asked that."

20 "Let it be asked: that, if a straight line falling on two straight lines make the interior angles on the same side less than two right angles, the two straight lines, if produced indefinitely, meet on that side on which are the angles less than the two right angles" (p. 155).

21 In the 1732 translation (London: D. Midwinter and A. Ward) of Barrow's edition. In the Latin original, they are "*Postuletur, ut à quovis puncto ad quodvis punctum rectam lineam ducere concedatur; Et rectam lineam terminatam in continuum recta producere; Item, quovis centro, & intervallo circulum describere.*"

and the mode in which established doctrine is presented to a learner. Barrow's concern is to explain what is required, on the part of the teacher and pupil, for the learning of science.

Barrow rejects the classical distinction between postulates and hypotheses. On Aristotle's view, as Barrow understands it, a postulate is "besides, or contrary to the Notion of the Learner" (p. 119), while a hypothesis is accepted by the pupil. Barrow objects that it "matters nothing" whether the pupil's "preconceived opinions" "agree or disagree with the Mind of the Teacher." They are nothing "to the Purpose of the Science" and must be discarded. But while it is the student's duty to throw off mere prejudices, "the Teacher of a Science is bound by his Office . . . to take away all reasonable Scruple from the Student." Principles "not only may, but ought to be" justified by "the Master, if the Scholar requires it" (p. 104). So what Barrow requires of the pupil is not baseless assent, but readiness to "be convinced by the Master's Arguments, or instructed by clear Examples of the Matter proposed" (p. 122).

Examples, in particular, are central to the learning of first principles. Definitions as Barrow conceives them "are no otherwise demonstrable, than the Hypotheses [*sc.* postulates] themselves are, from which they proceed, i.e. than by shewing that the Name is adapted to a Thing having a manifestly possible Condition." Since the name can be shown suitable "by arrogating to one's self that Right, which belongs to any Teacher" (if not "by appealing to the Authority of Lexicons, to vulgar Use, [or] to the Suffrages of the Learned" (p. 113)), the work is to show that the supposed condition is possible. Barrow maintains that postulates, and thus the suppositions they express, are "explained and demonstrated by some Example or Experiment more intelligible, and suitable to the Capacity of the Learner." Such examples are easy to come by, for mathematical reasoning is carried out on "the most simple and common" objects, "such as lie exposed to Senses." It is "seen and experienced to be done daily" that "*a Right Line may be either drawn, or conceived to be Drawn through any two points assignable*" (p. 57). More specifically, it can be shown "by the obvious Use of a Pair of Compasses, or the Revolution of a Wheel above its Center" that it "can be done or conceived" that the "*Rotation of a Right Line*" produces "the *plane Figure* (which is called a *Circle*)." Whoever "obstinately refuses" to admit possibility that is in such a way "made evident as far as it may be done" can be at best "admonished to apply himself to some other Study" or only "left to his own Liberty, as an unfit Learner" of geometry (p. 112).

On Barrow's view, the role of the senses in mathematical reasoning is to demonstrate the possibility of the processes by which objects come to exist and agree with geometrical descriptions (that is, to verify what is expressed by postulates). Some of these processes are just events in the physical world: "if not all, yet the most and chiefest Effects" of those studied in geometry are the figures obtained and motions executed by bodies themselves (p. 21). To reason about what follows if, for instance, "a Star or celestial Point in the Heavens be carried through a right, circular, elliptic, or any other Line, with an uniform or apparently equable Motion respecting some determinate Point, in a certain Time" is to do geometry (p. 27). Barrow argues that both the mixed sciences and physics itself are branches of geometry because the objects they study are just the concrete, extended quantities created by these processes (pp. 21–22). The required motions can also be brought about by human design, for instance "by the obvious Use of a Pair of Compasses, or the Revolution of a Wheel" (p. 112).

It is clear enough how the possibility of these events is proved by sense, but less clear that the traces they leave match geometrical descriptions. Barrow holds that the senses also prove the possibility of a more important class of processes, which bring sensible objects into agreement with geometry. Having conceded that we are "far from" finding the attributes of mathematical objects or gathering universal propositions by the senses, Barrow rejoins:

> By *Sensation* indeed may be deduced the Possibility of Mathematical Hypotheses: Thus *ex. gr.* we know that a Right Line can be drawn between two assigned Points; because we perceive by the Sense, how a Progress may be made from one Point to another, wherein if there be any Unevenness or Deflection, it can be so far rectified by the Hand as to make a Line *sensibly Right*; from whence we infer by our Reason, there being no Repugnance on the Part of the Thing, that all other Roughness and Exorbitances may be pared off and corrected, and so the Line become *perfectly Right*.
>
> (p. 75)

The ability to rectify unevenness or deflection "by the Hand" falls within the scope of what we "perceive by the Sense."[22] So the experience that proves the possibility expressed by the first postulate is not only of the motion that progresses from one point to the other, but also of the activity of "paring off" excess. Thus, the practice that removes "the Disguises and Covers of circumjacent Matter" is not alien to geometry; on the contrary, it is assumed at the foundation of the science, as the way in which its first principles win acceptance.

This practice also proves the "reality," i.e. the relevance to substances and their powers, of mathematics. The practice of "rectifying by hand" can be compared to the sculptor's "discovery" of Caesar's image: if it does not first create a concrete, extended object, yet it makes the underlying matter match some particular geometrical conception. In spelling out his negative condition on possibility, Barrow claims that "the Dominions of Reason do far exceed

22 As is indicated by the use of the subjunctive in both the first and second clauses: "*Sentimus enim ab uno ad alterum utcunque fieri posse progressum, in quo si quid inest asperum et a rectitudine deflectens, illud abradi possit, eatenus quidem manu nostrâ, donec ad sensum recta videatur effecta linea*" (Whewell 1860, p. 84).

the Limits of Nature; the intelligible World is vastly farther extended and more diffusive than the sensible World, and the Understanding contemplates many more Things than the Sense" (pp. 111–112). Through manual practice, we can know the extent to which these "dominions" are realized by concrete things. So Barrow claims that the comparison of geometrical objects requires "nothing . . . which savours of Mechanism,"

> but only as far as every [object] is involved in some sort of Matter, is exposed to the Senses, or is visible and palpable; so that what the Mind demands to be understood, the Hand can execute in Part, and the Praxis can in some Measure emulate the Theory. Which Imitation notwithstanding, is so far from weakening or depressing the Strength and Dignity of *Geometrical Demonstration*, that it affords it a much more strong Confirmation and an higher Advancement, by sensibly proving the Reality and Possibility of the assumed Supposition, which indeed (as we have often insinuated) is the genuine Foundation of all Science; and thus establishing the Authority of Reason by the Suffrage of Experience.
>
> (p. 188)

It is central to Barrow's argument that the constructions that confer existence and geometrical precision on objects are also the means by which theorems are proved. Since these processes are elements of mathematical inference, his account of them completes the case that mathematical inference is tantamount to causality.

But because Barrow fails to resolve an important ambiguity, his argument appears not to succeed. If the practice that is supposed to prove "reality and possibility" is understood as God's activity of making bodies move, of course it has the necessity that finite causes lack. But contrary to Barrow's suggestion, it does not seem that the extent of this power can be known to us through the senses. On the other hand, the practice can be taken to consist of *our* activities of manipulating instruments and paring off excess. On Barrow's view, the extent of these capacities is indeed known by the senses: he introduces the notion of "sensibly right" to mark the point at which our activities of fabricating and refining eliminate all discernible discrepancy between a material thing and its geometrical template. But what we thereby accomplish seems to fall short of what Barrow needs. It is not clear that "sensible" accord between geometry and nature constitutes true agreement, for there can still be discrepancy, even if we cannot perceive it. Barrow himself indicates that we succeed only "in part" or "in some measure" in bringing material things into accord with geometrical description. And even if our activity sometimes produces true agreement, there is no reason to suppose that it brings about its effects with any more necessity than the natural finite causes dismissed by Barrow. Thus, it remains unclear whether Barrow's appeal to practice yields a genuine alternative to efficient causation.

4.3.1 *"The foundation of geometry appertains to mechanics"*

There can be no question that Newton was familiar with Barrow's conception of mathematical objects as products of motion. For in a rare acknowledgement of intellectual debt, he reports that "its probable that Dr Barrow's Lectures might put me upon the consideration of figures by motion, tho I not now remember it."[23] I will now set out the echoes of Barrow's view in Newton's Preface.

In his Preface, Newton is out to overturn the distinction between mechanics and geometry "by the attribution of exactness to *geometry* and of anything less than exactness to *mechanics*" (p. 381). In an unfinished treatise on geometry from the 1690s (now called *Geometria*), Newton inveighs against this "common opinion." He maintains that it "defines mechanics" from "the ignorance and imperfection of mechanicians," in that "mechanics as it is commonly practiced is imperfect and without exact laws," while the operations of and mechanical practice required for geometry are "exact" (*NMP* VII, p. 289). In the Preface to the *Principia*, Newton likewise contends that the imperfection that seems to distinguish mechanics comes "not from the art" but "from those who practice the art," for "anyone who works with less exactness is a less perfect mechanic, and if anyone could work with the greatest exactness, he would be the most perfect mechanic of all" (p. 381).

Newton here takes his readers to understand mechanics as an art. As Alan Gabbey explains, mechanics was "a theoretical discipline in that it dealt mathematically with problems arising out of the construction and use of machines." But it was also a practice, dedicated to the construction and use of devices to "re-arrange things *contra naturam* and for human ends" (1993b, p. 134). It was an art both in the sense of producing outcomes not in accordance with nature, and in the sense of requiring ingenuity and skill. Newton highlights this conception of mechanics, reminding us in the Preface's second sentence that "the ancients divided *mechanics* into two parts," "rational" and "practical." According to Newton, the latter, which "comprises all the manual arts," has given its name to the whole (p. 381).

In *Geometria* (*NMP* VII, p. 289, quoted above), Newton makes the point that the standard for geometrical exactness is set (in that it is among the demands made) by geometry's postulates. In the *Principia*'s Preface, he stresses that mechanics meets this demand. Whoever can work with the greatest exactness is still a mechanic, because

> the description of straight lines and circles, which is the foundation of *geometry*, appertains to *mechanics*. *Geometry* does not teach how to describe these straight lines and circles, but postulates such a description.

23 *Add MS* 3968.41, f 86v. Quoted in Westfall (1980, p. 131). It should be noted that Newton may be referring to Barrow's "Geometrical," rather than his "Mathematical," lectures. See note 37 below.

> For *geometry* postulates that a beginner has learned to describe straight lines and circles exactly before he approaches the threshold of geometry . . . To describe straight lines and to describe circles are problems, but not problems in *geometry*. *Geometry* postulates the solution of these problems from *mechanics*, and teaches the use of the problems thus solved.
>
> (pp. 381–382)

In assigning the description of straight lines and circles to mechanics, Newton goes against the ancient tradition that counts these figures as "geometrical" solutions to problems, while disallowing more complex "mechanical" curves, on the grounds that they are generated through a use of instruments that precludes genuine understanding.[24] In *Geometria*, Newton explicitly challenges the basis of this distinction. He states unequivocally that all constructions are mechanical – and qualify as exact by the standards of the postulates:

> any description of curves by instruments, even that of the circle itself by compasses and of the straight line by a ruler, is mechanical, and [the ancients] consequently postulated the descriptions of those they received into geometry – not that these might, insofar as they are geometrical, be described by men (for who has seen a line without breadth?) but that, once their description is granted, all the rest of what geometers derive therefrom shall accurately follow from it.
>
> (*NMP* VII, p. 383)

Newton's rhetorical question is of the same sort as Barrow's "Who ever did see or perceive by Sense an *exact Right Line* . . . ?" (p. 75). His answer also takes the same form as Barrow's: the postulates license the ascription of exact attributes even to objects whose crudity is apparent to the senses.

Newton also follows Barrow in taking the postulates to secure the conditions under which alone geometry can be learned: in his words, to specify what the "beginner" must be able to achieve, with exactitude, "before he approaches the threshold of geometry."

Even as these passages demonstrate *that* Newton adopts Barrow's strategy for proving the relevance of geometry to natural phenomena, they raise the question of *why*. It does not profit Newton to argue that geometry qualifies as a science by Aristotelian standards, for by the 1670s the Aristotelian model of science had been supplanted by the quantitative theories of Galileo, Descartes, and Huygens.

One possibility, pursued to illuminating effect by Mary Domski and Niccoló Guicciardini, is that Newton's allegiance to ancient standards of demonstration bolsters his opposition to Cartesian method in the sciences.[25] In particular,

24 See Mancosu (1996, pp. 71–74).

25 In Domski (2010) and Guicciardini (2006). Kargon (1966) also argues that Barrow supplies Newton with an alternative to Cartesian methodology. But Kargon (following

they show how Descartes is a target of the argument of the Preface. Newton writes that the distinction between mechanics and geometry "by the attribution of exactness to *geometry*" "has now come to be usual" (*NMP* VII, p. 289). These scholars emphasize that it became prominent in Descartes's *Géométrie.* Shifting the classical boundary, Descartes relegates to "mechanics" those figures not "described by a continuous motion or by several successive motions, each motion being completely determined by those which precede" (Descartes 1954, p. 43). Newton, however, must defend geometry's prerogative to "treat an unbounded variety of curves in the plane" in order to extend it to natural philosophy, where "the curves represent motions in nature" and so make it possible to specify the forces required for them.[26]

I will show how Barrow's conception of geometrical reasoning also opens up an alternative to the way systems of bodies in motion came, under Descartes's influence, to be treated in natural philosophy. Perhaps the most obvious way in which Barrow's strategy suits the purposes of the *Principia* is in giving mathematical reasoning the authority to speak of causes, without supposing them to operate in any accepted modality. Barrow is under pressure to fit causes under the Aristotelian headings, and in response he argues that geometrical demonstration pertains to causes which are neither material, efficient, nor final, nor formal in any but an etiolated sense. I show (in Section 4.3.2) that unlike Barrow, Newton intends to leave it open that the causes established by mathematical reasoning can ultimately fit into the dominant scheme of modes of causation (which is no longer Aristotelian). But I argue (in Section 4.3.3) that in the *Principia* he takes himself to have laid celestial and terrestrial motion to the same cause, although he has not satisfied this demand.

4.3.2 The "mathematical/physical" distinction and the question of cause

The distinctive reasoning of the *Principia* has proved highly controversial. In Book III, Newton treats (sufficiently precise) observational results that stand in the mathematical relationships demonstrated in the first two Books as measurements of gravitational force. Important examples are that the times of orbits of satellites are as the 3/2 power of their distances from the center of the body they orbit and the proportionality, to distance from the center of the Earth, of the acceleration towards it of both the Moon and terrestrial bodies. At the close of Book III, Newton claims to have "explained [*exponere*] the phenomena of the heavens and of our sea by the force of gravity" (p. 943), which seems to mean nothing other than that gravity causes celestial and terrestrial motions. What

the lead of E. A. Burtt) takes Barrow's contribution to be the rejection of "hypotheses," and does not address the positive role of postulates in the development of science.

26 Domski (2003, p. 1123). Guicciardini (2004) gives examples of curves treated by Newton that are "mechanical" by Descartes's criteria.

has most struck his readers is that he claims to explain them without specifying any mode by which the force of gravity operates.

From the outset, critical discussion of the *Principia* has centered on the question of whether it really establishes the causes of these motions. For some of Newton's contemporaries, this was the question whether the *Principia* consists in mechanics in its traditional acceptation, namely the application of mathematics to phenomena of weight and motion, which lacks the power to determine their causes. Newton's critics insisted that the reality of a force could be established only by "physical" reasoning, not its "mathematical" or "mechanical" variant. For instance, the review of the *Principia* in the journal of the French Academy of Sciences claims that the demonstrations of Books I and II must be regarded "as only mechanical; indeed, the author recognizes himself... that he has not considered their principles as a Physicist, but as a mere geometer." The reviewer complains that the hypotheses by which Newton purports to explain the world system "serve as the foundation only for a treatise of pure mechanics" because they are "arbitrary for the most part." A "Physics as exact as [Newton's] mechanics" would have to "substitute true motions for those he has supposed."[27]

Newton's claim to explain the phenomena of nature was resisted by his more innovative peers as well. The preceding generation of motion theorists – the most qualified readers of the *Principia* – held that a natural phenomenon was not explained until it was shown to result from "mechanical" principles. Insofar as mechanical explanation takes as its model the working of machines, it shows phenomena to result from something like pushing or pulling, localized to the surface of a body.[28] In the *Principia*, Newton seeks to avoid this demand: to identify the cause of celestial and terrestrial motions, without specifying a mechanism by which gravity operates. Newton resists the demand because he does not accept the reasoning by which such explanations were purportedly established. Because they invoke configurations too small to be observed, they could only be regarded as hypotheses.[29] Newton insists that it is far from "rigid Consequence" to infer that an hypothesis is "thus because not otherwise or because it satisfies all phaenomena."[30] For without a complete enumeration of alternative accounts, "numerous hypotheses may be devised, which shall seem

27 *Journal des Sçavans* vol. 16 (August 2, 1688), 237–238. Translated in Koyré (1968, p. 115). I have altered the translation, following Gabbey (1992, p. 321).

28 The objection that attraction is unintelligible absent such a mechanism is most closely associated with Leibniz. For discussion, see Janiak (2007). It is also attributed to Huygens. See, for instance, Koyré (1968, p. 118) (where the relevant passage is quoted); Larmore (1987, p. 88); Smith (2002a, p. 150); and Maglo (2003, p. 147). But (as Maglo notes) Huygens also had empirical grounds for opposing attraction. See Schliesser and Smith (forthcoming).

29 See Nadler (1998, p. 520).

30 Letter to Oldenburg, 6 February 1671/2, in Newton (1959–1977 I, pp. 96–97).

to overcome new difficulties," making it appear that certainty can never be obtained.

But while Newton resists the demand, he sees justice in it. It is commonly observed that throughout his life, Newton continued to try to find a cause for gravitational attraction in the immediate operation of some "agent" on the interior or surface of a body.[31] Since he finds it repugnant to conceive gravity as the immediate action (at a distance) of passive matter,[32] he appears forced to demote it from a cause of motion to a mere phenomenological rendering of the tendency to move. A number of writers note that Newton's distinction between "mathematical" and "physical" treatment of attraction is intended to relieve this pressure.[33] The distinction is elaborated at a crucial juncture of Book I. As George Smith points out, Newton revisits it in a Scholium following a proposition (I.69) that "lays the groundwork for Newton's law of gravity by asserting that in the relevant inverse-square case the forces directed towards the various bodies must be proportional to the masses of those bodies" and preceding two Sections (12 and 13) that "lay the groundwork for Newton's claim that his law of gravity holds *universally* between individual particles of matter." Because "no hypothetical contact mechanism seems even imaginable to effect 'attractive' forces among particles of matter generally," this Scholium thus occurs just "where adherents to the mechanical philosophy would start" to reject Newton's reasoning (2002a, p. 141). At this point, Newton asserts the prerogative to consider "attraction, impulse, or any sort of propensity toward a center" both mathematically, without "defining a species or mode

31 John Henry shows (in 1994a) that Cohen, Koyré, McMullin, and A. Rupert Hall hold in common that because Newton could not accept that a quality internal to matter could exert attractive power at a distance, he continued to seek an alternative account of gravitation. Henry concludes that it is "almost canonical" to read Newton as denying that "gravity could be an inherent property of matter" and that "action at a distance [is] possible 'without mediation'". However, Henry challenges this interpretation, as does Schliesser. See the following note.

32 In his letter to Bentley of 25 February 1692/3, Newton insists that gravity *must* "be caused by an agent acting constantly according to certain laws," whether material or immaterial, for that "gravity should be innate, inherent, and essential to matter, so that one body may act upon another at a distance and through a vacuum without the mediation of anything else by and through which their action and force may be conveyed from one to another is to me so great an absurdity that I believe no man who has in philosophical matters any competent faculty of thinking can ever fall into it." (In Newton 1959–1977 III, pp. 253–254, and Newton 2004, pp. 102–103.) Henry (in 1994a) and Schliesser (2010b) stress that Newton's concern in this passage is to differentiate his view from Epicureanism.

33 Janiak in (2007); Smith in (2002a); McMullin in (1989), (2001, pp. 295–297), and (2002). These writers differ as to how much of the *Principia*'s argument is conceived by Newton as "physical." On Janiak and Smith, see note 50 below. According to McMullin, later portions of the *Principia* are "physical" by the criteria of the Scholium in Section 11, but not by those of Definition VIII. McMullin concludes that "Newton's use of these terms is not entirely consistent" throughout his works (2002, p. 291).

of action or a physical cause or reason" for them, and as forces (Definition VIII, p. 409).

From Newton's discussion, it is clear that "mathematical" treatment brackets at least some questions of causation. To his mechanist readers, he says

> I use the word "attraction" here in a general sense for any endeavor whatever of bodies to approach one another, whether that endeavor occurs as a result of the action of the bodies either drawn toward one another or acting on one another by means of spirits emitted or whether it arises from the action of [any medium] impelling toward one another the bodies floating therein. I use the word "impulse" in the same general sense, considering in this treatise not the species of forces and their physical qualities but their quantities and mathematical proportions, as I have explained in the definitions. Mathematics requires an investigation of those quantities of forces and their proportions that follow from any conditions that may be posited [*quae ex conditionibus quibus eunque positis consequentur*].
>
> (pp. 588–589)

"Then," Newton continues, "coming down to physics, these proportions must be compared with the phenomena, so that it may be found out which conditions of forces apply to each kind of attracting bodies." In leaving it to physics to determine which of these alternatives causes the endeavor, "mathematical" treatment leaves it open that a mechanism will be found.[34] But Newton does not make clear whether "mathematical" treatment suspends *all* consideration of cause. That he conceives attraction as a force does not settle the question, for it can still be taken (in the same way as what is now called "Coriolis force") as a mere calculating device.[35]

Thus, many mechanist natural philosophers take Newton to deny gravity the status of a cause. Like the early reviewer who takes Newton to "himself recognize" that he treats forces only under the "arbitrary" suppositions made in mechanics, they in effect put Newton's work in the tradition of the "mixed sciences," understood as mathematical descriptions intended only to "save" the phenomena (cf. Janiak 2007, p. 131). Thus, both the traditional understanding of mechanics, and the explanatory standards of the mechanical philosophy,

34 It is left open, specifically, that a "corporeal medium" will be found that "impels the bodies floating therein" by local action. As Janiak puts it, Newton's "use of the terms 'attraction' and 'attract' is intended to be compatible with any physical account of gravity in the sense that it is not intended to rule out any physical medium that 'pushes' or 'pulls' bodies" (2007, p. 135). Newton thus contrasts "mathematical" consideration of centripetal forces "as attractions" with "physical" consideration of them as "impulses" at the start of Book I's Section 11 (p. 561).

35 Janiak gives this as "a classic example" of a "fictitious force," such that "any phenomenon associated with" it "would in fact be caused by some combination of factors independent of the Coriolis force" (2007, n. 1).

encourage the view that Newton's "mathematical" reasoning is not intended to specify any cause.

4.3.3 *Postulates and Newton's mathematical/physical distinction*

In the context of the *Principia*, the interpretation of the "mathematical/physical" distinction just bruited is hard to sustain. As Janiak points out, Newton "does not dodge the implication" that gravity "bears causal relations": for instance, he proclaims dramatically that it is "that force by which the moon is kept in its orbit" (2007, p. 130). Koffi Maglo argues that in light of the "philosophy of mathematics" adumbrated at the start of the Preface, "mathematical" treatment of force cannot be held to lack "ontological content" or "assumptions about a referent in nature."[36] The Preface indeed seems to make explicit that the mathematical portion of the argument deals with causally efficacious features of the natural world. After declaring that geometry is, in virtue of its dependence on mechanical practice, a part of "universal mechanics," Newton allows that it can yet be distinguished from mechanics proper. For "*geometry* is commonly used in reference to magnitude, and *mechanics* in reference to motion". The mechanics on offer in the *Principia* is thus "the science, expressed in exact proportions and demonstrations," of the "forces that are required for any motions whatever" and of the motions that *result from* them [*Motuum qui ex viribus . . . resultant*] (p. 382).

Consideration of the precedent Newton has in Barrow's work casts further doubt on this interpretation of the "mathematical." Barrow makes a distinction between mathematical and physical treatments of "force or motive power," on which mathematics conceives it as both "a quantum" (and so "subject to calculation") and as "the efficient cause of motion," "whatever its nature or the origin whence it arises, for we leave this discussion to physicists."[37] More important, however, is the analogy between Newton's mathematical/physical distinction and the way geometry is related to other sciences by its postulates, on his and Barrow's conception of their role. According to this analogy, the questions deferred to physics for investigation concern only the modality and physical character of causation.[38]

36 (2007, pp. 581–582). Cf. de Gandt (1999, p. 13) and Jammer (1954, p. 95), who traces Newton's "realistic conception of mathematics" to Barrow.

37 Barrow's *Geometrical Lectures*, in Whewell (1860, p. 166). Translation in de Gandt (1995, p. 109). Newton probably attended these lectures, and later helped prepare the text for publication.

38 Andrew Janiak explains how the argument of the *Principia* can show that gravity is genuinely the cause of certain motions yet fall on the "mathematical" side of Newton's "mathematical"/"physical" contrast, namely by remaining neutral as to its "underlying physical basis" (2007, p. 130). See note 50 below.

Newton claims to use the term "attraction" in a mathematical sense for "any endeavor whatever," whether it "occurs as a result of the action of the bodies either drawn toward one another or acting on one another by means of spirits emitted or whether it arises from the action of aether or of air or any medium whatsoever" (p. 588). What he here describes as the "mathematical" consideration of force appears to be just the way in which geometry relates, through its postulates, to its objects. In *Geometria*, he claims that plane figures "executed by God, nature or any technician you will are measured by geometry in the hypothesis [*sc.* postulate] that they are exactly constructed," no matter how, in physical reality, they are formed:

> A technician is required and postulated to have learnt how to describe straight lines and circles before he may begin to be a geometer. And it consequently does not matter how they shall be described. Geometry does not posit modes of description: we are free to describe them by moving rulers around, using optical rays, taut threads, compasses, the angle given in a circumference, or finally any mechanical means whatsoever. Geometry makes the unique demand that they be described exactly.
>
> (*NMP* VII, p. 289)

Then, just as mathematics in general investigates the quantities and proportions that follow from whatever conditions are posited, so geometry finds the "properties and proportions" (*NMP* VIII, p. 179) of figures however described.

And just as mathematics leaves it to physics to find the conditions of forces, so geometry leaves it to another science to find modes of description. Newton makes the point succinctly in the *Principia*'s Preface: "To describe straight lines and circles are problems," because they present the reader with something *to do* rather than a result *to prove*,[39] "but not problems in geometry. *Geometry* postulates the solution of these problems from mechanics" (p. 382).

This remark recalls the traditional view that what is assumed as a first principle in a "mixed" or "subordinate" science is proved in a higher science.[40] Newton seems here to rank the sciences in just such a hierarchy, while reversing the traditional priority of geometry over mechanics. The traditional conception of "subordination" also seems to fit the relationship, as conceived by Newton,

39 Newton is extending the distinction between problems and theorems, propositions proved in Euclid's *Elements*, to its first principles. Problems are propositions demonstrated by showing "what it was required to do," while theorems are demonstrated by showing "what it was required to prove" (in Heath's translation). Problems and theorems differ in the *sumperasma*, which summarizes what has been shown, and in the *protaseis*, the initial statement of the proposition (Mueller 1981, pp. 11–13). Formally, "a problem is cast as an infinitive expression seeking the construction of a geometric term in a specified relation to other given terms," while a theorem "is typically set in the form of a conditional asserting a property of a specified geometric configuration" (Knorr 1986/1993, p. 348).

40 For discussion, see Dear (1995, chapter 2).

between mechanics and physics. To consider force mathematically, in the way that geometry relates to its objects through its postulates, is "rational mechanics" as Newton defines it in the *Principia*'s Preface: the science, made exact, "of the motions that result from any forces whatever and of the forces that are required for any motions whatever." As mathematical doctrine, mechanics assumes the possibility of the forces required for the motions it specifies, while physics finds the features of interacting bodies that give rise to these endeavors. Mechanics thus appears poised between geometry and physics, depended upon by the one and dependent on the other. For it makes real the operations assumed as possible in geometry: namely, the description and generation of lines "through the continuous motion of points," and of circles and other "surface-areas through the motion of lines," which "are daily witnessed in the motion of bodies" (*De Quadratura Curvarum*, *NMP* VIII, p. 123).

Geometry and rational mechanics are, however, safe in assuming *that* what they require is possible. They depend on mechanical practice and physics (respectively) only to work out *how* it happens in the material world.

The tradition that Barrow takes himself to continue from ancient geometry thus allows Newton to assign certain topics to physics, without denying causal relevance to mathematical reasoning. I will now explain how this tradition motivates elements of Newton's reasoning that might otherwise seem heroic.

4.3.4 *Turning discrepancy into measurement*

So far, I have shown how Newton's conception of mechanics helps to distinguish the "mathematical" argument of the *Principia* from a mere description of phenomena. Clearly, mathematical reasoning as Newton conceives it has causal implications. But more is required to see how it can have the relevance Newton intends.

It is hard to understand the mathematical portion of the *Principia* as anything other than a series of mathematical constructs, successively revised to agree more closely with the data of observation and experiment. The first question we face is how the positing and refinement of these constructs differs from the hypothetico-deductive (HD) reasoning that Newton so clearly opposes. The physical content of the constructs makes them models in a central sense of that term: they are stipulated to be identical to the actual world in certain respects. But they are still hypotheses (as the tendency to interchange the terms suggests)[41] in that they are always liable to be replaced by alternatives that better

41 Models are distinguished from hypotheses in terms of how each compares with experience on some standard (but not consistently followed) usages. In general, while a hypothesis is proposed as a conjecture subject to decisive refutation, a model is proposed to explain a set of data and can be refined (Glass and Hall 2008). In a Bayesian framework, a model can be understood as a disjunction of hypotheses, such that the latter predict specific

agree with the data.[42] For, to begin with, it is not clear what constrains the introduction of new conditions (as modifications to or replacements of a given construct). Consideration of arbitrary conditions is supposed to be blocked by the Rules of Reasoning (added in the second (1713) edition), especially the Fourth: "*In experimental philosophy, propositions gathered from phenomena by induction should be considered either exactly or very nearly true notwithstanding any contrary hypotheses, until yet other phenomena make such propositions either more exact or liable to exceptions*" (p. 796). But to apply it, we need to better understand what entitles a proposition to "be considered either exactly or very nearly true."

Secondly, there appears to be no end to the progression in which each mathematical construct is replaced by another that better fits the data. Newton sees it as an acute problem to prove a fit between nature and any one mathematical construct. In drafts of a predecessor to the *Principia*, he reasons that in the system comprised of the Sun and planets, the Sun must deviate from the center of gravity, because it is drawn by attractive forces toward the planets, while the center cannot move. So

> the centripetal force does not always tend to that immobile center, and hence the planets neither move exactly in ellipses nor revolve twice in the same orbit. There are as many orbits of a planet as it has revolutions . . . and the orbit of any one planet depends on the combined motion of all the planets, not to mention the action of these on each other. But to consider simultaneously all these causes of motion and to define these motions by exact laws admitting of easy calculation exceeds, if I am not mistaken, the force of any human mind.
>
> (*NMP* VI, p. 78).

values while predictions from the former require the adjustment of parameters (Sober 2006).

42 On I. Bernard Cohen's understanding of the Newtonian "style," it is especially hard to distinguish from HD method. According to Cohen, consequences deduced from mathematically specified "artificial worlds created in the mind" are "transferred" to the ontologically distinct "level" of physical nature by substituting "physical equivalents" for "mathematical entities" (e.g. bodies with mass for mass points) and comparing them with experimental and observational data (1982, pp. 50–51). A mathematical construct is assured to be "not fictive" just as a hypothesis is confirmed: to the extent that it retrodicts known phenomena and predicts new effects (confirmed by observation) (1980, p. 110). But as long as it is legitimate to introduce further conditions, to produce a system that better agrees with observation and experiment, there can be no confidence that "the system and its conditions" match "the realities of the external world" (p. 64). So for the same reasons that hypotheses lack "certainty," according to Newton, it remains unclear whether the "ultimate system" arrived at in the *Principia* (of many interacting bodies) is "so congruent with reality that its laws and principles" are those of the Universe. Cohen takes Newton to acknowledge the tentative character of his reasoning, or at least to leave open whether it matches reality (1980, pp. 66–67).

Whatever darker morals we might draw from this passage, it suggests at least that for any mathematical system that calculates the trajectories of the planets, there will be a more complex alternative that better captures their true motions. So if phenomena can somehow be made to select a true description from among the alternatives, there is still no prospect of establishing one with certainty.

Of course, the human mind's "force" is limited in the complexity of alternatives it can formulate, and observational accuracy limits our ability to test them. Newton's point may then be that we can hope, not to bring mathematical systems into "exact correspondence [with] physical reality," but to make them unfalsifiable: elaborate them to the point that the "aspects of the system" they disclose are "of so small a magnitude that they [can] be ignored within the limits of observation, even with the best telescopes of the time."[43] We would then have an analogue, in Newton's terms, of Barrow's notion of making a line "sensibly Right", i.e., making it the case that no discrepancy between construct and reality can be discerned by the senses. But if Newton can only bring constructs into "sensible" accord with reality, his claim to mathematically exact and certain knowledge of nature will not be secure. We thus confront a second question: how the appeal to practice (in conjunction with postulates) can take Newton further than it does Barrow.

The groundbreaking work of George Smith and William Harper reveals how Newton's aspirations can be fulfilled. They make it their goal to explain how Newton's method differs from HD. For Smith, it is "exactly right" to say "that Newton saw empirical science as progressing by successive approximations" or idealizations. But by identifying conditions that the idealized mathematical constructs must satisfy, Smith makes clear that they are not arbitrarily proposed. He emphasizes that each of them can be, and is, "used to draw conclusions from phenomena." These conclusions are far more than failures of the phenomena to falsify the idealizations. According to Smith, they often "take the form of inferred measures of quantities" (2001b, p. 250). Similarly, Harper shows how the mathematical portion of Newton's argument establishes "systematic dependencies that make [what is] inferred count as" a parameter value "measured by the phenomena from which [it is] inferred" (2002b, p. 78). In virtue of their role in inferring conclusions from phenomena, the constructs have undeniable physical content.

One important condition on the constructs is that it must be possible to infer consequences from them even when they agree only approximately with phenomena. The effect of this restriction is clear in Proposition 2 of Book III, which asserts that the planets (Mercury, Venus, Mars, Jupiter, and Saturn) are

43 As I. Bernard Cohen puts it (1980, p. 92). He observes that Newton writes in the continuation of the draft that the physical situation, "the simple orbit and the mean among all errors," can be identified with the construct, "the ellipse of which I have already treated," if "those minutiae" are "ignored" (p. 265).

kept in orbit by forces directed towards the Sun and inversely as the squares of their distances from it. Newton cites the phenomena that the planets obey Kepler's second and third rules: that they sweep out equal areas in equal times and that their periods are as the 3/2 power of their distances from the Sun. He does *not* take as a premise Kepler's first rule, that the planets move in ellipses with the Sun at a focus, although he shows in Book I (Proposition 11) that under these conditions their force toward it is inverse-square. Instead he uses a proof of the inverse-square relationship for bodies that move in concentric circles with uniform motion (Proposition 4). Smith makes clear that Newton did not believe, nor take his readers to believe, that the planets move uniformly in concentric circles. In fact, the ellipse was taken as the best approximation of their orbits by the leading orbital astronomers of the time – in sharp contrast to Kepler's second and third rules, which fit the data no better than alternative constructions.[44] Smith argues persuasively that Newton takes the second and third rules as premises because if they hold to a high degree of approximation, the force will be at least approximately centripetal and inverse-square. Newton proves as much in Book I.[45] Crucial evidence for Smith's interpretation is that Newton also treats orbits in which the force varies as a complicated function of the distance from the attracting center,[46] which can closely approximate Keplerian ellipses without the force being even approximately inverse-square.

Part of Newton's response to the challenge posed by the complexity of the true motions is thus to grant that generalizations such as Kepler's rules must be approximate to some degree, and to make his inferences from phenomena as secure as possible in light of the impossibility of eliminating equally precise alternatives.

But by proceeding in this way, Newton can infer only that the force law holds to some degree of approximation. His overarching aim is to show that *exact* mathematical generalizations are satisfied. Smith explains how Newton's way of dealing with the discrepancy between mathematical generalizations and actual phenomena serves this goal. According to Smith, the idealized representations are to be taken as approximations that *would* hold exactly under specified physical conditions. Newton's approach in Book III of the *Principia* is thus to seek physical conditions under which the phenomena taken as premises would hold exactly (and, in accordance with the Fourth Rule of Reasoning, to take the inferred force law as exactly true under these conditions). He proves, for

44 By Ismaël Boulliau, Nicholaus Mercator, and Vincent Wing. See Smith (2002b, p. 34) and the first of Smith's Suppes Lectures (2007a).

45 Specifically, in Corollary 2 to Proposition 3 and Corollary 7 to Proposition 4.

46 In the Scholium concluding Section 3, Newton considers ellipses in which the force tends to a point between the focus and (geometrical) center, and in Proposition 7, he considers circles in which the force is directed at, and Kepler's second rule is satisfied relative to, a point off the (geometrical) center. See Smith (2002b, pp. 37–41).

instance, that the planets *would* revolve in exactly the same orbits each time, and describe exactly equal areas in equal times, if they were subject to no force other than the gravitational force of the Sun.[47]

On this way of proceeding, mathematical generalizations are taken to express physical conditions, which may or may not be realized. Accordingly, discrepancies between the generalizations and actual motions are treated as indicative of physical factors. Smith claims that the "main purpose" of the idealizations is "to bring to light ways in which the observed world systematically deviates from the ideal" (2002b, p. 52). So the detail that a mathematical description fails to capture is no longer an "impediment limiting the quality of empirical evidence" (2002a, p. 155), but is instead harnessed to drive ongoing research. The rationale for treating a generalization in this way is precisely that deviations from it are capable of driving the process Smith calls "successive approximation." Here, then, is a further condition that mathematical systems must satisfy.

As Smith elaborates the condition, it is that it must be possible to construe "systematic differences between the idealized representations of the motions at any stage and observation themselves" as "evidence either for the original theory or for a refinement of it" (Smith 2002b, p. 47). For an idealization to be taken as a good approximation at a given stage, deviation from it must be small in magnitude. But for the process to succeed in the long run, there must be deviation that cannot be attributed to observational imprecision or inaccuracy. So residual discrepancy is not written off, as below the sensible threshhold, but made into information. When differences are systematic, they count as evidence that the world matches the mathematical system at least well enough that it can signal that the specified conditions are not met. Further evidence for the theory comes in the form of physical factors (sources of perturbing force) accounting for the discrepancy. Ultimately, continued iterations of what Smith calls "the process of successive approximations" constitute evidence of a third kind. When they yield "increasingly small residual deviations from current theory," they serve to more deeply entrench the presupposed force law as well to "tighten the range over which" it holds to a close approximation (2002a, p. 161).

This approach may well seem to beg the question of mathematics' relevance to nature (in the form that it confronts Newton, whether a mathematical result is also the specification of a cause). For it begins with the assumption that mathematical generalizations express physical conditions. According to Smith, the justification for "viewing every deviation from" a mathematical generalization "as physically significant" is just that the generalization "be well suited to initiate science by successive approximations" in the sense that "it

47 In Propositions 14 and 13, respectively, of Book III. See Smith (2005, p. 133) and the first of Smith's Suppes Lectures (2007a).

would hold exactly in certain identifiable circumstances."[48] Smith concedes that such an "hypothesis," which can be "established" only through the success of research "predicated on" it, must involve "an element of wishful thinking." But he holds that it is appropriate to venture one in a situation in which "if it does not [hold] – if the empirical world does not cooperate – then there is no apparent way to get beyond mere conjecture," while if it does, "there are prospects for empirically driven, sustained research" (2005, p. 151). To venture such assumptions is thus a condition under which alone a certain practice is possible. The risk they carry is "of being misled by the apparent high quality of the initial evidence obtained from the readily accessible data," when this evidence owes its promise to "accidental or parochial factors" (2005, p. 149).[49] Newton's response is to "immediately [push] the theory for all it is worth," in application to "problems that *prima facie* have nothing to do with the original evidence for it" (2002a, p. 165), and so expose limitations on the research strategy. The assumption is thus retrospectively justified by the success of the practice.

When mathematical generalizations drive research in this way, their content is causal in a sense that Janiak makes precise. First, they limit a range of ("previously disparate") phenomena that "have the *same* cause." Secondly, as they set topics for physical research they help to delineate the modalities of causation. To say, for instance, that "gravity is as the masses of the objects in question and inversely proportional to the square of the distance" is to say that "mass and distance are the only salient variables" in the "causal chain" from gravity from phenomena (2007, p. 142), leaving it to physics to fill in its links.[50]

This stepwise procedure for securing the physical significance of mathematical generalizations has an antecedent in the way postulates prove the reality of mathematical reasoning, according to Barrow. Barrow holds both that the possibility asserted by the postulates must be granted at the outset by a learner, and that it can in a way be proved by the senses, specifically by the experience

48 "[F]or then observed deviations from it would indeed reflect specific physical factors, and not just imprecision in a description" (2002b, pp. 50–51).

49 Smith gives as an example Newton's assumption (in later versions of Book II of the *Principia*) that when a body falls vertically in a fluid, the resistance is dominated by the inertia of the fluid, at least to such an extent that a law can be established for this force. The failure of his attempts to isolate species of resistance forces reveals the "'law'" to be "a mere curve-fit over a restricted domain." See (2005, p. 141) and (2002a, p. 164).

50 Smith gives a similar formulation of the conditions under which "a component of a mathematically characterized force can be considered" as a causal factor. He includes among them that the "respects in which [the force's] magnitude can vary must be given by a general law that is independent of the first two laws of motion." But Smith's point is that the force is characterized "physically" when this and further conditions are satisfied. In this respect, Smith differs from Janiak, who denies that the treatment of force in Book III is "physical" because Newton does not there "delineate the physical cause of the force", as would be "crucial to such a treatment" (Janiak 2007, n. 45).

of "correcting" roughness and excess. We have seen that for him, the activity that proves this possibility brings underlying matter into accord with geometrical conceptions (if it does not first create extended concrete objects). Barrow understands this activity concretely, as one of literally "paring away." But he emphasizes that the processes whose possibility is asserted by the postulates, and proved through experience of this activity, include the auxiliary constructions executed in the course of Euclidean proof. It is thus open to him (although not his stated view) to understand the practice that proves the possibility asserted in the postulates more abstractly, simply as *mathematical argument*. Then the "correcting" that vindicates the concession demanded at the outset could be understood as the derivation of results more adequate to physical reality, that is, such as to leave no unaccounted-for roughness or excess. As proof of the relevance of mathematics to nature, it would of course be less direct.

The way in which Newton's method is justified, on Smith's interpretation, is thus only a short reach beyond the work done by postulates on Barrow's conception. What remains is to explain how Newton can hold that postulates justify reasoning of the sort Smith outlines. I will suggest that not only can Newton draw on this understanding of postulates, but he can be seen to address one of its weaknesses, namely the unclarity in Barrow's conception of postulated activities.

4.3.5 Postulates and the practice of measurement

Newton agrees with Barrow that practice of a certain kind founds geometry. We can see this by revisiting Newton's conception of the relations between the sciences. It is only by having both theoretical and practical aspects that mechanics holds the place between geometry and physics. While mechanical theory and practice are merely two sides of a single discipline, this discipline's links to physics, on the one hand, and geometry on the other, are not of the same kind. As *doctrine*, mechanics is founded on physics.[51] But what suits it for its role in founding geometry, that of finding the required technique or craft, is its heritage as art. Because the operations assumed as possible in geometry are made real by making bodies move, to which "the manual arts are especially applied" [*praecipue versentur*], its foundation is mechanical *practice* (p. 382).

Newton maintains in his Preface that mechanics secures the possibility of, specifically, the operations *postulated* by geometry. So on his view, as on

51 Mechanics as practice does not depend on physics, or indeed on any other science. For it seems to have no principles of the kind that are proved in another science. At least, Newton claims at one point that mechanics "holds a place among the mathematical sciences through its axioms and demonstrations," but "not among the mathematical arts. Its practice is not founded on postulates but is purely manual" (*NMP* VIII, p. 179). Elsewhere, however, he lists postulates that are "lawful in mechanics" (177).

Barrow's, a certain practice proves the possibilities asserted by the postulates. We have seen that this practice can be understood in different ways: as human or divine; as operating on extant objects, or first bringing objects into being; abstractly or concretely. It is not clear how Barrow understands it, other than as concrete. Newton articulates the alternatives more precisely – without, however, seeming to choose among them.

In *Geometria*, Newton distinguishes between the genesis of geometry's subject-matter and the operations that comprise its practice. He makes clear that "the intention of postulates" is not to "teach the genesis" of figures (*NMP* VII, pp. 292–293). Postulates pertain, rather, to operations, and only the operations which geometry does *not* "teach how" to effect.

> Geometry neither teaches how to describe a plane nor postulates its description, though this is its whole foundation. To be sure, the planes of fields are not formed by the practitioner but merely measured. Geometry does not teach how to describe a straight line and a circle but postulates them; in other words, it postulates that the practitioner has learnt these operations before he attains the threshold of geometry. Once, however, these are previously understood and granted it teaches all the other operations of mensuration ...
>
> (*NMP* VII, pp. 288–289)

Newton insists that "both the genesis of the subject-matter of geometry" and "the effection of its postulates" are the concern of mechanics (pp. 288–289).[52] Though he refers to figures as "executed by God, nature or any technician you will" [*a Deo Natura Artifice quovis confectas*], he says little to explain how mechanics makes possible processes of the first sort.

The operations postulated by geometry, on the other hand, can be characterized more precisely. Newton extends the traditional division between practical and theoretical parts beyond mechanics into all of mathematics. Postulates, in general, pertain to the activity that makes mathematics art. Thus Newton writes that "the mathematical sciences" are "sciences inasmuch as they teach the truth by means of definitions, axioms, and theorems, but arts (skills) insofar as they deliver and exhibit its practice by means of postulates and constructions of problems" (*NMP* VIII, p. 179).[53] (And, as if to counter the implied distinction between art and science, he claims that "to set the very principles

52 A dimension of Newton's view is thus left out when he is taken to understand postulates as existence-claims (as in Garrison 1987, p. 611, and Guicciardini 2003, p. 417).

53 Newton writes in *Geometria* that the postulates of geometry ought to be "useful" "in that any practitioner should find them readily applicable in his measuring" (*NMP* VII, p. 291). Similarly, Barrow contends that "every Science is both Speculative and Practical: *Speculative*, as it *speculates*, *i.e.* seeks, investigates, and demonstrates Truths agreeable to its Object: and *Practical*, as those Truths when found and demonstrated, may be referred to Use, and reduced into Practice" (p. 50).

[of science] useless . . . would render the whole of science – and even its very name – empty and futile.") According to Newton, the practice that geometry "delivers and exhibits" is *measurement*: in it, problems are solved by applying postulates to "determine and set forth measures of all figures and magnitudes which are proposed by the definitions." Indeed, it can leave to others the problem of forming its objects precisely because its purpose "is neither to form nor move magnitudes, but merely to measure them" (*NMP* VII, p. 291). The operations it postulates are thus "those most necessary, useful, and expedient in the technique of measurement" (*NMP* VIII, pp. 179–181).

Newton suggests that the activity of measurement may have a divine practitioner. But because measuring is within our power in a way that creating bodies and initiating movement are not, on Newton's view the activity to which postulates pertain is shown by experience to be possible. Newton further improves on Barrow in that the practice whose possibility is asserted by the postulates, as he conceives it, more decisively proves the physical reality of mathematical conceptions. While the activity of making things "sensibly right" may leave discrepancy beyond our sensory threshold, which we can only ignore, Newton harnesses discrepancies to drive research that ultimately vindicates the mathematical theory as a description of nature. I believe what enables Newton to go further is his conception of the role of postulates in science. It is important to him that postulates should be acceptable to knowers facing constraints, because they are (on his view) key in overcoming those constraints. In particular, they secure the relevance of mathematical generalizations – to whose precision alone are our finite capacities suited – to causality in Nature.

Newton often speaks of measurement in concrete terms, as the laying-out of measuring sticks or rods,[54] so that its dependence on mechanics is obvious. When geometry is taken to deal with measures in this sense, its relevance to nature is likewise clear. But Newton also has a more abstract conception of measurement, under which even the theory of the *Principia* is included. Although he distinguishes between geometry and "rational mechanics" in the

54 In the conjectural history of geometry that opens *Geometria*, Newton clearly thinks of it this way. He reminds us that "geometry" "means the art of 'earth-measure'", and "a geometer was called by the Romans a 'ten-footer' from his measuring stick" (*NMP* VII, p. 287). He is thus led to qualify the distinction between the genesis of geometry's subject-matter and the operations that comprise its practice (which he presents as exclusive at e.g. *NMP* VII, p. 291). Newton charges geometry with the formation of those magnitudes used to measure others. His history begins with problems of measuring terrain that "were very speedily and accurately constructed by descriptions of straight and circular lines." While the practitioner "measured plane figures universally as he met with them," he "formed none by measuring other than rectilinear and circular ones. These, then, are the sole instruments of geometry" (*NMP* VII, p. 287; cf. p. 293: The description of "the sphere, the cylinder and the cone" was not postulated because they "were regarded not as measuring instruments but as magnitudes to be measured").

published Preface, he writes in the draft of a revision that the *Principia*'s contents are geometry in the "rather broad sense" of "that which instructs how to measure magnitudes described and defined not only by local motion but in any other manner whatever" (*NMP* VIII, p. 451). Since Newton counts the *Principia* as a mathematical science comprising a practice, he must conceive a role for postulates in it.

Their role can be understood as securing the presuppositions of the method Smith calls "successive approximation." The first step of this reasoning is to suppose of a mathematical generalization that under certain physical conditions it would hold exactly. This is to take it as an exact measure of physical factors. According to Newton, the assumption that physical items can be measured with exactitude is licensed by postulates. He writes in *Geometria* that plane figures "are measured by geometry in the hypothesis [*sc.* postulate] that they are exactly constructed," no matter how, in physical reality, they are formed (*NMP* VII, p. 289).

Newton goes on to say that geometry postulates "a technician who knows how to form straight lines and circles" in order to teach "how through their formation appointed magnitudes are to be measured" (*NMP* VII, p. 291). Postulates thus license the further assumption that the items dealt with in geometry can be not only objects, but *instruments*, of exact measurement. In this way, they secure the condition under which the method can continue into further stages. For the method iterates, not by measuring the deviation from a particular generalization, but by treating observed deviation as a measure of quantity. For example, in the argument of Book III the system comprised of Earth and the Moon is initially treated as isolated, as though it were subject to the force of no other bodies. In this way, Newton can establish the relationship between the weight of a satellite towards its central body and its "quantity of matter" (inertial mass).[55] The approximation is good, as Newton argues in Proposition 3, because the Moon is so much closer to Earth than to the Sun. But after asserting the universality of gravitational attraction, Newton returns to this system and indicates how mathematical results of Book I (Proposition 66 and its corollaries) can be employed to "find the forces of the sun that perturb the motions of the moon" given the deviations of the Moon's orbit from Kepler's first and second rules. Thus the discrepancies, like the original phenomena, are taken as measures of force, and the reasoning in progress is further justified by associating them with physical sources. Postulates thus ground a way of making mathematics ever more adequate to the infinitely complex totality of causes.

55 First, of course, by comparing the Moon's acceleration towards the Earth with the value for gravitational acceleration at the Earth's surface (measured by pendulum experiments). But also, by comparing the acceleration of the satellites towards the Sun with their acceleration towards their central bodies (Newton 1999, p. 808). See Harper (2002a, pp. 187–189).

Conclusion

I have contended that if we take Newton to address the relevance of mathematical reasoning to causal relationships in Nature in the opening remarks of his Preface, we see how they pertain to the "mathematical"/"physical" distinction drawn there, and the results that follow.

As Newton's manuscripts show, he holds (with Barrow) that postulates guarantee that bodies and motions in Nature can be found to agree with the exact descriptions of geometry. Like Barrow, Newton emphasizes that geometry comes to have objects by means of certain activities: thus he claims in the Preface that what geometry postulates (namely the finding of straight lines and circles) depends specifically on the *practice* of mechanics. Newton appears to resolve the ambiguity in Barrow's conception of this practice. As I interpret him, he takes it to consist of the reasoning carried out in the course of the *Principia.*

Barrow aims to reconcile geometry's precision with the observable crudity of Nature. He forswears the attempt to specify the causal powers of bodies and varieties of matter, on the grounds that there are none. In stark contrast, Newton is concerned with the unceasing multiplicity of causal factors, on account of which every mathematical description of a natural process is liable to be replaced by a more complex alternative. I have explained that when he takes as a postulate that magnitudes can be measured exactly, he can be taken to mean that the postulates license us to take some of these approximations as exact. By asserting the possibility of this practice of mathematical reasoning, they rationalize the "wishful thinking" it involves. They thus make it possible to cross the threshold of a science which, as Smith, Harper, and Janiak explain, can specify causes in Nature.

Note regarding citations

Newton's *Principia* is cited according to the Cohen and Whitman edition (1999). Citations from Newton abbreviated *NMP* are from Newton (1967–1981). Blancanus's *De Mathematicarum Natura Dissertatio* is cited according to the translation in Mancosu (1996). Barrow's *Lectiones Mathematicae* are cited according to the translation by John Kirkby (1734/1970). Euclid's *Elements* are cited according to the Heath edition (1956).

PART II

Philosophical themes in Newton

5

Cotes's queries

Newton's empiricism and conceptions of matter

ZVI BIENER AND CHRIS SMEENK

5.1 Introduction

The relation of Isaac Newton's natural philosophy to his method of inquiry is of central importance to Newtonian scholarship. In this chapter, we investigate this relation as it concerns Newton's ideas about the nature and measure of matter. We argue that a conflict between two conceptions of "quantity of matter" employed in a corollary to proposition III.6 illustrates a deeper conflict between Newton's view of the nature of extended bodies and the concept of mass appropriate for the *Principia*. The conflict was first noted by the editor of the *Principia*'s second edition, Roger Cotes. His "two globes" objection demonstrates that Newton employed two different measures of "quantity of matter," related to competing views on the nature of matter. On what we call the "dynamical conception of matter" – dominant in the *Principia* – quantity of matter is measured through a body's response to impressed force. On the "geometrical conception of matter," quantity of matter is measured by the volume a body impenetrably fills. The discussion with Cotes reveals Newton's commitment to the geometrical conception: he assumes all atoms have a uniform specific gravity; i.e., that the inertia of completely filled bodies is proportional to their volume. On the dynamical conception of matter, there is no reason for this proportionality to hold. A purely dynamical conception is consistent with the inertia of completely filled bodies varying in proportion to their volumes, or with bodies treated as non-extended point particles. By analyzing the exchange with Cotes (and related texts), we show that before Cotes's

Many thanks to J. E. McGuire, who along with B. R. Goldstein, was our first guide to Newton, for insight and encouragement. Additional thanks to others who provided us with comments and discussion: Katherine Brading, Tyler Burge, John Carriero, William Harper, Andrew Janiak, Eric Schliesser, George Smith, Sheldon Smith, Howard Stein, Wouter Valentin and other participants in the "Newton and/as Philosophy" conference. Our treatment of Newton's method of inquiry is indebted to Howard Stein's work.

prodding in 1712, Newton held both conceptions of matter and apparently saw no conflict between them.

We trace Newton's failure to recognize the conflict between the two conceptions to Newton's allowance for the justification of natural philosophical claims by two types of a posteriori, empiricist methodologies, which both turn away from the a priori Cartesian approach Newton deplored. Although both proceed "from the phenomena," we argue that there are important differences between them. The first, underlying the dynamical conception, is mathematical and relies on a nuanced interplay between specific phenomena and their theoretical descriptions. Recent work by George Smith, Bill Harper, and Howard Stein has shown how this methodology was used in the *Principia* to justify the laws of motion. Drawing on their analyses, we briefly characterize this method, using Newton's reply to Cotes's better-known "invisible hand" objection as an illustration. The second empiricist method, underlying the geometrical conception, also proceeds from the phenomena, but does not draw on the technical resources used in the first. Instead, its conclusions are intended to follow from general features of our experience, in a way articulated most clearly in *De gravitatione* (hereafter, *DG*) and through certain of Newton's examples in Rule III of the *Regulae Philosophandi.* We argue that although both methods of inquiry are based on empirical considerations, the relationship of theory to evidence in each is distinct. Centuries of debate attest to the difficulty in extracting from Newton's methodological discussions a clear account of evidential warrant that spans all of his work. We do not tackle this general project here. Instead, we highlight two different types of arguments from the phenomena endorsed by Newton, and argue that he failed to clearly distinguish them. He thus failed to recognize that one was not as secure as the other. In the *Principia* and *DG*, the two conceptions of matter are justified by these different types of arguments, yet prior to Cotes's "two globes" objection, Newton treated the two conceptions as if on equal footing, without recognizing their different sources of warrant. Cotes's objection forced Newton to reconsider the status of the geometrical conception. Although he never drew general conclusions regarding the relation between his two methods of inquiry, he came to side with the more sophisticated method of inquiry and, in revisions prompted by Cotes, stated the geometrical conception hypothetically. Given the deep-seated Cartesian and atomistic roots of the geometrical conception in Newton's thought, this was a profound shift.

We begin (in Section 5.2) by introducing the geometrical and dynamical conceptions of matter and the associated measures of quantity of matter. To do so, we review the reasons that led Newton to abandon aether theories of gravitation and accept the existence of void spaces. Although these forced Newton to reject much of the Cartesian analysis of space and body, we show that the geometrical conception of matter present in *DG* and later work betrays a lingering debt to Descartes. At the close of Section 5.2, we explicate the a posteriori method of inquiry underlying the geometrical method of quantifying matter. In Section 5.3 we turn to the a posteriori method underlying the dynamical conception of matter and the argument for universal gravitation (AUG). Drawing on these accounts of the contrasting methods, we investigate the conflict between the two conceptions of matter in Section 5.4.

5.2 Newton's conceptions of matter

5.2.1 Against the mechanical aether and Cartesian body

In the years leading up to the *Principia*, Newton's natural philosophy underwent a "radical conversion" (borrowing Westfall's phrase); he abandoned the fundamentals of Cartesian natural philosophy and replaced them with novel conceptions of space, motion, and body. This radical conversion was motivated in large part by Newton's rejection of the idea that the planets are carried in their orbits by an aetherial vortex. We begin by elucidating two empirical reasons that led Newton to abandon a gravitational aether and clarifying how this a posteriori line of reasoning undermined Cartesian philosophy.

As early as 1664, Newton took the cause of terrestrial gravitation to be mechanical and formulated a mechanical aether theory akin to other contemporary theories. The central idea of these theories – that a body's weight could be explained in terms of an aetherial fluid exerting pressure on a body's inner surfaces – appears to have persisted in Newton's accounts through the 1670s, even as the gravitational aether became more intricately tied to active principles inspired by his alchemical studies.[1] The central role played by the gravitational aether in the 1670s makes its nearly complete (if temporary) disappearance from Newton's natural philosophy in the period preceding the composition of the *Principia* remarkable.[2]

In *E1* Newton gave two decisive empirical reasons for abandoning the aether. (*Principia* editions are abbreviated as *E1*, *E2*, and *E3*.) First, he became convinced that the planets and comets encounter negligible resistance to their motion. In the first two theorems of the *De motu corporum in gyrum*, Newton derived Kepler's area law and the harmonic law for a central force *with no resistance.* The accuracy of Kepler's law in describing planetary motions implied that there was no need to introduce a resisting force alongside the centripetal force holding the planets in their orbits.[3] Newton strengthened his case in later

1 See "Of Gravity and Levity," in Newton (1983, pp. 362–365, 426–431), and Wilson (1976, pp. 192–195) regarding Newton's early views, and Dobbs (1991, chapter 4) regarding the development of Newton's views through the 1670s.

2 Dobbs (1988) details Newton's abandonment of the aether at the time of *E1*; McGuire (1966) discusses later shifts in Newton's views.

3 Newton was aware that even if there is no resistance Kepler's laws fail to hold *exactly* for universal gravity due to the perturbing effects of each planet on the other planets'

drafts and in the *Principia.* The persistence of planetary motion over thousands of years is also incompatible even with slight aetherial resistance, which would lead to a steady decrease in quantity of motion (Herivel 1965a, p. 302). Although the negligible resistance encountered by the planets is compatible with an aetherial vortex in which the planets move with the aether, reconciling the motion of comets, especially retrograde and highly eccentric ones, with an aetherial vortex is difficult.

Second, Newton failed to detect aether resistance in a series of pendulum experiments reported in Book II. Based on the realization that a gravitational aether must penetrate to the inner surfaces of bodies – without such penetration, the aether's action could only depend on a body's surface area – Newton designed experiments to measure the *internal* resistance due to the aether.[4] Newton constructed a pendulum consisting of a "round firwood box" suspended from a cord. He measured the oscillations of the empty box, then filled the box with various metals, adjusting the cord to the same length. The metal-filled box weighed 78 times as much as the empty one; in the absence of internal resistance Newton expected the oscillations of the full pendulum bob (from the bob's increased inertia) would take 78 times as long to decay. Newton initially assumed that filling the box would not change its *external* resistance. Because the decay only took 77 times as long, Newton concluded that the internal resistance must be over 5,000 times less than the external resistance. In *E2* and *E3*, Newton interpreted this to mean that *aether resistance* caused the damping:

> This argument depends on the hypothesis that the greater resistance encountered by the full box does not arise from some other hidden cause but only from the action of some subtle fluid upon the enclosed metal.
>
> (Newton 1999, p. 723)

However, in *E1* (in a passage subsequently omitted), Newton proposed a different cause:

> But I suppose that the cause is very different [than the aether acting on the internal surfaces of the box]. For the times of the oscillation of the full box are less than those of the empty one, and therefore the resistance to the *external surface* of the full box is greater, by virtue of its velocity and the length of its oscillations, than to the empty box. From which it follows that the resistance [due] to the internal parts of the box is either zero or entirely insensible.
>
> (translated in Kuhn 1970b, pp. 106–107)

orbits (Herivel 1965a, p. 301). However, these departures from Keplerian motion differ in character from the departures Newton expected for a resisting medium.

4 See Newton (1999, pp. 722–723). Note that *external* resistance may arise due to the air alone or the air and the aether conjointly, but the experiment is designed so that external resistance (as well as buoyancy of the air) is held constant. Cf. Kuhn (1970b, pp. 106–108).

Newton's conclusion is phrased cautiously. He did not claim that the aether does not exist; instead, he inferred only that if there is an aether, then its *resistance* is either nil or negligible. However, he had also concluded from other experiments (reported in the same scholium) that the primary contribution to resistance is proportional to the *material density* of the fluid through which an object moves. Thus, the experiments gave Newton grounds to reject a *mechanical* aether, although with his usual care he did not claim that they rule out an aether altogether (cf. Smith 2001b).

Although these considerations triggered Newton's radical conversion, they were not decisive for his contemporaries and successors. In Newton's later treatment in the *Principia*, fluid resistance arises primarily from the inertia of the fluid, and the dominant component of the force of resistance is proportional to ρv^2 (ρ is the density of the fluid, and v is the relative velocity). Leibniz demurred to Clarke regarding this assumption, arguing that "it is not so much the quantity of matter as its difficulty in giving place that makes resistance" (Alexander 1956, p. 65). Leibniz had earlier distinguished between two sources of resistance, viscosity and density, and argued that they make distinct contributions to the overall resistance for different types of fluids. Drawing this distinction between different types of resistance opens up the possibility of an aetherial fluid with resistance not proportional to density – which would avoid Newton's arguments. In fact, the possibility is much easier to realize than Newton had anticipated (Smith 2001b). In 1752, d'Alembert showed that a fluid without viscosity has exactly zero resistance, undercutting Newton's proposal that the dominant contribution to fluid resistance arose from the fluid's inertia. This error does not detract from Newton's insight that a single force law was sufficient to account for planetary motions, but it does undermine Newton's empirical case against the aether. Yet the problem was not solely with *Newton*'s arguments against the aether. Contemporary versions of aether theory were also based on misconceptions regarding fluids and the nature of resistance, and any aether theorist faced the challenge of elucidating how the aether produced gravitational effects without causing appreciable resistance (cf. Aiton 1972).

Newton's rejection of a mechanical aether left him without a mechanical explanation of gravitation, along with an awareness of the obstacles to providing one. Here we focus on one fundamental consequence of this awareness for Newton's thought: he was forced to reconsider Descartes's doctrines regarding the nature of body and space, and replace them with ones compatible with the existence of void spaces.

Newton's most sustained critical discussion of Descartes appears in *DG*. The stated aim of the manuscript is the study of the gravitation and equilibrium of fluids. Written in the geometrical style, it begins with a series of definitions and closes with two theorems. Newton makes room for his own definitions of space, body, and motion with a long philosophical discussion expressly devoted to undermining the corresponding Cartesian definitions – to "dispos[ing] of [Descartes's] *figmenta*." The main thrust of this digression is that an adequate definition of motion requires an appropriate structure relating locations over time.[5] Descartes's plenum lacked the necessary structure, leaving Descartes with a definition of motion that failed to support distinctions fundamental to his physics. Newton overcame this defect by introducing space as a distinct entity with a sufficiently strong structure, albeit an entity that did not fit neatly into traditional ontological categories.[6]

Even in this overtly philosophical context Newton supported his arguments against Descartes with empirical evidence in favor of void space. On the basis of pendulum experiments (that may have been either the experiments discussed above or precursors), Newton asserted that the resistance of the aether is "over ten or a hundred thousand times *less*" than the resistance of quicksilver (Newton 2004, p. 35, emphasis added).[7] Newton also took resistance to moving through a medium as a consequence of the material nature of the medium's parts. As he put it, "if we set aside altogether every resistance to the passage of bodies, we must also set aside the corporeal nature [of the medium] utterly and completely" (Newton 2004, p. 34). Because two bodies cannot simultaneously occupy the same region of space, one body resists the passage of another body through the region it occupies. Though controversial, if this view is accepted then the failure to detect resistance is decisive evidence against the Cartesian plenum.

Rejecting the plenum posed a clear challenge to the Cartesian identification of extension as the principal attribute of body. In *DG*, Newton singled out Descartes's so-called "elimination argument" (*Principles of Philosophy*, II.4 & II.11) as the main argument for this thesis. According to Newton, Descartes argued that various sensory properties such as hardness, weight, and color can be abstracted from a body without endangering its status *as a body*. Only the elimination of extension can destroy a body's corporeality, and so extension alone constitutes body's principal attribute, or, as Newton put it, "pertain[s] to [body's] essence" (Newton 2004, p. 21). To Descartes's argument, Newton countered that to be recognized as body, a body had to possess not only extension, but "faculties," particularly the ability to stimulate perceptions and

5 What is actually required in the *Principia* is the distinction between inertial and non-inertial motion; this only requires an affine connection and not the stronger structure that would be provided by identifying the "same position" over time; see Stein (1967). However, Newton was apparently unclear on this issue in *DG*; some of his criticisms of Descartes presume a stronger structure than necessary.

6 See Stein (1967), Rynasiewicz (1995a,b), DiSalle (2002).

7 Based on this brief allusion it is unclear how these experiments relate to discussions in other texts. See Dobbs (1991, pp. 134–143), and Westfall (1971a, pp. 341, 375–77) for further discussion.

"transfer action" to other bodies.[8] The core of Newton's critique was the claim that "although philosophers do not define substance as an entity that can act upon things, yet everyone tacitly understands this" (Newton 2004, p. 21). Newton – *contra* Descartes – held that what we should primarily care about is not what a substance is, but what it does.

This difference of orientation is also evident in the stated aim of Newton's speculation regarding body. By contrast with Descartes, Newton's goal in *DG* was the development of an account of body *sufficient* to serve as a basis for physical theory and *sufficient* to capture the phenomenal properties of bodies, the properties of "a kind of being similar in every way to bodies, and whose creation we cannot deny to be within the power of God, so that we can hardly say that it is not body" (Newton 2004, p. 27). Newton was clear, however, that he could not establish more than the sufficiency of his account. In particular, he made no claims to reveal the necessary essence or nature of body.[9]

5.2.2 The geometrical and dynamical conceptions of matter

But we must not stress only the *differences* between Newton and Descartes. Although Newton's conception of body in *DG* differed from Descartes's both in its content and metaphysical pretensions, it still possessed vestiges of Cartesianism. While defining body in terms of regions of space endowed with additional attributes – attributes foreign to Descartes's account – Newton still followed Descartes by treating bodies as regions of space, as extended geometrical structures, albeit not geometrical structures *simpliciter*. In *DG*, the character of bodies is partially dependent on the character of space. Space, in turn, has geometrical structure – it is full of "all kinds of figures, everywhere spheres, cubes, triangles, straight lines, everywhere circular, elliptical, parabolical, and all other kinds of figures, and those of all shapes and sizes, even though they are not disclosed to sight" (Newton 2004, p. 22 ff.). Bodies, as regions of space, are consequently geometrical, although they admit non-geometrical properties as well.

8 Newton was also familiar with the predecessor of the argument of *Principles* II.4 in the *Second Meditation*, but does not address it explicitly in *DG*; see Harrison (1978, p. 132) and Newton (1983, p. 23).

9 The epistemological status of Newton's accounts of space and body differs. While Newton emphasized the tentative status of his account of body – as one possible account compatible with experience – he did not treat his account of space as similarly conjectural and tentative, as Stein (2002) emphasizes. Consequently, when we speak of the "nature" of body according to Newton, we do not mean to impute to him any form of essentialism or a conception of natural philosophy according to which the goal of philosophizing is to draw observable consequences from the natures of ontological primitives.

In *DG*, Newton treated bodies as "*determined quantities of extension which omnipresent God endows with certain conditions*," namely:

> (1) that they be mobile; and therefore I did not say that they are numerical parts of space which are absolutely immobile, but only definite quantities which may be transferred from space to space;
>
> (2) that two of this kind cannot coincide anywhere; that is, that they may be impenetrable, and hence that oppositions obstruct their mutual motions and they are reflected in accord with certain laws;
>
> (3) that they can excite various perceptions of the senses and the imagination in created minds.
>
> (Newton 2004, pp. 28–29)

Central to this account of body is the notion that bodies are primarily "determined quantities of extension." The reliance on a determinate spatial substratum as a precondition for the existence of bodies is one of the main features of Newton's account. After providing the above definition of body, Newton emphasized one of its main anti-Aristotelian implications; namely, that it does away with the need for a property-less substratum as the metaphysical support for properties and forms and instead makes do with space itself:

> [F]or the existence of [bodies] it is not necessary that we suppose some unintelligible substance to exist in which as subject there may be an inherent substantial form; extension and an act of the divine will are enough. Extension takes the place of the substantial subject in which the form of the body is conserved by the divine will; and that product of the divine will is the form or formal reason of the body denoting every dimension of space in which the body is to be produced.
>
> (Newton 2004, p. 29)

Newton's analogy between his account and hylomorphism demonstrates that extension was as central to his conception of body as the substantial subject was for the conception of body of his Aristotelian adversaries. On Newton's account, extension is necessary for the application of so-called "form" and thus for the existence of body. "Body," as defined in *DG*, "is that which fills space" (Newton 2004, p. 13). It is not necessarily that which gravitates, nor that which moves, nor that which is tangible and visible (although it may also be any of those).

The conception of body as impenetrable extension takes precedence in *DG* over the nascent conception of body defined by laws of motion. In *DG* Newton held that bodies must move "in accord to certain laws," but the phrase does not acquire special significance without the juxtaposition of *DG* against later texts. Taken by itself, *DG* defines body primarily as a region of filled-in extension and only secondarily in terms of laws governing motion. On Newton's account, tangibility, visibility, and other traits that constitute the "corporeality" of matter according to our senses all depend, primarily, upon the impenetrability of regions of space (Newton 2004, pp. 27–28). Motion has a secondary role in constituting that corporeality because motion only makes impenetrability manifest to our senses. Newton only introduced motion once regions of space were rendered impenetrable:

> [W]e may suppose that there are empty spaces scattered through the world, one of which, defined by certain [spatial] limits, happens by divine power to be impervious to bodies, and by hypothesis it is manifest that this would resist the motions of bodies and perhaps reflect them, and assume all the properties of a corporeal particle, except that it will be regarded as motionless. If we should suppose that that impenetrability is not always maintained in the same part of space but can be transferred here and there according to certain laws, yet so that the quantity and shape of that impenetrable space are not changed, there will be no property of body which it does not possess.
>
> (Newton 2004, p. 28)

Even if mobility has only a secondary status in this passage, it is still essential to Newton's account both here and in the "determined quantities of extension" passage. However, our point is that mobility *of impenetrable regions* is essential, not mobility taken by itself. The centrality of the impenetrability of the extensional substratum reveals Newton's residual Cartesianism: in *DG* he considers bodies to be essentially extended geometrical structures – geometrical structures made real by further conditions, but geometrical structures nevertheless.

Newton's manner of quantifying body in *DG* further illustrates his residual Cartesianism. Newton measured quantity of matter through a body's geometrical rather than dynamical properties. After defining the absolute quantity of force as a product of the force's intension ("the degree of its quality") and extension ("the amount of space or time in which it operates") Newton wrote:

> [M]otion is either more intense or more remiss, as the space traversed in the same time is greater or less, for which reason a body is usually said to move more swiftly or more slowly. Again, motion is more or less extended as the body moved is greater or less, or as it is diffused through a larger or smaller body. And the absolute quantity of motion is composed of both the velocity and the magnitude of the moving body.
>
> (Newton 2004, p. 37)

In modern terminology, Newton equated momentum (the "force of motion") to the product of the velocity (intension) and the "magnitude of the moving body" (extension). The latter is measured by the body's volume ("the amount of space in which [the force of motion] operates") rather than by the body's resistance to impressed forces (i.e., inertia). Given that (from the *Waste Book* onward) Newton equated force of motion with the product of velocity and *quantity of matter*, here Newton measures quantity of matter through its volume or quantity of extension (see Herivel 1965a, p. 26). We call this method of quantification, along with Newton's account of the nature of bodies as primarily determined quantities of extension, Newton's *geometrical conception of matter*.

Two caveats must be made regarding this geometrical conception. First, it is Cartesian in inspiration without being wholly Cartesian. Newton did not attempt to reduce all of a body's properties to geometrical properties, nor to treat any single property as a body's principal attribute. However, Newton did follow Descartes in considering extension as essential to our understanding of body *and* to the practice of physics vis-á-vis the measure of the quantity of matter associated with body. Second, although we have highlighted the geometrical conception's indebtedness to Newton's Cartesianism, the conception is also closely tied to Newton's atomism, particularly his belief in the uniformity of nature. This is an important aspect of Newton's thought, but we can only touch on it briefly in Section 5.4 below.

In *DG*, the geometrical measure of quantity of matter is not supplemented with a precise dynamical measure, as it is in the *De Motu* drafts and the *Principia*. According to what we call the *dynamical conception of matter*, a quantity of matter is measured by its response to impressed force, not by the volume of space which it impenetrably fills. As with the geometrical conception, the dynamical conception also incorporates a view regarding the nature of bodies, which we consider shortly. Newton introduced the dynamical measure of quantity of matter in Definitions I and III of the *Principia*. In Definition III, Newton states that the internal force of a body (its *vis insita*) "is always proportional to the body and does not differ in any way from the inertia [*vis inertia*] of the mass except in the manner in which it is conceived" (Newton 1999, p. 404). We are to understand that *vis insita* is also proportional to a body's *quantity of matter* since Definition I states that:

> I mean this quantity whenever I use the term "body" or "mass" in the following pages.
>
> (Newton 1999, p. 404)

Together with Law II, these two definitions establish a proportionality between a body's quantity of matter and the force responsible for the body's dynamical properties.[10]

Although Definition I also states that "Quantity of matter is a measure of matter that arises from its density and volume jointly," the quantification

10 See McGuire (1994) on the nature of *vis insita* and Bertoloni Meli (2006a) on its connection to *vis centrifuga*.

method implied by Definition III is used throughout the *Principia* almost exclusively. Even in Definition I Newton made explicit that quantity of matter "can always be known from a body's weight for – by making very accurate experiment with pendulums – I have found it to be proportional to the weight."

This is a far cry from *DG*. Of course, Newton did define force in *DG* as either "external" – "one that generates, destroys, or otherwise changes impressed motion in some body" – or "internal" – "by which existing motion or rest is conserved in a body, and by which any body endeavors to continue in its state and opposes resistance" (Newton 2004, p. 36). But, unlike in the *Principia*, the dynamical method is not used to *quantify* body.[11] Lacking a clear statement of Law II, the dynamical measure of a quantity of matter remains vague in *DG* and intertwined with the conception of body as that which fills space. *De gravitatione* adumbrates the dynamical conception of matter, but does not contain it fully and certainly does not contain its central element, the measurement of quantity of matter by a body's response to impressed force. In the *Principia*, the two methods of quantifying matter co-exist, but the geometrical conception is relegated to the wings while the dynamical conception takes center stage.

What account of the nature of bodies accompanies the dynamical measure of matter in the *Principia*? The *term* "body" appears in the definitions and laws, but Newton does not explicitly define "body" or provide an account of body's possible nature like the one provided in *DG*.[12] Nevertheless, in contrast to *DG*, in the *Principia* Newton characterizes material bodies almost exclusively by their dynamical properties. This suggests a transformation in Newton's view: the *Principia* provides clear formulations of the concept of force and the laws of motion, but bodies are defined only derivatively – as the entities subject to forces and for which the laws of motion hold. The nature of body in the *Principia* thus depends upon whatever constraints are implied by satisfaction of the laws. Furthermore, empirical support for this view of bodies derives from the empirical support for the laws of motion and the physical theory based on them.

Yet the dynamics of the *Principia* place surprisingly weak constraints on the nature of body. In particular, "bodies" satisfying the laws of motion need

11 In *DG* Newton used the term *vis inertia* for the internal principle of motion (Newton 2004, p. 36) in much the same sense as he used the term in an excised portion of Definition 1 in the Lucasian lectures (1685, Newton 1967–1981: Vol. V). This contrasts with his use of *vis insita* in the early *De Motu* drafts (see Herivel 1965a, pp. 26–28). We thank George Smith for this point. Yet despite the appearance of the term *vis inertia*, *DG* is a transitional text which only hints at the concept of *vis inertia* developed in the *Principia*. This is not surprising, since in *DG* Newton is working out the metaphysics of "natural power" on which *vis insita* depends; see Stein (1990).

12 Newton did define "body" in unpublished definitions intended for *E3* of the *Principia*, but these only bolster the argument below, cf. McGuire (1966). We discuss them briefly in Section 5.4.

not have any geometrical properties whatsoever. This may appear to conflict with Newton's various theorems regarding extended bodies, such as the famous proofs to the effect that a spherical body can be treated as if the mass were concentrated at a point (Propositions I.71–75). However, even these proofs only require that the force acting on or produced by the whole body is the sum over forces related to its constituent parts. They do not require the attribution of geometrical properties to the parts of the spherical bodies, and are compatible with bodies treated as Boscovichian point-particles characterized by parameters such as "quantity of matter" that have no geometrical basis. The austerity of the dynamical conception of matter stems from the limited mathematical framework of the *Principia.* The generalization of Newtonian theory to continuum mechanics leads to a richer notion of body that does have implications for the geometrical properties of bodies.[13]

That said, *DG*'s geometrical conception of matter did not disappear from Newton's thought following the elaboration of the dynamical conception in the *Principia.* In drafts of corollaries to III.6 written in the 1690s, Newton assesses the connections between gravitational aethers, matter theory, and the existence of void (McGuire 1967). In doing so, he assumes that the appropriate measure of quantity of matter is the volume of the basic particulate constituents of matter. These manuscripts indicate that Newton continued to take the geometrical conception of matter seriously post-*Principia.* Finally, Newton's reply to Cotes's "two globes" objection – also concerning III.6 – relies on the geometrical conception, and took place over 20 years later, in 1711. We return to the two globes objection in §IV. We now turn to the two types of a posteriori, empiricist arguments we believe are associated with the two conceptions of matter.

5.2.3 The a posteriori character of the geometrical conception of matter

How did Newton establish his geometrical conception of body a posteriori? Two distinct a posteriori contributions can be discerned. First, the results of pendulum experiments and the accuracy of Kepler's "laws" pushed Newton to reject the Cartesian identification of body with extension. In this regard, his path towards a new conception of body is similar to his path towards a new conception of light in his optical work (Shapiro 2004; Stein ms.) In both, Newton took a narrow set of experimental results to be sufficiently crucial to warrant revision to a fundamental concept of natural philosophy.

But there is an important difference: whereas the prism experiments, the crucial experiments in Newton's research on colors, were used to both refute

13 For example, Cauchy's generalization of Newton's laws involves contact forces and the outward normal defined over the contact boundary; Boscovichian point particles lack boundaries and Cauchy's formulation does not apply, cf. Truesdell (1968) and Smith (2007).

the extant conception of light and suggest a new conception (i.e., that white light is not a natural kind but is composed of individually homogeneous rays of differing refrangibilities), the pendulum experiments were used only to refute the Cartesian doctrine. On our view, Newton rejected an account of gravity based on the results of pendulum experiments along with his success in modeling planetary motions using a single force law. This rejection undercut the associated Cartesian accounts of body and space. However, the constructive element of Newton's geometrical conception of body was not secured by an *experimentum crucis*; rather, it seems to have been secured by a different type of argument from the phenomena.

This argument proceeds from the experience of any body whatsoever. Newton attempted in *DG* to provide an account of body that is sufficient for capturing the "evidence of our senses" (Newton 2004, p. 28). The traits of body he aimed to save were all quite generic and are reflected in the overall character of our experience; for example, that body is visible, tangible, audible, etc. Newton's account of body as mobile, impenetrable, and sensible extension is able to save these traits because it is set against a framework of natural philosophical presuppositions – e.g., that an object is visible because it reflects light and audible because it can move adjacent air – but, given this framework, the evidential basis for Newton's account includes any and all experiences of body. Importantly, success within this framework does not rely on any quantitative notion of "strength of evidence" that can help arbitrate between Newton's account and possible competitors – where by "strength of evidence" we mean any measure of the fit between a given theory and its evidential basis that allows discrimination among competing theories according to degree of evidential warrant. Rather, it relies on a notion of warrant akin to the one mechanical philosophers used to justify their mechanical models, but one that does not appeal to first principles or privileged modes of explanation. Strikingly, it does *not* rely on the sophisticated notion of warrant used in the AUG. Note Newton's explicit reference to the underdetermination of *DG*'s account of body:

> [I]t is hardly given to us to know . . . whether matter could be created in one way only, or whether there are several ways by which different beings similar to bodies could be produced . . . [H]ence I am reluctant to say positively what the nature of bodies is, but I would rather describe a certain kind of being similar in every way to bodies, and whose creation we cannot deny to be within the power of God, so that we can hardly say that it is not body.
>
> (Newton 2004, p. 27)

Newton is explicitly open to the possibility that another hypothesis regarding the nature of body can save the phenomena equally well.

This a posteriori method of arriving at claims regarding the nature of matter resembles the one offered in Rule III and its drafts (McGuire 1968; 1970). In Rule

III, Newton claimed that certain qualities of bodies are "universal," qualities that can be attributed to any body whatsoever and so constitute the core of our understanding of body (Newton 1999, p. 796). Often, Newton referred to such claims of universality as being "deduced from phenomena" (e.g., Newton 1999, p. 943). Body's universal qualities are extension, hardness, impenetrability, mobility, and inertia. However, Newton's evidence for their universality is not homogeneous. One of our theses is that deducing or gathering propositions "from phenomena" does not have a univocal meaning for Newton, and so the resemblance of Rule III to *DG* concerns some universal qualities – more will be said about the others in Section 5.3.3.

First, Rule III, like *DG*, appeals to our general experience of bodies as the evidential basis from which claims regarding the extension, hardness, and impenetrability of bodies ought to be drawn. Newton wrote, echoing *DG*, that:

> The extension of bodies is known to us only through our senses . . . [and] because extension is found in all sensible bodies, it is ascribed to all bodies universally. We know by experience that some bodies are hard . . . [and] justly infer from this not only the hardness of the undivided particles of bodies that are accessible to our senses, but also of all other bodies. That all bodies are impenetrable we gather not by reason but by our senses. We find those bodies that we handle to be impenetrable, and hence we conclude that impenetrability is a property of all bodies universally.
>
> (Newton 1999, p. 795)

In each case, our experience of bodies broadly conceived forms the evidential basis of the generalization. Still, the evidential basis recommended by Rule III is more restrictive than the one used in *DG*. According to Rule III, only those qualities found in "all bodies on which experiments can be made" *and* passing the intension and remission criterion may be "taken as qualities of all bodies universally" (Newton 1999, p. 795). Thus only *some* features of our experience of bodies remain relevant to generalization about body; visibility and audibility, for example, are eliminated. Nevertheless, the remaining features are those that are truly general and are present in *all* experiences of body. Achieving this generality is precisely the point of Newton's application of the intension and remission criterion. Any quality that is not *always* present in our experience of bodies – i.e., one that can be remitted to zero and thus disappear, or one that is not present in some bodies – is not universal.

Second, regarding the first four qualities mentioned, Rule III, like *DG*, does not invoke a notion of evidential warrant similar in complexity to the one used throughout the *Principia*. This is because, while the intension and remission criterion can be made precise, it is unclear when in the course of empirical investigation we can be content that it is satisfied for "all bodies on which experiments can be made." Newton's examples do not help. According to Newton, the extension of bodies is made manifest in *all sensible bodies*. However, we know by experience that hardness is only found in *some* bodies while impenetrability is only found in "those bodies that we *handle*" – presumably a smaller class than "all *sensible* bodies." Is the judgment of universality regarding one of these better than the others? Newton suggested, but did not elaborate, a notion of strength of evidence: "the argument from phenomena will be *even stronger* for universal gravity than for the impenetrability of bodies, for which . . . we have not . . . even an observation, in the case of the heavenly bodies" (Newton 1999, p. 796, emphasis added). Something like simple enumerative strength seems to be at work here: the more instances of a quality we have, the stronger the judgment of its universality. This is a far cry, however, from the sophisticated and more robust relation between theory and evidence implicit in the *Principia.*

The lack of a robust notion of evidential warrant or strength would not be bothersome by itself, but we argue in Section 5.4 that, on at least one occasion, Newton overstated the evidence in favor of the geometrical conception of matter. The reason, we argue, is that Newton failed to distinguish the type of argument given in *DG* for the geometrical conception of matter from the type of argument used in the *Principia.* In order to clarify the latter type of argument and its notion of evidential strength, we will use Cotes's invisible hand objection.

5.3 The invisible hand

As the editing of *E2* neared completion in 1713, Cotes began writing a preface contrasting Newton's "experimental philosophy" with Cartesian and Aristotelian approaches. To exemplify Newton's method he intended to present a "short deduction of the Principle of Gravity from the Phænomena of Nature, in a popular way" (Newton 1959–1977, V, p. 391). However, he encountered a difficulty.

Cotes accepted the first two steps of Newton's AUG: (1) the planets are held in their orbits by an inverse-square centripetal force directed towards the Sun, and (2) this force can be identified with terrestrial gravity, via the Moon test. What troubled him was the next step, discussed in III.5.c3 and III.7. In this step, Newton applied the third law to the centripetal force holding planets in their orbits, and concluded that a given planet also attracts the Sun. Newton argued that gravity is a *mutual interaction* between the Sun and planet. Cotes objected that this step requires further hypotheses about the nature of gravitation:

> ye Force by which they [the planets] are continually diverted from the Tangents of their Orbits is directed & tends towards their Central Bodies. Which Force (from what cause whatever it proceeds) may therefore not improperly be call'd Centripetal in respect of ye revolving Body & Attractive in respect of the Central. . . . But in the first Corollary of the 5th

> [proposition of Book III] I meet with a difficulty, it lyes in these words *Et cum attractio omnis mutua sit.* I am persuaded they are then true when the Attraction may properly be so call'd, otherwise they may be false. You will understand my meaning by an Example. Suppose two Globes A & B placed at a distance from each other upon a Table, & that whilst A remains at rest B is moved towards it by an invisible Hand. A by-stander who observes this motion but not the cause of it, will say that B does certainly tend to the centre of A, & thereupon he may call the force of the invisible Hand the Centripetal force of B & the Attraction of A since ye effect appeares the same as if it did truly proceed from a proper & real Attraction of A. But then I think he cannot by virtue of this Axiom [*Attractio omni mutua est*] conclude contrary to his Sense & Observation that the Globe A does also move towards the Globe B & will meet it at the common centre of Gravity of both Bodies. This is what stops me in the train of reasoning by which I would make out as I said in a popular way the 7th Prop. Lib. III. I shall be glad to have Your resolution of the difficulty, for such I take it to be.... For 'till this objection be cleared I would not undertake to answer one who should assert that You do *Hypothesim fingere.* I think You seem tacitly to make this Supposition that the Attractive force resides in the Central Body.
>
> (Newton 1959–1977, V, p. 392)

There are two ways of reading Cotes. On the first, there is a stark *empirical* contrast between the "invisible hand" scenario and Newton's account of gravitation. According to Cotes, the invisible hand moves Globe B without moving Globe A. According to Newton, however, true interactions are mutual, and so the central body of a gravitational system (Globe A) is predicted to move, however slightly. The mismatch between prediction and observed motion is Cotes's problem: "I think [an observer] cannot by virtue of this Axiom [*Attractio omni mutua est*] conclude *contrary to his Sense & Observation* that the Globe A does also move towards the Globe B" (Newton 1959–1977, V, p. 392, emphasis added). Cotes presumed that in a gravitational system, like in the invisible hand case, "Sense & Observation" will show that the central body does not move. Determining the motion of a central body in a real-world case is not straightforward, but it is possible. For truly mutual interactions between two bodies there is a "two-body" correction to Kepler's third law (see, e.g., Smith 2002b, p. 44). Neither Newton nor Cotes could have made an empirical case in favor of this correction based on contemporary observations, but the question was open to empirical resolution.

Newton shifted the terms of the debate rather than treating this as an unresolved empirical question. To account for the subsequent exchanges we focus on a second reading. On this reading, we take Cotes to suggest that *two* invisible hands are acting jointly to move the globes in a way identical with the predictions of Newton's theory. This scenario is not empirically distinguishable from Newton's description of planetary motions; instead, it suggests that Newton's argument for extending the third law to the central body rested on an unacknowledged assumption about the nature of gravity.[14] Although Cotes claimed that Globe A does *not* move, Newton seems to have responded to Cotes as if he had posed this more telling objection.

Newton's replies to Cotes focused on two points, both ultimately reflected in changes to the *Principia*. First, Newton defended the third law as a crucial feature of his conception of force by showing that it is necessary for extending the first law to systems of interacting bodies. This reply missed the point (of the second reading). Cotes did not challenge the third law itself, but rather the identification of the bodies involved in its application. Nonetheless, Newton's discussion reveals the third law's importance in going from a mathematical characterization of force based on the first two laws to a physical characterization of forces treated as mutual interactions among bodies. Second, Newton responded to the charge of feigning hypotheses: he clarified the nature of hypotheses in his method and argued that objections of a certain kind, exemplified by Cotes's invisible hand, should simply be set aside. By following Newton's response to Cotes and considering the status of the laws of motion, we argue that Newton properly answered Cotes here. We contrast the limited sense in which the laws are "hypothetical" on Newton's account with the role of hypotheses for his contemporaries (and the a posteriori character of the geometrical conception of matter), and by doing so describe the sophisticated a posteriori reasoning of the *Principia*, the basis for the dynamical conception of matter.

5.3.1 Applying the third law

The invisible hand highlights the ambiguity in Newton's application of the third law in the third step of the AUG. Suppose that invisible hands and an attractive force produce indistinguishable motions of an orbiting and a central body. The third law implies that there is an equal and opposite force corresponding to the force holding the body in its orbit, but it does not specify the nature and location of this force. Should it be a reaction force acting on the central body, or a force pushing against the invisible hand?

Newton's argument depends on one of two assumptions. Either, first, gravity is a force of attraction causally residing in the interacting bodies alone, an attractive force "properly so call'd", as Cotes put it; or, second, whatever underlying mechanism is responsible for gravitation must itself produce a reaction force on the central body. Suppose, for example, that an aether mediates the gravitational interaction. Newton's application of the third law would be appropriate only if there is no net momentum transfer from the two gravitating bodies

14 See Densmore (1996), Koyré (1968), Harper (2002b), Stein (1990a), and Harper in this volume. Kant rediscovered the problem, apparently independently of Cotes; see Kant (2002, pp. 225–226) and Friedman (1992, pp. 149–159).

to the aether. In this case, even though the third law *properly applied* yields a reaction force on the aether pressing against the planet, the aether interacts with the planet and Sun in precisely the right way to produce a reaction force on the Sun. Either option conflicts with Newton's claim that the validity of the AUG does not depend upon "hypotheses" regarding the nature of gravity.[15]

Newton responded to Cotes by defending the validity of the third law of motion itself. He asked Cotes to consider two bodies *A* and *B* acted on by no net external forces, such that the forces *between* *A* and *B* do not satisfy the third law. Say, for example, that *A* exerts a greater force on *B* than vice versa. Newton emphasized that the resulting imbalance of forces would cause the bodies to accelerate off to infinity, a result that conflicts both with experience and the first law of motion (Newton 1959–1977, V, p. 397). Newton added text to the *Principia* to the same effect. In the scholium following the Laws, Newton considered sections of the Earth cut off by parallel planes equidistant from the center. As before, an imbalance of the gravitational forces felt by these two parts of the Earth would lead to the Earth accelerating off to infinity with no net external force.[16] These examples reveal the intimate connection between the third and first laws. In order for the first law to hold for the center of mass of a closed system of interacting bodies, the third law must hold for the interactions among the bodies, although Newton's examples only involve *contiguous* bodies pressing against one another.

This line of response highlights the importance of the "mutuality" of force, Newton's crucial novelty. As Stein (2002) emphasizes, speaking of *separate* forces acting on two bodies, which happen to come in an action–reaction pair, is misleading. In Newton's usage, the "force" corresponds to an interaction between bodies that is *not* broken down into separate "actions" and "reactions," except in our descriptions of it. Newton's own "popular" version of the third book, the *System of the World*, included a clear statement to this effect:

> It is true that we may consider one body as attracting, another as attracted; but this distinction is more mathematical than natural. The attraction resides in each body towards the other, and is therefore of the same kind in both. . . . In this sense it is that we are to conceive one single action to be exerted between two planets arising from the conspiring natures of both.
>
> (Newton 1934, p. 568)

15 There is a second, distinct objection that Cotes did not raise (discussed in Harper 2002b): are the motions sufficient to establish that the motive forces of two interacting bodies are equal in magnitude? The equality can fail if, in modern terms, one allows the gravitational constant *G* to vary rather than treating it as a universal constant. The assumption that *G* is a universal constant amounts to treating the various acceleration fields produced by the celestial bodies as instances of the same type of force.

16 Applying this line of reasoning to orbital motion is less straightforward; see Harper (2002b) for discussion.

This conception of force as an interaction manifested by equal and opposite impressed forces is built into the Laws of Motion. It also plays a crucial role in distinguishing apparent forces from real forces. Given a body in motion, the first two laws allow one to infer the existence of a force producing the motion that may be well defined quantitatively (given a definite magnitude and direction), without considering the question of what produces the force. But the third law further requires that the force results from an interaction between the body in motion and some other body. (Coriolis forces illustrate this distinction: the force is well defined quantitatively and can be inferred from observed motions, but there is no "interacting body" responsible for the force.) The first two laws figure primarily in treating forces from a mathematical point of view, whereas the introduction of the third law marks an important physical constraint. Although Newton famously abstained from requiring a full account of the "physical cause or reason" of a force as a precondition for establishing its existence, any further account of the physical nature of the force would have to satisfy the constraint imposed by the third law.[17]

But even granted this conception of force as mutual interaction, Cotes's query prompted justification for Newton's identification of the second "conspiring" body. The first two steps of the AUG established that the force producing orbits of the celestial bodies is closely related to their respective central bodies: it is directed toward the central body, and it varies as the inverse square of the distance from the central body. These features make plausible Newton's identification of the central body as the second body whose "conspiring nature" produces the interaction. If the list of candidates is limited to other *known* bodies, there are few plausible choices other than the central body. Cotes accepted the first two steps of the AUG, and so accepted those features of the force law apparently related to the central body. However, Cotes was correct to insist that this plausibility argument is inadequate. Newton's rivals pursuing a vortex theory of planetary motion aimed to recover both these aspects of the force without introducing a truly mutual interaction with the central body. They did so by introducing an analog of the "invisible hand" – an aether that was unobservable except for its gravitational effects.

In sum, although this part of Newton's response emphasizes the viability of the third law, this clarification is not sufficient to answer Cotes without further stipulations regarding the bodies referred to by the law. However plausible these further stipulations may seem, they beg the question as a reply to Newton's contemporary critics. One may charge that these stipulations are inconsistent with

17 Smith (2002a: 150) argues that Newton imposes five further conditions that must be satisfied for a component of a mathematically characterized force to qualify as physical; cf. Janiak (2007, 2008), regarding Newton's mathematical and physical characterizations of force.

Newton's own method, that they are feigned hypotheses. Newton responded to this charge directly.

5.3.2 *Status of the laws and the dynamical conception*

Since Cotes concluded that Newton did indeed "feign hypotheses" in the AUG, Newton offered two clarifications. First, Newton reprimanded Cotes for applying the term hypothesis too broadly:

> as in Geometry the word Hypothesis is not taken in so large a sense as to include the Axioms & Postulates, so in experimental Philosophy it is not to be taken in so large a sense as to include the first Principles or Axiomes wch I call the laws of motion. These Principles are deduced from Phænomena & made general by Induction: wch is the highest evidence that a Proposition can have in this philosophy.
>
> (Newton 1959–1977, V, pp. 396–397)

To make this clear in the *Principia*, Newton added the well-known passage immediately following "hypotheses non fingo" to the General Scholium (Newton 1999, p. 943). Thus, apart from defending his application of the third law, Newton argued that the laws of motion had a distinctive status.

Characterizing this status is a delicate matter, but the exchange with Cotes and a comparison of Newton with his contemporaries sheds light on Newton's position. Newton regarded the laws of motion as having a more secure status than the hypothetical models pursued by "mechanical philosophers" such as Huygens. Huygens characterized the aim of physics as the construction of mechanical models that rendered various phenomena intelligible. Confidence in a hypothetical model was based on the caliber of explanations it offered, its ability to predict novel phenomena, and other theoretical virtues such as simplicity. The well-recognized problem with this approach was the possibility of alternative, yet equally satisfactory, models. Mechanical philosophers often addressed it by insisting that their models satisfied further constraints, such as the compatibility of the models with privileged first principles, particularly with the ontology of matter in motion. These constraints, however, did not eliminate underdetermination worries; they merely limited their scope.

Newton, however, did *not* conclude that certainty was unattainable in natural philosophy.[18] His criticisms of mechanical theories were combined with the assertion that his own method could establish results with as much certainty "as the nature of things admit," a certainty guaranteed by a criterion of evidential warrant distinct from that of the mechanical philosophers (cf. Harper and Smith 1995). Propositions that met this more stringent criterion qualified as

18 See, e.g., Newton's evaluation of his optical work; Newton (1978, p. 106).

"deduced from the phenomena" or "proved by experiments," and Newton claimed that they were not susceptible to the problems faced by mechanical hypotheses. Specifically (cf. Section 5.2.3), confidence in Newton's reasoning about natural phenomena did not depend on its conformity with first principles regarding fundamental natures.

But there is more to "deduction from phenomena" or "proof by experiment" than a disregard for first principles, particularly in the context of the *Principia*. Specifying the difference is no mean feat, but we are able to draw on prior work on the implicit methodology of the *Principia* by Howard Stein, Bill Harper, and George Smith.[19] Despite disagreements on several finer points, this line of work highlights two general contrasts between Newton and the mechanical philosophers. First, Newton's predecessors – such as Galileo – did not deal with the complexity of actual motions as Newton did. Although the consequences of Galileo's theory of uniformly accelerated motion were not taken to apply exactly to actual motions, a rough conformity between actual and theoretically described motions was taken as evidence in favor of the theory. More refined judgments of conformity, however, require an assessment of factors such as air resistance and measurement imprecision, and these are problematic precisely because they are not treated in the original theory.[20] By way of contrast, Newton had an elegant way of handling the complexity of actual motions. He took the care to prove theorems that could underwrite "robust inferences," that is, inferences whose conclusions (usually claims regarding forces) hold approximately if their antecedents (usually observational claims) hold approximately (Smith 2002b). For example, Newton's use of the precession theorem in the first step of the AUG makes it possible to infer properties of the gravitational force from actual motions even if they only *approximately* satisfy a simple mathematical description, such as Kepler's laws. An initial theoretical description is not blocked by the complexity of actual motions. The argumentative structure of the *Principia* further illustrates that Newton approached the complexity of actual motions piecemeal, building up to increasingly complicated descriptions in what Cohen called the "Newtonian Style" (Cohen 1980). As we shall see below, this style is also crucial for establishing the epistemological warrant of Newtonian mechanics.

Second, the Newtonian laws of motion by themselves do not entail specific predictions about directly observed motions – falling bodies, for example – but must be supplemented with assumptions regarding the forces at play. The laws

19 See Harper (1990, 2002b), Harper and Smith (1995), Smith (2001b, 2002a,b), Stein (1991, ms), and Harper and Smith in this volume.

20 Many mechanical philosophers expected that such effects could not be incorporated into theories of motion; Galileo, for example, doubted whether air resistance could ever be handled theoretically, but defended mathematical idealization nevetheless, see McMullin (1985).

are thus not *directly* "deduced from phenomena" on the basis of successful predictions. This claim apparently runs counter to Newton's defense of the laws in the scholium following the Laws and Corollaries (Newton 1999, pp. 424–430). There, Newton discussed phenomena (such as the ballistic pendulum) which could plausibly be taken as the basis for a "deduction" of each of the laws. For example, Newton defended the third law as a natural extension of the static treatment of forces to cases wherein the mutually balanced forces apply to different bodies. Although the successful treatment of these phenomena provides evidence in favor of the laws of motion, this is not a case of simple predictive success. For each, further assumptions regarding the forces at play are required. The scholium persuasively establishes that a variety of phenomena are compatible with the laws of motion when motion is characterized dynamically using Newton's definitions, but it does not uniquely entail the laws. The challenge in giving an account of the status of the laws – and the status of Newtonian mechanics more generally – is to clarify the sense in which *indirect* empirical support accrues to the laws, and thus the dynamical conception of matter, that is nonetheless stronger than that offered by mere predictive success of hypothetical models.

Harper's and Smith's reconstructions of the status of Newton's laws help here. Harper holds that empirical success is judged according to whether observed motions provide multiple agreeing measurements of theoretical parameters used to describe them (Harper 1990). This approach shifts the focus from the predictive success of a single model to the stability of parameter values across a set of theoretical descriptions. Smith emphasizes the importance of approaching actual motions by a series of approximations (Smith 2002b). An initial inference establishes the *approximate* validity of the gravitational force law as applied to actual motions, but one can further calculate trajectories on the assumption that the gravitational force holds *exactly* in a precisely specified situation – such as two point-masses interacting solely via the gravitational force. Discrepancies between this initial theoretical account and the actual motions may indicate that some idealizing assumptions are flawed, and the next step is to drop these assumptions and provide a more elaborate theoretical description. On Smith's account, the laws of motion accrue empirical support with each stage in a series of approximations when discrepancies between actual motions and a particular stage can be explained by relaxing idealizing assumptions in a way that is self-consistent and that identifies further physical details of the system.

This brief sketch is sufficient to contrast the methods of inquiry associated with the dynamical and geometrical conceptions of matter. Clearly, both Harper's and Smith's accounts of Newton's method depend crucially on the exactness provided by the mathematical framework of Book I. But Newton claimed to have established "from the phenomena" not only the laws of motion and gravity, but also the impenetrability and extension of bodies (see Sections 5.2.2–5.2.3). Can similar accounts be given for these claims?

Certainly, Newton does *not* introduce parameters characterizing impenetrability and show how various phenomena give agreeing measurements of them. Nor does he give controlled idealizations that can be utilized as first approximations in order to derive the properties of impenetrability and extension from observed motions, and then proceed to develop successively more detailed approximations. The evidential warrant for such inferences "from the phenomena" relies on a less sophisticated chain of reasoning than does the warrant provided for the laws of motion. We return to this issue shortly, when we consider a case in which a deduction from the phenomena that uses a sophisticated mathematical framework is pitted against one that does not.

Before doing so, however, we must note a feature that is *shared* by both types of inferences "from the phenomena." Newton also warned Cotes against overstating the certainty of any *a posteriori* deductions:

> Experimental philosophy proceeds only upon Phenomena & deduces general Propositions from them only by induction. And such is the proof of mutual attraction. And the arguments for ye impenetrability, mobility & force of all bodies & for the laws of motion are no better. And he that in experimental Philosophy would except against any of these must draw his objection from some experiment or phænomena & not from a mere Hypothesis, if the Induction be of any force.
>
> (Newton 1959–1977, V, p. 400)

Newton acknowledged that the laws of motion were "hypothetical" in the sense of being open to revision, but limited in how they may be revised. In modern terminology, "provisional" or "corrigible" are more apt for capturing Newton's meaning. For Newton, the laws of motion are not hypothetical due to the threat of underdetermination and alternative models. Rather, they are hypothetical – provisional, corrigible – because in establishing them one must generalize from a limited set of phenomena, and this necessarily inductive step may be overturned by new evidence. In an unsent draft of the letter above, Newton elaborated:

> One may suppose that God can create a penetrable body & so reject the impenetrability of matter. But to admitt of such Hypotheses in opposition to rational Propositions founded upon Phænomena by Induction is to destroy all arguments taken from Phænomena by Induction & all Principles founded upon such arguments. And therefore as I regard not Hypotheses in explaining the Phenomena of nature so I regard them not in opposition to arguments founded upon Phænomena by Induction or to Principles setled upon such arguments... This Argument holds good by the third Rule of philosophizing. And if we break that Rule, we cannot

> affirm any one general law of nature: we cannot so much as affirm that all matter is impenetrable.
>
> (Newton 1959–1977, V, p. 398)

By the time of this exchange, the earlier portions of Book III had already been printed and new material could not be added. In *E3*, however, Newton added Rule IV, a claim much to the same effect but now no longer treated as a consequence of Rule III (Newton 1999, p. 796). Rule IV clarifies that the uncertainty Newton associated with deductions from the phenomena was quite different than that associated with mechanical models. Taking the results of such a deduction to apply without exception introduced uncertainty, but merely the uncertainty of any inductive generalization. Newton further acknowledged the possibility that the results of a deduction may only be approximations to further, more exact theoretical descriptions. But in both cases, Newton held that the way to handle the associated uncertainty was to continue to compare observations and their theoretical descriptions, with the hope of turning up contrary phenomena indicating error. Pursuing "hypotheses" in the sense of the mechanical philosophy had no part in this effort.

5.3.3 Gravity as an essential property

How did Cotes respond to Newton's elaboration of his method? Cotes was tempted to bite the bullet and assert that the matter of the central body actively produces the gravitational force felt by the orbiting body, that it is the physical seat of the force of gravitation.[21] The third law applies in this instance because the central body, rather than some intermediary, is *directly* responsible for the force felt by the orbiting body. However, this suggests an intimate connection between matter and gravitation, and so a question arises about how to characterize this connection. In writing the preface to *E2* Cotes initially called gravitation an essential property of matter – a property "without which no others belonging to the same substance can exist" (Newton 1959–1977, V, pp. 412–413) – but was reprimanded by Clarke. In response, Cotes substituted "primary" for "essential," but still treated gravitation as on par with impenetrability, extension, and mobility; it has, he wrote, "as fair a claim to that title" as the other properties.[22]

21 There are two senses in which gravitation can be ascribed to matter (McMullin 1978, pp. 59–61). First, gravity causes deviations from inertial motion in accordance with the second law, and matter plays a *passive* role by responding to the impressed force (gravity). But a body must also produce the impressed force felt by other bodies, and this second, *active* sense is more problematic for Newton. For the third law to apply to an attractive force between two bodies, without any mediation, each body must respond to and also produce the force.

22 See also Newton (1999, p. 392).

Cotes did not elaborate, but he might have defended himself as follows. Inertia is taken to be essential to material bodies because the laws of motion – the laws detailing the relations between inertia, impressed force, and motion in bodies – require it. To be a body subject to the laws of motion is necessarily to be a body with inertial properties. Likewise, gravity has a "fair claim" to the title of an essential property because the understanding of attractive forces at work in the *Principia* requires it. The *Principia* demonstrates that all bodies attract one another according to a single force law, and so, taking this force law as his guide in determining the essential properties of matter, and having no indication that this force law could be explained by some deeper mechanism, Cotes is ready to claim that gravity is essential to material bodies. For Cotes, physical theory itself is the guide to determining essential properties. As promised in Section 5.2.3, we can now also see why the list of qualities generalized by Rule III of the *Regulae Philosophandi* is heterogeneous. The force of inertia, for example, is essential for the Newtonian theoretical description of actual motions. But extension, hardness, and impenetrability are not.[23] Moreover, gravity and inertia are established by the complex method outlined in the previous section, but extension, hardness, and impenetrability are not. As we shall see in Section 5.4, Cotes recognized that the qualities treated by Rule III are not on an equal footing and thus that not all "deductions from phenomena" generalized by Rule III are equally meritorious.

For Newton, however, responding to the objection by taking gravity as an essential property was a misstep. The physical characterization of gravity as a real rather than merely apparent force requires at least that it is a mutual interaction satisfying the third law. This is an important constraint on the nature of the force and it runs deeper than might be expected, but Newton does not follow Cotes in taking this to have direct implications for the ultimate cause of gravitation or the essential properties of matter. Newton's original reprimand – that Cotes applied the term "Hypothesis" too broadly – is instructive. For Newton, the application of the third law to the orbiting and central body is a crucial step in moving from a mathematical characterization of a force, as a well-defined quantity inferred from observed motions, to a characterization of the physical causes, species, and proportions of real forces. But taking this step does not require determining the cause of gravity or the relation of gravity to the essential properties of matter. The application of the third law has a "hypothetical" or provisional character, in the limited sense in which the overall framework of the laws of motion is "hypothetical." However, this sense is not analogous to the hypothetical character of mechanical models. The true nature of the gravitational force – i.e., whether or not it acts *immediately* as a force of interaction between the orbiting and central bodies – is a separate question,

23 Mobility has a curious status as an object of the intension/remission criterion, so we leave it aside here.

not directly related to the status of laws of motion, and Newton reserved judgment regarding it.[24] To speculate, as Cotes did, that the application of the third law is inconsistent with the true, yet unknown, cause of gravitation is to repeat a common mistake of the mechanical philosophers, namely to judge an experimentally established proposition on the basis of its compatibility with claims regarding the fundamental nature of bodies. Given his skepticism regarding such claims, Newton rejected the need for such a compatibility check, and this was one of the most distinctive aspects of his method.

In sum, in our opinion Newton's answers to Cotes only *seem* to fail to recognize the question of whether gravity is mutual *per se* because Newton *purposely* rejected any discussion of what gravity is, *per se*. Newton's reference to the conspiring nature of both orbiting and central bodies should not be taken to mean that gravitational attraction resides essentially in either. Had Newton explicated his own methodological tenets with enough clarity, he could have made it clearer to Cotes that he chose to remain agnostic about the implications of his own theory regarding the essential natures of bodies. However, his lack of explicitness on this occasion, the fact that he often entertained deeper explanations (albeit with sufficient caveats), *and* the fact that he was the sole natural philosopher endorsing this approach, all contributed to Cotes's confusion and willingness to consider such implications. The same pattern of misunderstanding recurs in Cotes's query about the proportionalities that hold between weight, inertia, and quantity of matter. There, however, Cotes shows Newton to be mistaken about the claims warranted by his own method.

5.4 Proportionalities

In III.6, Newton demonstrated that:

> All bodies gravitate toward each of the planets, and at a given distance from the center of any one planet the weight of any body whatever toward that planet is proportional to the quantity of matter which the body contains.
>
> (Newton 1999, p. 806)

The weight of a body does *not* depend on properties such as form or texture. This distinguishes gravity from forces such as magnetism, and also sets Newton's view apart from several contemporary accounts that left open the possibility that gravity could depend upon a wide variety of a body's properties.[25] In the text of the proposition, Newton described a pendulum experiment meant to establish that near the surface of the Earth the weight of a body is proportional

24 Newton famously denied that his characterization of gravitational force implied that brute matter could act directly at a distance; see, for example, the oft-quoted letter to Bentley (Newton 1959–1977, III, pp. 240–44).

25 See Westfall (1967, pp. 246–251), Koyré (1965, pp. 173ff., 185ff.).

to its quantity of matter, and further that the weight of Jupiter's moons is proportional to their quantities of matter.[26] The experiment was first mentioned in two manuscripts which follow the initial *De Motu* drafts.

Newton constructed two equal-length pendulums with wooden boxes as bobs and filled the wooden boxes with equal weights of gold, silver, lead, glass, sand, common salt, wood, water, and wheat. For each pair of materials, he measured the periods of oscillation. According to II.24, the mass of a pendulum bob is proportional to the product of its weight and the square of its period, $m \propto wp^2$. This proposition is based on two basic assumptions. First, $f_m \propto \frac{m\Delta v}{\Delta t}$, where f_m is the motive force, v is the velocity and t the time – a restatement of definition 8 (motive force). Second, $f_m \propto w$, motive force is proportional to the weight of the pendulum bob.[27] For a simple pendulum near the Earth's surface, the period depends upon both the length of the pendulum and the acceleration due to gravity. Since Newton used pendulums of equal length, the pendulums would only have different periods if the gravitational acceleration varied for different materials. Newton reported that the periods of two pendulums containing different materials were in fact the same, to within an accuracy of 1/1000, and so concluded that $m \propto f_m \propto w$ for all materials tested. Citing Rule III, he then generalized to "all bodies universally," even those composed of materials not tested in the experiment (Newton 1999, p. 809).

In corollary 3 of *E1*, Newton highlighted an important implication of this proportionality for matter theory; namely, that a vacuum exists:

> And thus a vacuum is necessary. For if all spaces were full, the specific gravity of the fluid with which the region of the air would be filled, because of the extreme density of its matter, would not be less than the specific gravity of quicksilver or gold or of any other body with the greatest density, and therefore neither gold nor any other body could descend in air. For bodies do not ever descend in fluids unless they have a greater specific gravity.
>
> (Newton 1999, p. 810)

Cotes objected that this argument implicitly assumes that completely filled regions of space possess identical specific gravities, which can be the case if and only if those regions contain identical quantities of matter. He illustrated the objection with a thought-experiment:[28]

26 See Harper's contribution to this volume.

27 As Newton notes in II.24.c5, the result also holds with "relative" (or buoyant) weight of the pendulum bob in place of w, because for a body immersed in a medium the motive force is the relative weight.

28 Cotes's Cambridge contemporary Robert Greene lodged essentially the same objection in Chapter VI of Greene (1712), albeit not nearly as perspicaciously as Cotes.

> Let us suppose two globes A & B of equal magnitudes to be perfectly fill'd with matter without any interstices of void Space; I would ask the question whether it be impossible that God should give different *vires inertia* to these Globes. I think it cannot be said that they must necessarily have the same or an equal *Vis Inertia*. Now You do all along in Your Philosophy, & I think very rightly, estimate the quantity of matter by the *Vis Inertia* & particularly in this VIth Proposition in which no more is strictly proved than that the Gravitys of all Bodys are proportionable to their *Vires Inertia*. Tis possible then, that ye equal spaces possess'd by ye Globes A & B may be both perfectly fill'd with matter, so no void interstices remain, & *yet* that the quantity of matter in each space shall not be the same. Therefore when You define or assume the quantity of Matter to be proportionable to its *Vis Inertia*, You must not at the same time define or assume it to be proportionable to ye space which it may perfectly fill without any void interstices; unless you hold it impossible for the 2 Globes A & B to have different *Vires Inertia*. Now in the 3rd Corollary I think You do in effect assume both these things at once.
>
> (Newton 1959–1977, V, p. 228)

Cotes emphasized that contrary to Newton's assumption in the third corollary, the two ways of quantifying matter – based on response to impressed force (*vis inertiae)* and volume filled – need not agree. If they do not, one can account for differences in specific gravity without postulating a vacuum. The implications for Newton's anti-Cartesian, anti-plenum arguments are clear.[29]

But Cotes's objection also has broader implications, implications that tie together our treatment of the geometrical and dynamical conceptions of matter. Cotes's objection shows that he recognized the possibility of measuring "quantity of matter" in the two distinct, but possibly conflicting, ways. If both the dynamical and geometrical measures are correct, i.e. if both *vis insita* and extension are proportional to quantity of matter, it should follow that both are proportional to one another. However, a proportionality between the dynamical and geometrical measures can be justified neither a priori nor empirically. First, nothing in the concepts of spatial impenetrability or force of inertia necessitates a determinate proportionality between them. Second, although the pendulum experiments are intended to prove that gravitation depends upon the quantity of matter, as Cotes indicated to Newton, "no more is strictly proved [in them] than that the Gravities of all Bodys are proportionable to the *Vires Inertiae*." Whether the gravities of bodies are further proportional to their *quantities of*

29 As with the invisible hand objection, Kant also criticized Newton on precisely this point. See Proposition XII of the *Physical Monadology* (Kant 1992, p. 64). A passage from the *Critique* (Kant 1998: (A173/B215–A174/B216)) more closely parallels Cotes's argument (we thank Kent Baldner for bringing it to our attention).

matter depends on how one defines "quantity of matter."[30] If one defines it to be proportional to the inertia of a body, then the experiments support the desired conclusion. But if one defines it to be proportional to the extension a body impenetrably fills, they do not. Cotes's objection reveals, although he does not say so directly, that the choice of an appropriate definition is crucial for the AUG: Assume that quantity of matter, defined geometrically, can vary in relation to *vis inertia*, as in Cotes's two globes. We can replace *vis inertia* with weight in the conclusion to III.6, since the pendulum experiments show that they are proportional at a given distance; thus, quantity of matter geometrically defined is not proportional to weight at a given distance. That is, if quantity of matter is defined to be proportional to quantity of extension, even at a given distance, *the quantity of matter of a body is not proportional to its weight.* Cotes's objection undermines not just the III.6.c3, but III.6 itself, and thus the AUG. If Newton wants to maintain that quantity of matter can be defined by *either* quantity of extension or quantity of inertia, he must assume that the two are determinately proportional, a claim for which he can offer no justification. This was Cotes's point.

Newton attempted to rebut Cotes by claiming that matter has inertial properties proportional to its quantity and geometrical properties due to its impenetrability, and that these two entail a fixed proportionality of inertia to extension. Yet this missed Cotes's point. The point was that these two facts, which Cotes did not dispute, do *not* entail the proportionality of inertia to extension.[31] Newton's second response to Cotes (after Cotes reiterated his reasoning) illustrates his misunderstanding and his continued commitment to *both* the dynamical and geometrical conceptions of matter and their a posteriori character. He wrote:

> I have reconsidered the third Corollary of the VIth Proposition. And for preventing the cavils of those who are ready to put two or more sorts of matter you may add these word[s] to the end of the Corollary: [1] From pendulum experiments it is established that the force of inertia is proportional to the gravity of a body. [2] The force of inertia arises from the quantity of matter in a body and so is proportional to its massiness [*massa*]. [3] A body is condensed by the contraction of the pores in it, and when it has no more pores (because of the impenetrability of matter) it can be condensed no more; and so in [completely] full spaces [the force

30 One might object that Newton and Cotes are conflating inertial and gravitational mass, see Densmore (1996, pp. 313–330). The problem is distinct from the objection under consideration.

31 Cotes's position shifted slightly during this exchange: whereas initially he objected to the implicit assumption of the proportionality of inertia to quantity of extension ("You *must not* at the same time define . . ."), he later allowed that the proportionality could be invoked as an unproved assumption. In either case, his objection is that Newton's explicit commitments do not entail that the proportionality holds.

> of inertia] is as the size of the space. Granted these three principles the corollary is valid.
>
> (Newton 1959–1977, V, p. 240)

Since Newton and Cotes explicitly agreed on [1], the source of their disagreement lies in [2] or [3]. In [2], Newton implicitly defined quantity of matter to be proportional to the force of inertia. Since Cotes had already written to Newton that "all along in your Philosophy, & I think very rightly, you estimate the quantity of matter by the *Vis Inertiae*," the source of conflict must be [3]. In [3], Newton deduced from [2] and the impenetrability of matter that the inertia of matter is proportional to the extension it solidly fills. Clearly, Newton took this to be a valid inference. According to Cotes, however, Newton's reasoning is circular: he implicitly assumed that the force of inertia is determinately proportional to the extension solidly filled by matter in order to deduce that, after condensation, the force of inertia would be determinately proportional to the extension filled by matter. Cotes wrote in his subsequent response:

> I am not yet satisfied as to the difficulty unless You will be pleased to add, That it is true upon this concession, that the Primigenial particles . . . have all the same *Vis Inertiae* in respect to their magnitude or extension in *Spatio pleno*. I call this a concession because I cannot see how it may be certainly proved either a Priori by bare abstracted reasoning; or be inferr'd from Experiments.
>
> (Newton 1959–1977, V, p. 242)

Cotes took Newton to be putting a uniformity constraint on the fundamental, "Primigenial" particles of matter, particles that are inaccessible to direct experimental investigation. The uniformity constraint is the claim that all fundamental particles have identical specific gravities; or, equally, that their quantity of matter is uniformly proportional to their extension. Newton had appealed to the uniformity constraint from his student days: in the *Certain Philosophical Questions*, in his draft and final revisions to Hypothesis III of *E1*, and in his considered arguments against the vacuum. It is a fixture of Newton's thought that had gone unchallenged until this exchange with Cotes, although Newton appears to have justified the constraint by subtly different means at different points in his career (McGuire 1970). At the beginning of this exchange with Cotes, Newton believed that the constraint could be justified a posteriori. His initial responses demonstrate that, by his own lights, the uniformity of primigenial particles followed from observable facts regarding the extension, impenetrability, and inertia of matter. Cotes's objection pointed to the conflict between the geometrical and dynamical measures of matter, both of which, according to Newton, were derived a posteriori.

However, the geometrical definition (and the conception of matter underlying it) is derived by a different sort of a posteriori argument than the dynamical definition (and the conception of matter underlying it). The geometrical definition is derived from the claim that, as Newton puts it in Rule III and as he articulated more elaborately in *DG*, "extension is found in all sensible bodies." This derivation is in some sense immediate – it rests on no sophisticated mathematical chain of reasoning, no process of approximation, and no fixing of causal parameters. Within Newton's broadly mechanical account of perception, it simply follows from our experience of any body whatsoever. The dynamical definition is a crucial part of Newton's account of force, developed and used to account for a variety of motions in the *Principia*.

Ultimately, Newton backed down. In III.6.c4 of *E2*, he rephrased the anti-plenum argument in the form of a conditional, acknowledging the assumption Cotes insisted on:

> If all the solid particles of all bodies have the same density and cannot be rarefied without pores, there must be a vacuum. I say particles have the same density when their respective forces of inertia [or masses] are as their sizes.

If the fundamental particles have a fixed ratio between inertia and volume, *then* a vacuum must be granted. Yet the interchange with Cotes shows that Newton's initial inclination was to positively maintain that all primigenial particles are uniformly extended in proportion to their quantities of matter, despite the fact that his pendulum experiments and the mathematical structure used to interpret them recommended no such steadfastness. In fact, the dynamical conception supported by the results of the *Principia* is compatible with treating matter as constituted by Boscovichian point-particles, with the quantity of matter appearing solely as a parameter of these points. The geometrical properties of matter play no role in physical explanations in this schema, since such explanations depend solely on the laws of motion and the further specification of inter-particle forces.

Newton's initial failure to see this point reflects, on our view, a failure to clearly distinguish the distinctive a posteriori methods described above. As we saw in the treatment of Rule III, Newton conceived of properties like extension and impenetrability as having the same status as inertia, despite the fact that they were supported by a distinctive line of argument that was not intertwined with the AUG or the deduction of the laws of motion from phenomena. Cotes, to his credit, was quite clear that Newtonian mechanics does not support a geometrical conception of matter. He even pointed out the precarious status of extension to Clarke:

> I understand by Essential propertys such propertys without which no others belonging to the same substance can exist: and I would not undertake to prove that it were impossible for any of the other Properties of Bodies to exist without even Extension.
>
> (Newton 1959–1977, V, pp. 412–413)

Cotes obliquely entertained the possibility that Boscovichian non-extended point-particles can constitute bodies and that our *experience* of bodies – even our experience of those qualities that seem immutable and invariably present – is no guide in questions of essentiality. For Cotes, to repeat a point made in Section 5.3.3 regarding gravity, physical theory itself is the guide to determining essential properties. It just so happens that within Newtonian mechanics inertia plays a central role in giving an account of observed motions whereas extension does not. Insofar as Cotes is concerned, so much the worse for extension.

Newton was not far behind. After Cotes's objection highlighted the incongruity between the two conceptions of matter, Newton began to doubt the geometrical conception more thoroughly. The change of mind for an astute and tenacious figure such as Newton is significant: Newton did not back down in response to the invisible hand objection because he was certain of his correctness. In response to the two globes objection, however, Newton modified his views. In a series of draft definitions intended for Book III of *E3* (dated by McGuire (1966) to 1716), Newton explicitly addressed his now-changed conception of body. He wrote:

> *Definition II* Body I call everything which can be moved and touched, in which there is resistance to tangible things, and its resistance, if it is great enough, can be perceived.
>
> (p. 115)

Lacking from this definition is any mention of the extension of bodies. The *only* definitional property of body here is its inertial resistance. A far cry indeed from *DG*'s definition:

> Definition 4. Body is that which fills space.[32]
>
> (Newton 2004, p. 13)

5.5 Conclusion

We have stressed two aspects of Newton's thought. The first is Newton's empiricist method, and the two approaches he took to justifying claims in natural philosophy. The approach exemplified by the AUG contrasts sharply with the method of the mechanical philosophers. Unlike the mechanical philosophers, Newton did not allow the satisfaction of intelligibility constraints (e.g., that only contact action is comprehensible) to serve as justification, even if partial, for a physical theory; the justification for the laws of motion and universal

32 We take the development of Newton's views on body to show that *DG* cannot be automatically taken to reflect Newton's mature metaphysical views. Rather, it is best taken as Newton's relatively early attempt to explicate the philosophical infrastructure in which his physics is embedded, but by no means the last word.

gravitation is their ability to serve successfully as a framework for describing motions. Newton's response to Cotes's invisible hand objection reflected this methodological stance. Cotes objected that Newton had inappropriately assumed that gravitational force must be produced by the orbiting and central bodies, despite his professed agnosticism regarding the underlying cause of gravity. Newton responded by clarifying that his characterization of gravity as a force obeying the three laws was hypothetical in the same limited sense that the laws of motion are hypothetical and did not entail further assumptions regarding the essential properties of matter or the underlying cause of gravity. The second approach to establishing results a posteriori is exemplified by the account of body in *DG* and some of Newton's statements in Rule III. It involves a more direct argument, essentially reading off the properties of matter from the general experience of bodies. It does not draw on a precise mathematical framework like that of the *Principia*, and so the ways of clarifying evidential warrant within the first approach apply. It is consequently unclear how to assess the strength of the conclusions derived from this type of reasoning.

Second, there is an uncomfortable union in Newton's thought between two competing conceptions of matter. The geometrical conception reflects Newton's Cartesian roots and was linked to the possibility of an aetherial explanation of gravitation. Although Newton decisively rejected several aspects of Cartesian thought in *DG*, he retained an account of bodies that took their geometrical properties to be fundamental. Consequently, he took a body's quantity of matter to be proportional to the volume it impenetrably fills. At the same time, Newton developed the distinctive dynamical conception of matter of the *Principia*, which measures quantity of matter by a body's response to impressed force. Newton apparently treated the two measures as aspects of a single, coherent account of matter. Cotes's second objection brought out the tension between these two conceptions. Cotes argued that Newton's claims could not be sustained without an explicit assumption regarding the fundamental constituents of matter, betraying Newton's professed agnosticism on such matters. Although Newton's response to Cotes reflects his failure to clearly distinguish the two approaches to a posteriori reasoning characterized above, there is evidence he took Cotes's criticism to heart and attempted to dispense with *DG*'s geometrical conception.

6

Newton's scientific method and the universal law of gravitation

ORI BELKIND

6.1 Introduction

When Newton first presented his argument for the law of gravitation not all were convinced. Huygens, while admiring Newton's achievement in the *Principia*, was skeptical. Since gravitational attraction contradicts the principles of mechanical philosophy, Newton's theory seemed counterintuitive and even absurd.[1] Leibniz was also critical of Newton's argument.[2] After all, how is one to accept action at a distance without relying on any contact forces or whirling fluids? Newton, on his part, argued that his theory was based on impeccable reasoning. Even if his gravitational force violates the scientific sensibilities of the day, one still has to accept it as fact.

Newton explains his attitude towards hypotheses in the General Scholium to Book III of the *Principia*:

> For whatever is not deduced from the phenomena must be called a hypothesis; and hypotheses, whether metaphysical or physical, or based on occult qualities, or mechanical, have no place in experimental philosophy. In this experimental philosophy, propositions are deduced from the phenomena, and are made general by induction. The impenetrability, mobility, and impetus of bodies, and the laws of motion and the law of gravity have been found by this method.
>
> (Newton 1999, p. 943)[3]

I would like to thank Arthur Fine, Eric Schliesser, Andrew Janiak, George Smith, Bill Harper, Steffen Ducheyne, Geoff Goddu, and the anonymous reader for CUP for their helpful comments and suggestions on various versions of this paper. I would also like to thank the University of Richmond for providing research support and Michele Bedsaul for help with editing.

1 See Maglo (2003) for an account of the reception of Newton's gravitational theory.

2 See Leibniz's letters to Newton from March 7, 1692/3 (Newton 2004, p. 106).

3 It is important to note that the Scholium to Book III was added to the second edition of the *Principia* in 1713, some 26 years after the first edition in 1687. In these long years between the first and second edition, Newton received a lot of criticism for failing to include a

According to Newton the nature of a hypothesis does not matter; hypotheses have no place in experimental philosophy. The historical context here is that Newton is primarily concerned with undermining the Cartesian explanation for gravitation as a legitimate alternative to his attraction force. According to Newton, mechanical explanations are no more scientific than occult qualities if we cannot deduce them from the phenomena. The empiricist rhetoric gives Newton an important advantage over his Cartesian opponents.

In Rule 4 for the Study of Natural Philosophy, Newton again codifies his approach:

> In experimental philosophy propositions gathered from phenomena by induction should be considered either exactly or very nearly true notwithstanding any contrary hypotheses, until yet other phenomena make such propositions either more exact or liable to exceptions.
>
> (Newton 1999, p. 796)[4]

Thus even if our scientific intuitions – our hypotheses about how the Universe works – contradict the propositions we derive from the phenomena, we do not have good reasons to reject them. The phenomena should dictate what we take as true. But even if we derive the proposition from the phenomena, we should not consider it completely safe from refutation. Scientists may extend the investigation into new domains or discover new phenomena that would show the proposition to be false. It seems as if according to Newton we must *take* the scientific proposition deduced from the phenomena to be true, without denying the possibility of it being refuted in the future.[5]

However, it is not exactly clear what Newton means by "propositions gathered from phenomena by induction." Is there a rule of induction that tells us how to produce general statements from our observations? What does this rule look like? According to a long tradition in the philosophy of science, formal rules

proper explanation of gravity, and we can see the Scholium as an attempt to answer critics. See I. B. Cohen's introduction to the *Principia* (Newton 1999, pp. 274–280).

4 This rule was only added in the third edition of the *Principia* from 1726, almost 40 years after the initial publication of the *Principia*! It is possible that this rule reflects only Newton's latest thought on methodology. See I. B. Cohen's introduction to the *Principia* (Newton 1999, p. 200).

5 In the preface to the second edition of the *Principia*, Cotes defends Newton from the charge that he treats gravity as an occult force:

> occult causes are not those causes whose existence is very clearly demonstrated by observations, but only those whose existence is occult, imagined, and not yet proved. Therefore gravity is not an occult cause of celestial motions, since it has been shown from phenomena that this force really exists.
>
> (Newton 1999, p. 393)

While Cotes may not have the same philosophical views as Newton, the defense Cotes provides here is very much in line with the wording of Rule 4.

of induction are either extremely hard or impossible to formulate. Perhaps by "experimental demonstration" Newton means simply that causal laws are derived from the phenomena through deductive reasoning. Duhem (1982, pp. 190–195) famously argued that it is impossible for Newton to have used logical deduction in deriving the law of gravitation from the phenomena. Newton started with the elliptical orbits of the planets and deduced from them the inverse-square nature of the law of gravitation.[6] He then used the law of gravitation to calculate corrections in the planets' orbits, using the law of gravitation to determine the planets' deviations from pure elliptical orbits. Following a strict deductive method cannot reach conclusions that demonstrate a premise to be false or show it to be only approximately true. Thus, Newton must have followed some other strategy in deriving his law of gravitation.

Proponents of the Hypothetico-Deductive (HD) method worry that hypotheses always extend what can be shown with a few observations or experiments. Moreover, they argue that for any favored hypothesis, there may be others that are consistent with the phenomena. Thus there is no foolproof procedure which can be given for generating hypotheses. Newton warned us in Rule 4 that we ought to be careful when deriving a proposition from the phenomena. In case we discover new phenomena that deviate from the scientific proposition, it should be rejected. Newton was therefore well aware that induction is fallible. However, while he admitted that induction is fallible because of future discoveries, he seemed to think that there is a rule of induction that safely charts the course from a given set of phenomena to the scientific proposition derived from it.

It is quite tempting to be swayed by the HD method when considering the fallibility of induction. If there were a mechanical rule of induction, which would allow us to directly infer the scientific proposition from the phenomena, this rule of induction would resemble deduction in its strength. Inferences that start from true premises would be infallible. However, it is difficult to find a rule of induction that operates in the same way, in all contexts. If the procedure for generating scientific propositions from the phenomena is not foolproof, can it even be formulated? It seems plausible to conclude that the *process* by which a scientific proposition is generated could never be universalized and made into a rule, and so any inductive procedure followed must be merely a contingent one with no universal validity. We ought to confine the process of generating a scientific proposition to the "context of discovery." But in case there is no rule of induction, all scientific propositions are essentially conjectures since there is

6 A superficial reading demonstrates that the claim that the law of gravitation is derived from the elliptical orbits of the planets is a mischaracterization of Newton's argument in Book III of the *Principia*. However, Duhem's point can be made using the actual argument in Book III. See Smith (2002b) and section 3 for a full discussion.

an ineliminable gap between the phenomena and the scientific proposition.[7] To find a gray area between these two choices, i.e., between there being a rule of induction and there being no rule, we need to show how an inductive form of reasoning, starting from a *particular* set of phenomena, follows a specifiable procedure while still being fallible.

If proponents of the HD method are right, they must explain Newton's self-description as following an inductive method. Either Newton's methodological assertions were misinterpreted, and he in fact was endorsing the HD method despite his explicit assertions; or, the method Newton endorsed explicitly is not the one he followed as a matter of fact.[8] According to Hanson, for example, when Newton expresses his commitment to inductivism, he is actually endorsing the

7 This is the reasoning that led Karl Popper to his falsificationism:

> ... it is obvious that this rule or craft of "valid induction" is not even metaphysical: it simply doesn't exist. No rule can ever guarantee that a generalization inferred from true observations, however often repeated, is true.... And the success of science is not based upon rules of induction, but depends upon luck, ingenuity, and the purely deductive rules of critical arguments.
>
> (Popper 2003, p. 70)

This leads Popper to articulate his methodological rule. First a scientist, in virtue of some leap of the imagination, formulates a hypothesis. Then, he or she derives a testable implication from the hypothesis. Finally, if the testable implication is shown to be false when compared with observations, the scientist concludes by *modus tollens* that the hypothesis is refuted. Otherwise, we have reason to take the hypothesis seriously. Popper articulates a methodological rule that encodes this approach:

> Once a hypothesis has been proposed and tested, and has proved its mettle, it may not be allowed to drop out without "good reason". A "good reason" may be, for instance: replacement of the hypothesis by another which is better testable; or the falsification of one of the consequences of the hypothesis.
>
> (Popper 2003, p. 53)

Popper's HD method therefore suggests that no valid distinction can be made between a mere hypothesis and the propositions that are deduced from the phenomena. We have to start with hypotheses. We then use deductive rules to see if they cohere with our observations.

8 Some commentators doubted that Newton ever meant to espouse the inductive method. According to Hanson (1970), for example, we should be careful to interpret correctly Newton's use of the word "hypothesis," since Newton did not use this word very consistently. Some occurrences of the word "hypothesis" in the first edition of the *Principia* were replaced by the word "phenomenon" in the second edition. The "hypothesis" that the solar system is at rest is explored by Newton in both the first and the second editions of the *Principia* to settle the controversy between the geocentric and heliocentric systems. Hanson differentiates between four different kinds of scientific propositions: a supposed observational claim, which functions like the initial conditions we specify when solving a physical problem; a confirmed observational claim; a supposed theoretical claim and a confirmed theoretical claim. All of these may be referred to as hypotheses, with varying meanings, depending on the context.

HD method. Newton's use of the word "hypothesis" is simply meant as "an expression of some philosophical or metaphysical prejudice" (Hanson 1970, p. 32). This pejorative use of "hypothesis" merely describes impeding metaphysical stances or prejudices that have no testable consequences and therefore cannot be confirmed or refuted. We should keep in mind that Newton was constantly being pestered by the Cartesians about Newtonian gravity being a force that attracts at a distance. The metaphysical prejudice of his day was that all physical forces are reducible to mechanical forces of push-and-shove. But the Cartesian whirling fluids hypothesis, used to explain gravity, has no testable implications. So Newton's methodological remarks do not suggest that hypotheses have no place in experimental philosophy (Hanson 1970, p. 31); only those hypotheses that have no testable implications should be ignored. However, as Worrall (2000, p. 47) argued, Hanson's attempt to mitigate Newton's inductivism renders Newton's methodological remarks inconsistent with his scientific practice. There are examples of hypotheses Newton excludes solely on the basis that they were not derived from the phenomena, even though they had empirical implications consistent with the phenomena.[9]

According to another reading of Newton's methodological remarks, we should distinguish between Newton the scientist and Newton the rhetorician. Newton the scientist made conjectures and hypothesized that the motion of the planets are governed by a force of gravitation obeying the inverse-square law. Newton the rhetorician claimed to have followed a strict inductive method.[10]

During the last third of the twentieth century, some philosophers of science started rehabilitating Newton's inductive method. Jon Dorling (1973 and 1990) proposed a method he coined 'Demonstrative Induction' (DI). Dorling argued that general propositions may be inferred deductively from the phenomena if additional background assumptions are used. According to Dorling, the history of science demonstrates that the DI method was used in many cases to derive new theoretical claims from the phenomena. These deductions conferred scientific credibility on their conclusions that led to the exclusion of other

9 Worrall claims that

> Newton's famous attitude toward material emission ("corpuscular") theory of light would be irreducibly mysterious if this Hanson-style view were correct. As is well known, Newton many times and very heatedly insisted that this emission theory was a *mere* hypothesis because it could not be deduced from the phenomena; and yet the theory is clearly testable.
>
> (Worrall 2000, p. 47)

10 This is how Imre Lakatos put it:

> The schizophrenic combination of the mad Newtonian methodology, resting on the *credo quid absurdum* of "experimental proof" and the wonderful Newtonian method strikes one now as a joke. But from the rout of Cartesians to 1905 nobody laughed.
>
> (Lakatos 1978, p. 212)

hypotheses consistent with the phenomena. Norton (1993, 1994, and 1995) and Harper (1990 and 2002a) also demonstrated the importance of the DI method and of the closely related method of Eliminative Induction.[11] According to Norton, the background assumptions used in these demonstrations belong to the very core of the theories held by the scientific community. They constitute the most basic and general assumptions about systems. Thus we may take these background assumptions to be nearly certain, given that it would take a scientific revolution to show the falsity of this core set of beliefs. Indeed, as Norton argues, some of these assumptions are so general and weak that they survive scientific revolutions. It is not that these assumptions do not carry some inductive risk, but we should not forget their (almost) universal validity in the eyes of the scientific community.

Obviously the strength of the DI method depends on the plausibility of the background assumptions. Some of the assumptions may be justified on a priori grounds, since, for example, much could be learned from the phenomena through mathematical reasoning. Some of the assumptions may themselves be phenomenal laws. However, both these cases are not interesting from a methodological perspective, since the propositions derived in such cases are themselves merely phenomenal laws. Moreover, the conclusions of scientific arguments frequently involve additional theoretical entities that were not present in the phenomena. From the observed motions of the planets, for example, Newton derived the existence of a gravitational force. This force goes "beyond" or "behind" the phenomena, and can be taken as the cause that generates the phenomena.[12]

There is also a sense in which the conclusions of scientific arguments often count as "laws of nature" that carry some necessity that extends the regularity described in phenomenal laws.[13] If the DI method is to describe our scientific

11 According to Eliminative Induction, the additional background assumption is a disjunctive statement (with the "exclusive or" connective). One requires only a few observations to confirm one of the disjuncts and eliminate all the other disjuncts. The methods of Demonstrative and Eliminative Induction are equivalent from a logical point of view.

12 Granted, Newton did not think that his account of the gravitational force was the whole story. His search for a mechanical explanation of this force suggests that further elaboration of the mechanism which leads to the inverse-square law may still be discovered. But this does not undermine the fact that the notion of force by itself is not present in the phenomena.

13 I should note here that by treating laws of nature as "necessarily true," I do not mean propositions whose negation is inconceivable. It is perfectly conceivable for Newton that bodies could have been created without impenetrability, inertia, or gravity; we can imagine God creating material bodies without those properties. However, there is a sense of "necessity" that is relevant in this context, which is the applicability of laws of nature to both actual and counterfactual scenarios. It is obvious that Newton thinks his laws of nature apply "universally," in the sense that we can compare the actual trajectories of bodies to their counterfactual ones. Laws of nature dictate what the counterfactual trajectories would have been.

practice, then we must examine how this method manages to confer on its conclusions their requisite necessity. A natural assumption is that the origin of the conclusions' necessity can be traced to the background assumptions. However, by doing so we have merely pushed the problem one step back. We can immediately ask what renders the background assumptions necessary. To avoid a possible regress, one needs to find a scientific proposition that naturally presents itself as universally valid.

If we follow Norton's account, it seems as if the necessity of background assumptions stems from their central role in a scientific paradigm. The core assumptions of a scientific paradigm function as near certain propositions in scientific inferences. However, if the near certainty of the background assumptions stems from their role as core assumptions of the scientific paradigm, then it is not clear to what extent the DI method is distinct from the HD method. As Worrall (2000 pp. 69–76) argued, the background assumptions of DI arguments can be overturned during scientific revolutions. Thus, even though they hold a significant place in a paradigm, these propositions function as hypotheses. Scientists presume these assumptions are true and rely on their truth whenever they investigate the phenomena.

On the surface of things, the HD and DI methods conceive of empirical confirmation differently. According to the DI method, a new scientific proposition is deductively inferred from the phenomenal laws together with background assumptions. In the HD method, the observable implication is logically derived from the hypothesis together with the background assumptions. Thus, it seems as if the methods differ in how they think of scientific inferences; either the scientific proposition is the conclusion of a deductive inference (the DI method) or one of its premises (the HD method). But where the inference begins is not really essential, since at the end what we are concerned with is the logical consistency of a set of propositions, some of them theoretical, some of them empirical. In both methods background assumptions pose constraints on the new propositions one incorporates into the accepted system of beliefs.[14]

14 The distinction between the DI and the HD methods is blurred even further when we recognize that a conclusion of a DI argument may get additional empirical support from other domains of the phenomena. When evidence gathered from various domains converges in support of a theoretical claim, we take this convergence as increasing the plausibility of the claim. Someone endorsing the DI method would justify the added credibility of converging evidence with the help of a common cause principle. That is, all else being equal, we would prefer a theory which does not posit multiple causes for the various phenomena, over a theory which posits separate causes. In fact, the purpose of Newton's Rules 1 and 2 for the Study of Natural Philosophy is to guarantee that evidence gathered from different phenomena in favor of a scientific proposition will bolster its plausibility. But the common cause principle would be redundant, if we simply assume a hypothesis to be true, and then try to get confirmation for it from as many domains of the phenomena as possible. The process of confirmation seems to boil down to the same procedure (Worrall 2000, pp. 66–67).

The DI method is caught between two horns of a dilemma. If, on the one hand, the background assumptions constitute mathematical reasoning or other phenomenal laws, then we are not in a position to go "beyond" the phenomena to the causal laws that explain them, nor can we explain why the conclusions of DI arguments feel as if they are valid necessarily. If, on the other hand, the background assumptions incorporate more than mathematical reasoning or phenomenal laws, then the DI method becomes unstable and runs the risk of collapsing into the HD method. The first horn of the dilemma fails to capture our intuition that scientific arguments manage to go beyond the phenomena, and that the conclusion of deductions from the phenomena are not mere representations of phenomena but explanations of them. In the second horn of the dilemma the DI method fails in what it purports to do; i.e., to distance itself from the somewhat arbitrary or contingent status of hypotheses.

In what follows, I will provide an account of Newton's argument for the law of gravitation which shows it as following the DI method. An important aim, however, is to show that Newton's method does not collapse into the HD method. In deriving the universal law of gravitation, Newton is following what is mostly a deductive argument, which begins with Kepler's phenomenal laws and ends with the universal law of gravitation. The additional background assumptions Newton brings to bear are not hypotheses. These additional assumptions are grounded in experience, but they take on universal validity; i.e., they are taken to be true in both actual and counterfactual cases, once they are elevated into general assumptions about the structure of physical systems.[15] The near certainty attributed to these background assumptions stems not from their widespread acceptance, but from their unique role in making the structure of physical systems intelligible.

In one crucial step of the argument, Newton elevates an empirical claim into a structural assumption before reaching his final conclusion. This additional structural assumption was not known to his fellow scientists before the publication of the *Principia*, so in no way can the universal validity of this assumption stem from its historical role as a core assumption in a scientific paradigm. This assumption receives its universal validity from its particular role in making apparent the structure of physical systems.

15 Some may object to my use of the notion of "universal validity." Ordinarily, the notion of validity is taken to be a property of arguments. A valid argument is one where the conclusion must be true given that the premises are. According to this view, only the notion of truth is a property of propositions. But I need a notion that would describe the difference between the generalization, "All A's are B's," which is only true for actual cases, and the necessary statement, "All A's are B's," which is true in both actual and counterfactual cases. Kant describes the latter kind of propositions as a priori, but his use of it carries the prejudice that necessary propositions can be arrived at independently of experience. I will therefore use "universally valid" to describe a necessarily valid proposition, without prejudging what sort of necessity is involved.

We shall describe the process through which a claim that was initially considered as empirical in nature becomes universally valid. First Newton replaces the empirical claim with an assumption which relates the properties of composite systems to properties of their parts. Once this assumption has the original empirical claim as its logical consequence, Newton takes it to be universally valid; i.e., he takes it to be valid in both actual and counterfactual circumstances. Thus when Newton elevates an empirical claim to the status of a structural assumption, he takes it to have natural necessity.

After a structural assumption is taken to be universally valid, it provides a reliable Archimedean point for turning other phenomenal laws into causal laws of nature. The upshot is that Newton followed carefully constructed inferences throughout the derivation; i.e., he deduced his universal law of gravitation from the phenomena. However, this deduction relies on the process of turning empirical claims into structural assumptions. As we shall see, this process of elevation is not foolproof. Structural assumptions can still be revised if other more encompassing or accurate assumptions are found. Thus, that structural assumptions have natural necessity does not imply that they are metaphysically necessary. Newton would not claim that it is impossible for him to have stumbled upon the wrong structural assumptions. Nevertheless, even though the process of elevation is not foolproof, it is still governed by a particular type of reasoning, and so we may think of it – and derivations based on it – as some form of inductive reasoning.

In Section 6.2, I will develop the notion of structural assumption and will show that structural assumptions are expressed in the law of momentum conservation and in a rule of composition governing the gravitational force. In Section 6.3, I will reconstruct Newton's argument in Book III of the *Principia.* I shall follow William Harper's (2002a) division of the argument into three relatively independent parts and will then demonstrate the role of structural assumptions in each of these parts. I will then conclude by analyzing the significance of Newton's distinction between a scientific proposition and a mere hypothesis.

6.2 Structural assumptions and their role in inductive reasoning

Throughout his argument in Book III of the *Principia*, Newton relies heavily on his three laws of motion as background assumptions. The laws of motion are partly justified through experiments and observations. Nevertheless, Newton applies these laws in domains far exceeding their empirical support. Also, applying these laws in the context of an attraction force outsteps the conceptual paradigm in which these laws were introduced. Stein (1991) argued that Newton's application of the third law of motion (equality of action and reaction) to the system of a central body and a rotating satellite exceeds the empirical basis of this law. The third law of motion was confirmed for collisions performed on the surface of the Earth where there is contact between the bodies. In collisions it is reasonable to assume that bodies act on one another equally and in opposite directions. Moreover, when Newton explicates the third law of motion in the *Principia*, he gives various examples where this law holds; when pressing a stone with a finger, when a horse draws a stone tied to a rope, and when a body impinges on another body. All these examples are ones where contact occurs. In the Scholium to the Laws of Motion, Newton also mentions pendulum experiments he performed to test the third law (Newton 1999, p. 426). Newton does argue as well that the third law of motion applies to forces of attraction. However, this is a conceptual argument, which presupposes the conservation of momentum, and not an empirical argument. Thus, according to Stein (1991, p. 217) it is clear that Newton's third law of motion functions as a hypothesis. Newton assumed the law to be true in all circumstances and in all contexts, beyond the domains in which it was empirically tested.

Many of Newton's contemporaries were astounded with Newton's bold application of laws of motion to the solar system. Leibniz, for example, thought it was absurd to apply these laws *without* presupposing they were caused by interplanetary fluids or some other mechanical cause. In the context of contact forces, it is reasonable to assume that action equals reaction. But without supposing forces are grounded in mechanical explanations, how are we to *explain* the validity of the third law of motion? Huygens, while accepting Newton's inverse-square result for celestial forces, did not believe that Newton adequately showed the universal nature of the force of gravitation. Applying the third law of motion to a central body with a distant satellite seemed nonsensical to him.[16]

One may think Newton's three laws of motion are universally valid due to their foundational role in Book I of the *Principia*. At least according to Newton's presentation of these laws it appears that they are not "deduced" from the phenomena. Newton asserts these laws before any phenomena are mentioned. A common strategy during the first half of the twentieth century was to treat the laws of motion as implicitly defining the meaning of the terms used in them.[17] According to this approach, we may think of the truth of these laws as stipulated. The law of inertia implicitly defines the state of being force-free as uniform rectilinear motion. Similarly, part of a definition of "force" implies the equality of action and reaction. Given that the use of the notion of "force" presupposes the stipulated truth of the axioms, Newton may legitimately apply the laws of motion to every place it is appropriate to use the notion of "force."

A proper critique of the above conventionalist approach is beyond the scope of this chapter. I shall only say that the main problem with this view is that the laws seem to have been produced by some arbitrary process. In effect the

16 He also thought he had good empirical reasons to reject Newton's reasoning (see Schliesser and Smith 1996).

17 See for example, Poincaré (1905, p. 97).

stipulated laws have the same epistemic status as a conjecture or a guess; only one cannot get any confirmation or refutation of these laws, because the terms are used to interpret the evidence. According to this view, the laws of motion are the "free creations" of the scientists who came up with them, and their treatment as axioms is not grounded in any reasoning process that could be reconstructed or analyzed according to scientific principles. A scientific inference that relies on such stipulations is essentially a version of the HD method.

Stein does not argue that the laws of motion are conventional. But he does argue that Newton's application of the third law of motion carries with it some irreducible element of stipulation. By affirming the universal validity of the law of equality of action and reaction, particularly in the context of action at a distance, Newton simply presupposed the law to be valid in all circumstances. This stipulation renders the third law a hypothesis – other propositions are compatible with the empirical evidence and the law is not dictated by the phenomena.

It seems as if Stein is presupposing that any stipulation of universality is *arbitrary*, at least in the sense that it is not dictated by the phenomena, and hence any proposition which extends its empirical and explanatory basis is a hypothesis. To be sure, the stipulation becomes less and less arbitrary the more the hypothesis is tested. One can even see the argument in Book III as an overall justification of the initially stipulated hypothesis. However, a close reading of Newton's methodological remarks suggests that he articulated a criterion for elevating empirical statements into universally valid propositions that *precedes* subjecting these propositions to further empirical tests. If such a criterion can be reconstructed, then much of the arbitrariness of stipulated hypotheses can be shown to have been eliminated prior to additional tests the hypothesis might be subjected to. As we shall demonstrate, this criterion for elevating empirical claims does carry some inductive risks, and it cannot be completely formal. However, it does place severe limitations on the type of propositions that are accepted as universally valid.

Newton's inductive procedure proceeds as follows. To receive their status as universal laws of nature, empirical claims have to be reconceptualized as assumptions about the structure of physical systems. A "structural assumption" is a rule governing the relation between parts of a physical system and their composite. For example, the property of "extension" is a structural assumption. Part of the meaning of "extension" is the relation it describes between parts of a physical system and the system as a whole. By definition, the extension of a composite body is the sum of the extensions of the parts, so we may say that the extension of the whole "arises" from the extensions of the parts.[18]

18 The notion of "part" and "whole" is used in many contexts with varying meanings (see Nagel 1961, pp. 381–383). But here the notion of part means something *like* the spatial part of a physical system. I do not identify part with a spatial part since Newton takes

When we move in our thought from parts of a physical system to the composite we are guided by assumptions that inform us how to combine the various descriptions of the parts in forming the description of the whole. At first glance it may not be clear how Newton's laws of motion express structural assumptions. After all, the first law describes the trajectory of a force-free particle, the second law describes the equality between the impressed force and the deflection of a body from its rectilinear motion, and the third law describes the equality of action and reaction. No parts and wholes of physical systems are being discussed by the laws of motion. However, it is clear that there is a strong conceptual connection between the laws of motion and the conservation of quantity of motion (the seventeenth-century term for linear momentum). Throughout the *Principia* Newton analyzes the action of forces by taking them to be operating through instantaneous impulses. Whenever such a force operates, it is essentially identified through a change of quantity of motion so that $f = \Delta p$. A continuous application of a force requires the identification of a force with instantaneous changes in quantity of motion, or $f = dp/dt$. In any case, it is quite clear that the three laws of motion are logically entailed by conservation of quantity of motion, since force is identified with the change in quantity of motion.[19]

"mass" to be the quantity of matter. Thus, for Newton an ultimate part of a physical system is given by an infinitesimal small volume of unit mass. This is obviously an idealization.

19 I am here ignoring a recent controversy about how to interpret the conceptual relationship between force and quantity of motion. The standard view is that Newton takes force to be the change in quantity of motion. However, the formulation of the second law in the *Principia* does not equate impressed force with the *rate* of change in quantity of motion, it simply claims that force is change in motion. Many commentators argue that Newton takes these changes to be instantaneous, and the force is thought of as impulsive. (See Cohen's introduction, Newton 1999, pp. 111–117.) Recently Pourciau (2006) argued that the standard interpretation of Newton's second law is mistaken. Pourciau's argument is largely based on a revision of the second law that Newton never published (Newton, 1967–1981, vol. VI, pp. 538–543). In his revision to the law, Newton explains that the force is proportional to the deflection of a body from the place it would have arrived at without the force (if it had continued moving in a straight line) to the place the body occupies after the force operated on it. Pourciau uses this unpublished text to argue that by "change in motion" Newton means something like $M\frac{\overrightarrow{PQ}}{h}$ where $\overrightarrow{PQ}$ is the deflection a body experiences as a result of the force, and h is the time elapsed. This quantity is change in "motion," and should be distinguished from change in quantity of motion given that this quantity is not equated with the mass times the *instantaneous* velocity of the body. This enables Pourciau to argue that Newton considered both continuous and instantaneous applications of force to be legitimate representations of "force." However, contrary to Pourciau's claims, there seems no reason to suppose that the quantity $M\frac{\overrightarrow{PQ}}{h}$ is not what Newton would consider as the change in quantity of motion that results from the impressed force operating on a body, whether the force is continuous or not. The quantity seems to be exactly the difference in quantity of motion between the end and

The conservation of quantity of motion can be thought of as a structural assumption. Newton defines quantity of motion as "a measure of motion that arises from the velocity and the quantity of matter jointly" (Newton 1999, p. 404). This definition does not look like a structural assumption. However, in his explication of the definition the structural aspect of the term becomes clear:

> The motion of the whole is the sum of the motions of the individual parts, and thus if a body is twice as large as another and has equal velocity there is twice as much motion, and if it has twice the velocity there is four times as much motion.
>
> (Newton 1999, p. 404)

The definition of quantity of motion as the product of mass and velocity is actually derived from the idea that the quantity of motion of the composite is the 'sum' of the quantities of motion of the parts. That is, if the motion of each part of a body is the same (i.e., each part has the same velocity), we take the quantity of motion of the composite body as the direct numerical sum of the motions of the parts. Thus, the structural assumption in this case is very simple, since it is represented with a the rule of addition. The property of the composite system is proportional to the number of parts in the system.

The conservation of quantity of motion first presupposes that the quantity of motion of the composite system arises from the "motions" we find in each part. If, in addition, we accept that a closed composite system doesn't gain or lose quantity of motion, then the conservation of this quantity follows. Thus, to arrive at the three laws of motion we only need to presuppose the structural assumption about quantities of motion and that a composite system can only gain or lose quantity of motion by exchanging quantity of motion with another system.

Our account of momentum conservation as presupposing a kind of structural assumption faces some interpretive obstacles. First, according to our modern understanding of Newtonian theory, even dimensionless particles have mass and momentum. A dimensionless particle doesn't have parts, and the mass it possesses is simply an inherent property that resists impressed forces. Thus, Newton's description of momentum as a property of a composite system arising from the properties of the parts does not make sense. Another difficulty in our interpretation is that in the above quote Newton is not distinguishing carefully enough between the size of a body and its mass. Newton says that if

beginning of the application of force, even if the application of force is not instantaneous. Regardless of whether or not Newton admits continuous applications of force, we ignore this interpretive issue and simply identify force with the *rate of* change in quantity of motion, in case the force is continuous, and with change in quantity of motion in case it is not.

a body is "twice as large" than another body with the same velocity, it would have twice as much quantity of motion. So Newton's explication of momentum "clarifies" the definition only by being careless in distinguishing between size and mass. Newton might have done better with the simple definition of quantity of motion as the product of mass and velocity.

Some investigation into the conceptual prehistory of the concept of mass and quantity of motion may help us alleviate the above two worries. We can make sense of Newton's obscure explication of quantity of motion when we juxtapose this definition over his writing in the *De Gravitatione* (Newton 2004, pp. 12–36). Scholars may not agree on the exact date of this text, but almost all commentators agree that it predates the *Principia* and provides a window into the evolution of Newton's thinking about mechanics. In this text Newton clearly thinks of material bodies as impenetrable places:

> Thus suppose that there are empty spaces scattered through the world, one of which, defined by certain limits, happens by divine power to be impervious to bodies, and by hypothesis it is manifest that this would resist the motion of bodies and perhaps reflect them, and assume all the properties of a corporeal particle, except that it will be regarded as motionless. If we should suppose that impenetrability is not always maintained in the same part of space but can be transferred here and there according to certain laws, yet so that the quantity and shape of that impenetrable space are not changed, there will be no property of body which it does not possess.
>
> (Newton 2004, p. 28)

Thus Newton's initial thinking identified bodies with impenetrable parts of space; that is, places. To understand the origin of the concept of mass, consider a thought experiment where all material bodies are simply identified as impenetrable places that move about. At first, a body's volume and its mass are the same in this imaginary world. If we think of the quantity of motion we find in each rigid body, it is essentially the product of its size and velocity as the size of the body is a good measure of the "number of parts" the body has. Thus, quantity of motion can be thought of as a structural assumption. Given that the quantity of motion of each part of the body is the same as its velocity, the quantity of motion of the composite body is the "sum" of all the quantities of motions of the parts, i.e., quantity of motion is the product of size and velocity. Now, imagine each body experiencing some contraction or expansion. To be able to describe the process of contracting or expanding a body uniformly by a factor of α, we give this body after the transformation a density of $\rho = 1/\alpha$. The mass of the body, which is the same as its original size before contracting or expanding is $m = \rho \times V$, where V is the volume of the body. Thus, it is natural to think of m as giving us the "number of parts" in a body, if we imagine this number to be preserved throughout the process of expansion or contraction. The mass is therefore the quantity of matter. But another way to think of mass is to identify it with the original volume of an impenetrable place that undergoes expansion and contraction. This thought experiment makes it clear why the equation $\vec{P} = m\vec{v}$ can be thought of as a structural assumption. Once we take the total quantity of motion of a composite body as arising from the motion that is found in each of its parts, the definition $\vec{P} = m\vec{v}$ follows as a consequence.

While it is not difficult to see why quantity of motion is a structural assumption for solid bodies, it is quite another claim to suggest that the conservation of this quantity of motion in dynamic cases reflects a structural assumption. While writing the *Principia*, Newton had good reasons to think that the conservation of quantity of motion for collisions is empirically well confirmed. Huygens, Wren, and Wallis were able to show that collisions are well described by the assertion that the quantity of motion of a closed system is conserved. Moreover, Newton himself conducted pendulum experiments to show that the equality of action and reaction does not depend on the material nature of the object. In these experiments, Newton let the bobs of two pendulums collide, and then measured the change in quantity of motion in each bob. It did not seem to matter whether the bobs were made out of gold, silver, string, or iron, the change in quantity of motion in one bob corresponded to the exact opposite change in quantity of motion in the other bob. It is therefore reasonable using enumerative induction to conclude that conservation of quantity of motion applies in collisions. However, Newton applied the third law of motion audaciously to regions far removed from the domain of experiments, and in the context of an attraction force. Thus either he stipulated the third law to be universally valid, i.e., as applying in all actual and counterfactual interactions, or he had some inductive argument for it.

An important inductive step takes place when Newton universalizes structural properties. His argument is that since the property of a composite body is reducible via a clear and unambiguous rule to the same property attributed to the ultimate parts, then it must be a universal property. To understand why we can universalize structural properties, compare this process with enumerative induction. Assume we observe that all examined crows are black. We would be tempted to think that the proposition, "All crows are black," is universally valid; i.e., unexamined crows would be black in the same way that examined crows are. However, this assumption carries great inductive risk, since both the property "crow" and the property "black" are properties of composite systems. If we cannot reconstruct these properties from the properties describing the ultimate parts of the object, there is no guarantee that the property "crow" cannot be instantiated without the property "black." In contradistinction, consider the claim that the quantity of motion of a closed system is conserved. In all examined cases this quantity was conserved. Do we have reason to believe that it would be conserved in unexamined cases? We do if we can reconstruct the quantity of motion of any physical system from quantities of motion belonging to the ultimate parts. If we assume that all ultimate parts of matter are alike, and can demonstrate that properties of composite systems are reducible to properties of ultimate parts by a known rule, then we have reason to universalize structural assumptions. Because the quantity of motion of any observed system arises only from the quantities of motion of the parts, it is reasonable to assume that the quantities of motion of all closed systems arise from the quantities of motion of their parts. Moreover, if each isolated part does not change its motion over time unless it transfers quantity of motion to another part, we may conclude that quantity of motion is universally conserved over time.

That structural assumptions are significant for taking propositions to be universally valid is evident in Newton's Rule 3 for the Study of Natural Philosophy. The rule states as follows:

> Those qualities of bodies that cannot be intended and remitted [i.e., qualities that cannot be increased and diminished] and that belong to all bodies on which experiments can be made should be taken as qualities of all bodies universally.
>
> (Newton 1999, p. 795)

In his explication of this rule Newton begins by insisting that the qualities that can be universalized must have a basis in experiments. But he also asserts that "qualities that cannot be diminished cannot be taken away from bodies." It is difficult to make sense of this criterion, since all properties, including extension, hardness and mobility, seem to be capable of being increased or diminished in magnitude. Newton explicates what he means in the following:

> The extension, hardness, impenetrability, mobility and force of inertia of the whole arise from the extension, hardness, impenetrability, mobility and force of inertia of each of the parts; and thus we conclude that every one of the least parts of all bodies is extended, hard, impenetrable, movable, and endowed with a force of inertia. And this is the foundation of all natural philosophy.
>
> (Newton 1999, p. 795)

Thus for Newton, a quality cannot be intended or remitted when we recognize it as being governed by a structural assumption. If properties divide neatly into the parts of a system whenever a process of division occurs, we have to assume that the ultimate parts of matter have these properties.

The reason for taking extension, hardness, etc. as universal properties is that for each of them the property of the composite object is compounded from the same property attributed to the parts. The property of "extension," for example, arises from the same property of extension that each of the parts has. The extension of a composite system is the sum of the extensions of the parts. Thus, because the property of extension divides neatly into the extension of the parts every time we divide the body, we can assume that every division is also a division of the composite property into the properties of the components. This implies that the ultimate parts of matter are extended. Also, the extension that the ultimate part takes up cannot be intended and remitted. A further conclusion is that we cannot separate the property of extension from material bodies; it must be a universal property belonging to all material bodies. It requires only a few observations to confirm that a property survives the division of an object into parts. Once this structural assumption is confirmed, we may take it to apply universally.

In Rule 3, Newton argues that the properties of impenetrability, mobility, and force of inertia apply universally because of their role as structural assumptions. The impenetrability of a composite body arises from the impenetrability of the parts, and so this property has to apply universally. The mobility and force of inertia of the composite body also arises from the parts, as is clear from the concept of quantity of motion.

We can now see the analogy between quantity of motion and the extension of bodies. The force of inertia of a composite body is reducible to the force of inertia of its ultimate parts. And because the tendency to continue moving in a straight line is reducible in such a way, we find that the quantity of motion is proportional to the quantity of matter, i.e., we find the motivation for the equation $P = mv$. Since the force of inertia of a composite body is governed by a compositional rule and can be reconstructed from the motion of microscopical parts, we may take it to be a universal property. This implies that even two remote bodies that interact have a conjoined force of inertia that is reducible to the force of inertia of each body (in just the same way that two remote bodies have a conjoined extension). The two remote bodies, unless disturbed, would continue to move in a straight line (taking the center of mass as their common trajectory). Because the conjoined force of inertia of the composite system is comprised of the forces of inertia of each part, an increase in quantity of motion in one body must imply a decrease in quantity of motion in the other body, so as to conserve the tendency of the composite system to move in a straight line.

The consequence of universalizing the force of inertia and taking it to be governed by a composition rule implies that the conservation of quantity of motion must be valid in both actual and counterfactual scenarios. As a consequence, the three laws of motion, since they are a logical consequence of momentum conservation, are presumed to be valid in every possible circumstance. For Newton, therefore, the laws of motion have natural necessity justified through Rule 3. Nevertheless, that the conservation law applies to all physical systems universally does not imply that his reasoning is infallible, and that it is not possible that he has derived the wrong structural properties. It is quite possible that newly found phenomenal laws or exceptions to known phenomenal laws would "dictate" alternative structural assumptions. Structural assumptions are not metaphysically necessary, even if they are taken to have natural necessity.[20]

Newton also utilizes Rule 3 to argue for the universal nature of the gravitational force:

> Finally, if it is universally established by experiments and astronomical observations that all bodies on or near the earth gravitate toward the earth, and do so in proportion to the quantity of matter in each body, and that the moon gravitates toward the earth in proportion to the quantity of its matter, and that our sea in turn gravitates toward the moon, and that all planets gravitate toward one another, and that there is a similar gravity of comets toward the sun, it will be concluded by this rule that all bodies gravitate toward one another.
>
> (Newton 1999, p. 796)

This explication of gravity shows the significance of the structural assumption governing the force of gravitation. Newton derives from the phenomena the empirical claim that gravitational acceleration does not depend on the mass of a body. The distance between two bodies, no matter what their shape, mass or chemical constitution, is enough to determine their rate of gravitational acceleration. The empirical fact regarding gravitational acceleration is then redescribed by Newton as a rule of composition governing the force of gravitation, which asserts that the total gravitational force is the sum of the gravitational forces operating on each part. The rule of composition is expressed in the formula $f_m = mf_a$ where f_m is the overall gravitational force operating on the body, m is the body's mass, and f_a is the gravitational force operating on each part. Since the gravitational force operating on a composite body survives the division of the body into parts, we may conclude that the gravitational force exhibits a structural assumption. The gravitational force operating on a composite body arises from the gravitational force operating on the ultimate parts.

Newton uses Rule 3 to argue that the susceptibility to the gravitational force cannot be separated from any physical body, and thus the gravitational force is

20 In a recent paper, Ducheyne (2005) argued that Newton used autonomous models to investigate the various properties of forces and interactions. These models are based on the laws of motion, the definitions in the beginning of the *Principia*, and various initial conditions and force laws. Only after these models were developed to various degrees of complexity were they compared with the phenomena. On the one hand, Ducheyne argues that the models are developed independently of the phenomena. On the other hand, he argues that the laws of motion on which the models were based were deduced from the phenomena. Ducheyne's account is problematic since it is not clear how the laws of motion can be derived from the phenomena and then used to construct counterfactual models.

shown to have universal validity. The inductive step involved therefore depends on Rule 3 and on the universalizable nature of structural assumptions.

6.3 Newton's argument for the universal law of gravitation

Once empirical claims are reconceptualized as structural assumptions, they function in Newton's argument as background assumptions in DI arguments. To show this, we will follow Bill Harper's (2002) division of Newton's argument for the universal law of gravitation into three significant parts. The first step of the argument relies on Kepler's *Area Law*, which asserts that the radius from the Sun to the planets, or from a planet to one of the moons, sweeps equal areas in equal times. The second step utilizes Kepler's *Harmonic Rule* which asserts that for all gravitating satellites, the period of rotation T is related to the radius of rotation as $T^2 \propto R^3$. The *Area Law* and the *Harmonic Rule* are provided as the Phenomena at the beginning of Book III. Phenomenon 1 describes the motion of Jupiter's moons relative to Jupiter. Phenomenon 2 describes the motion of Saturn's satellites relative to Saturn. Phenomena 3–5 describe the motion of the planets relative to the Sun. Finally, Phenomenon 6 describes the *Area Law* applied to the Earth's moon. Newton's first step of the argument deduces from Kepler's *Area Law* the centripetal nature of the gravitational force. In the second step of the argument, Newton deduces from the centripetal nature of the gravitation force and the *Harmonic Rule* the inverse-square nature of the force of gravitation ($f \propto 1/R^2$). In the third step of his argument, Newton deduces the universal nature of the force of gravitation, and the formula:

$$f = G\frac{m_1 m_2}{R_{12}^2}. \tag{6.1}$$

We shall demonstrate the role of structural assumptions in each part of the derivation.

6.3.1 *The first step: deriving the centripetal nature of the force of gravity from the* Area Law

The first step of Newton's argument infers the centripetal nature of the force of gravitation from the *Area Law*. The equivalence between the *Area Law* and the centripetal nature of the law is proven in Book I, Propositions 2 and 3.

Figure 6.1 describes Newton's idealized model for a body traversing equal areas in equal amounts of time. Newton's idealization consists of taking the motion of such a body as governed by an instantaneous force operating at points B, C, D and E at equal intervals of time. The distances AB, BC, etc. represent the velocity of the object if we take the force to act at integral multiples of the unit of time. The law of inertia implies that, had the force not acted on the body at point B, it would have traveled uniformly and would have reached the point c at

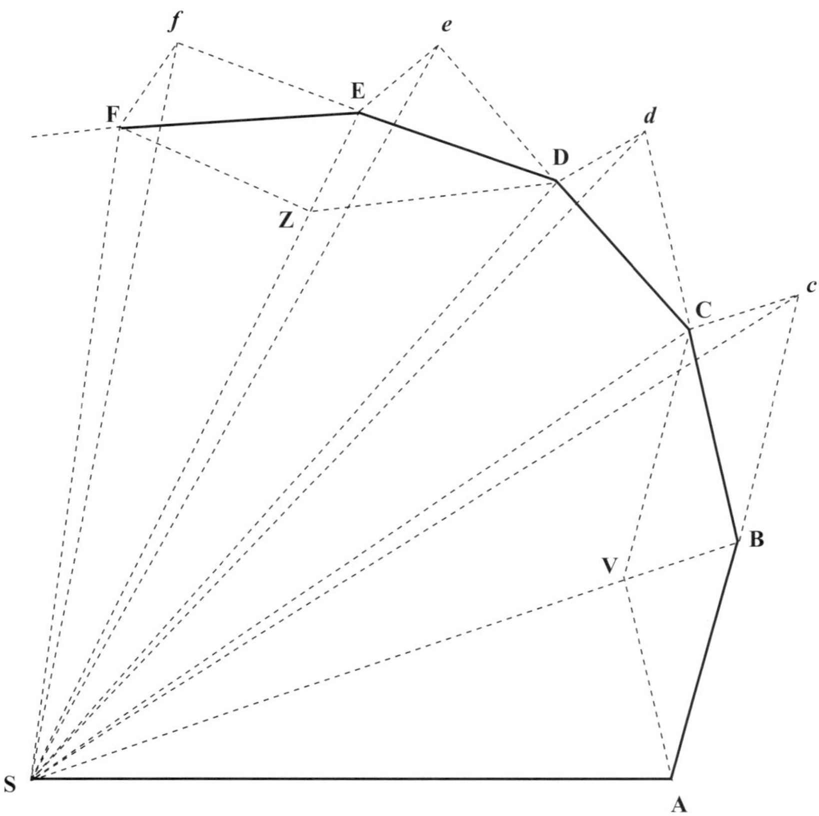

Figure 6.1 From the *Area Law* to the centripetal nature of the gravitational force

the same time the body has reached *C* in its actual motion. The conceptual link between the *Area Law* and the law of inertia is established when one compares the areas of triangles *SAB* and *SBc*. Because these triangles are of equal heights and bases, they are of equal areas. And since the model was constructed to retrieve the *Area Law*, by definition the area of triangle *SBC* equals the area of *SAB*. It follows that the area of *SBC* equals the area of *SBc*. The following equivalence holds: the area of the triangle *SBC* is equal to the area of *SBc* if and only if the change in motion *Cc* is parallel to (i.e., in the direction of) *BS* (Book I, Propositions 2 and 3). Thus, Kepler's *Area Law* is equivalent to the claim that change of motion is always directed at the immobile center *S*.

We may take the *Area Law* to be "measuring" the direction of the force. If the area traversed by the radius from *S* increases or decreases at *B*, then we know the direction of the force would have been off the line *SB*. Thus, it is clear that Newton was able in this part of the argument to translate a phenomenal law into a statement about the force generating the phenomena.

However, notice that Newton presupposes here the universal validity of momentum conservation in both actual and counterfactual cases. First, the law of inertia is taken to apply counterfactually. The trajectory *Bc* is taken as the trajectory the body would experience had a force not operated on it. Second, Newton presupposes that the direction of the force is identified with the change in momentum. The universal validity of momentum conservation far extends the empirical support of this claim, especially in the case of celestial bodies. But Newton argues in Rule 3 that mobility and the force of inertia are structural assumptions. The law of inertia must apply in both actual and counterfactual cases. Thus crucial to the argument is the natural necessity Newton attributes to the laws of motion.

We can summarize the structure of the first step of the argument as follows:

Argument I		
Premise 1	Kepler's *Area Law*	(*Phenomenal law*)
Premise 2	Momentum conservation	(*Structural assumption*)
2.1	The law of inertia applies counterfactually	
2.2	The force equals the change in momentum	
Premise 3	Euclid's geometry	(*Background assumption*)
Conclusion I	The gravitational force operates in the direction of an immobile center	

The first inductive step, therefore, follows the method of Demonstrative Induction. At this stage the background assumption used as a premise has already gone through the process of being elevated into a structural assumption. The natural necessity we attribute to the law of inertia enables Newton to "measure" the direction of the gravitational force. By comparing the actual trajectory of the body relative to its counterfactual one, Newton is able to conclude that the gravitational force operates towards the center of rotation. However, without applying Rule 3 Newton would not be able to justify his claim that the law of inertia applies in counterfactual cases.

6.3.2 *The* Harmonic Rule *and the inverse-squared distance nature of the gravitational force*

The second step in Newton's argument derives the inverse-squared nature of the gravitational force from the *Harmonic Rule*, which is:

$$T^2 \propto R^3 \tag{6.2}$$

where T is the period of the rotation around the center, and R is the radius. To carry out this part of the derivation, Newton makes an approximating assumption by taking the planets to be moving in perfect circular motion instead of ellipses. Newton proves in Book I, Proposition 4 that the centripetal acceleration of the body that rotates in a perfect circle is the following:

$$a = \frac{v^2}{R} \tag{6.3}$$

where the velocity v is the instantaneous velocity of the body and R is the radius of rotation. The proof relies on taking the polygon described in Figure 6.1 and reducing the length of the segments until they are indistinguishable from motion in a circle.[21] We can follow Newton's reasoning by examining Figure 6.1 again taking the radii *SA*, *SB*, etc. to be all the same since the body is now taken to be moving in a circle. The triangle *SBC* is again compared with the triangle *SBc*. Since *AB* and *BC* represent the velocity of the body, *BV* represents the change in velocity, and *SB* the radius of rotation, one can deduce the centripetal acceleration with the help of Euclidean theorems.[22] Since the instantaneous velocity of the body is related to the period of rotation through the equation $vT = 2\pi R$, we know from (6.2) and (6.3) and the conclusion of Argument I the centripetal acceleration of the body:

$$a = \frac{v^2}{R} = \frac{(2\pi)^2 R}{T^2} \propto \frac{1}{R^2}. \tag{6.4}$$

Thus Newton utilizes the *Harmonic Rule* to "measure" the gravitational acceleration. The assumptions that were employed in the first step of the derivation were employed in this step as well. In deriving the centripetal acceleration Newton relied on the law of inertia and on the identification of force with the change of momentum. However, here an important approximating assumption was used in the derivation, namely that the bodies are moving with perfect circular motion. This part of the deduction could be summarized as follows:

Argument II		
Premise 1	Kepler's *Harmonic Rule*	(*Phenomenal law*)
Premise 2	The gravitational force is centripetal	(*Conclusion I*)
Premise 3	Momentum conservation	(*Structural assumption*)
Premise 4	Satellites move with circular motion	(*Approximating assumption*)
4.1	$a \propto \frac{v^2}{R}$	
Conclusion II	$a \propto \frac{1}{R^2}$	

21 See the Scholium to Proposition 4 (Newton 1999, p. 452).

22 See Brackenridge and Nauenberg (2002) for a history of these calculations.

A complication in the derivation is the approximating assumption. According to Kepler's Laws, the planets are moving with ellipses around the Sun, and so the above argument applies to the motion of these planets only crudely. This complication is compounded by the fact that the phenomenal laws too hold only approximately, and one may find different curves to describe the data. The premises of Newton's derivation are known to hold only approximately, or in Newton's words, they hold *quam proxime* (very nearly). George E. Smith (2002a, 2002b) explicated the seemingly perplexing Newtonian procedure of beginning with phenomenal laws that hold only approximately, and then using the result of DI arguments to assess the origin of possible discrepancies between the observed phenomena and the ones predicted by the theory together with various idealized conditions.

Part of the story has to do with various calculations carried out by Newton to show that the approximations carried out in the premises of DI arguments cannot lead us too much astray in deriving the conclusions. As we have seen, Newton showed the biconditional between small deviations in the *Area Law* leading to small deviations of the force from the central gravitating body. Even if the *Area Law* does not hold precisely, Newton showed that small deviations would not have produced significant deviations in the conclusion. The DI argument is "stable" under small perturbations. Another example is the calculation Newton carried out to show that small deviations from the inverse-square law lead to a precession of orbits (i.e., he calculated that the apsidal angle θ is related to the index n of the exponent of centripetal acceleration – where $a = r^{(n-3)}$, and r is the radius of acceleration – as $n = (180/\theta)^2$). With this calculation Newton demonstrated that no precession implies the inverse-square law precisely. Because no precession in the planets' orbits are observed, we have indirect confirmation of the inverse-square law. We should therefore be careful in our evaluation of DI arguments. Newton's confidence in the conclusions of DI arguments is not only based on the relation between premises and the conclusion of the argument. It also relies on estimating the approximating assumptions made in the premises. Here, one needs to gauge the extent to which errors might have crept in. For example, Newton could have deduced from the elliptical orbits the inverse-square nature of the gravitational force. However, if small deviations from the elliptical orbits yield significant errors in the calculation of the force's power, such a derivation cannot be trusted. There must be some mechanism for mitigating the potential for errors.

Newton's detailed calculations show that his DI derivations are not sensitive to small deviations and inaccuracies. However, in order to assess such deviations, he relied on background assumptions that were taken to be valid in both the actual case and the idealized conditions presupposed in the argument. For example, the centripetal nature of the law of gravitation is deduced from the *Area Law* on the assumption that the first law of motion is valid universally. The law of inertia helps Newton bolster the validity of his conclusion by considering counterfactual scenarios and comparing these scenarios to the actual ones. But only by assuming that the law of inertia holds *universally* and *exactly*, can Newton show that small deviations from the *Area Law* correlate with small deviations from the centripetal direction of the force. We cannot estimate the inductive risk unless we take certain laws to be valid necessarily.

Smith is well aware of the paradoxical nature of approximating procedures, which seem to presuppose certain scientific claims as exact in order to calculate the errors that may arise from approximating assumptions (Smith 2002b, p. 45). Smith calls this common scientific approach the *exact-approximate duality*. He argues that the procedure is that of providing successive approximations, where each DI argument leads to results that yield further detail that help in evaluating deviations. What criterion do we use to designate certain scientific propositions as unquestionable background assumptions? Smith comes close to our account of structural assumptions in the following passage:

> Not just any old first approximation will permit such a sequence of successive refinements, as it might if this were merely tantamount to curve-fitting. The theoretical claim for which Newton requires the first-approximation phenomena to provide crucial evidence is a generic force law – the law of gravity in the case of orbital motions and his law for the resistance force arising from the inertia of the fluid in book 2. Moreover, when the force in question is a net force acting on a macroscopic body, he requires a compositional account of it in terms of forces acting on the individual parts of the body – in terms of microgravitational forces or, in his resistance case, in terms of the forces of impact of fluid particles on parts of the body. Finally, once inductively generalized, the force law ought to have as its consequences a host of idealized phenomena reaching beyond those providing the original evidence for the law . . . These idealized consequences are expected to agree with observation to increasingly high approximation, and to the extent they do, they of course provide further evidence of the law of force.
>
> (Smith 2002b, p. 48)

Smith identifies in passing Newton's attempt to provide a "compositional" account of the forces operating on a composite body. Thus when Newton provides an account of how composite forces are reduced to microphysical forces arising from interactions between individual parts of the body, he is safer in assuming that force-laws hold universally. The compositional account thus bolsters the robustness of scientific propositions we take to hold universally and exactly. In this Smith comes very close to our account of Newton's reliance on structural assumptions. The problem with Smith's account is that he doesn't recognize the methodological connection of the compositional accounts to Rule 3, and Newton's philosophical attempt to provide a methodological justification for taking such compositional accounts as evidence that laws hold universally and exactly. He also does not recognize how Rule 3 was intended to secure exactly those background assumptions Newton took as universally valid in his derivations and to distance Newton from an HD account of justification.

The scientific process of taking the conclusions of a DI argument and using them to create models of increasing accuracy is crucial and must supplement the initial derivation. We concur with Smith's claim that there is a risk that the conclusion will lead us down a "garden path," since not all the inductive risk is located within the initial derivation. But this process of finding more accurate models depends on finding laws that are valid in all actual and counterfactual cases. We cannot estimate the errors that creep into our observations without presupposing some rule that is valid in all circumstances.

6.3.3 Deriving the universal nature of the law of gravitation

In his third step of the derivation, Newton concludes that the gravitational force operates between any pair of masses, and is proportional to the product of the masses according to the following equation:

$$f = G\frac{m_1 m_2}{R_{12}^2}. \tag{6.5}$$

Newton's argument for the universal nature of gravitation occurs in Book III, Propositions 6 and 7. The argument in Proposition 6 begins with the observation that all earthly material objects move with the same gravitational acceleration. This fact was first discovered by Galileo. Newton describes pendulum experiments he conducted to show that all earthly matter gravitates towards the center of the Earth with an acceleration of $g = 9.8\ \mathrm{m/s^2}$. Moreover, since all of Jupiter's moons obey the *Area Law* and the *Harmonic Rule* relative to Jupiter, it follows that their acceleration toward Jupiter depends only on their distance from the planet. By the same argument, the acceleration of the planets toward the Sun depends only on their distance from the Sun. Moreover, the motions of Jupiter's moons relative to Jupiter are very regular, which implies that their acceleration toward the Sun is the same as that of Jupiter and independent of their relative mass. Thus, an important empirical claim to be deduced from all observations is that gravitational acceleration is independent of the mass of the body or its chemical constitution.

An important question remains. How does Newton derive the universal nature of the force of gravitation from this empirical claim? Some form of reasoning has enabled Newton to move from a claim that is valid for all observed bodies, to a universal law of nature asserting that a force of gravitation would operate between any pair of masses. According to Howard Stein (1970b, 1991) Newton utilizes the concept of field to make the inductive leap. Even though Newton does not explicitly use the notion of field, there are indications that he invented a very similar concept. In the beginning of the *Principia*, Newton makes an important distinction between the "absolute," "accelerative," and "motive" forces. The various forces are given in Definitions 6–8 of the *Principia*:

> The quantities of forces, for the sake of brevity, may be called motive, accelerative, and absolute forces, and, for the sake of differentiation, may be referred to bodies seeking a center, to the places of the bodies, and to the center of the forces: that is, motive force may be referred to a body as an endeavor of the whole directed toward a center and compounded of the endeavors of all the parts; accelerative force, to the place of the body as a certain efficacy diffused from the center through each of the surrounding places in order to move the bodies that are in those places; and absolute force, to the center as having some cause without which the motive forces are not propagated through the surrounding regions, whether this cause is some central body . . . or whether it is some other cause which is not apparent.
>
> (Newton 1999, p. 407)

The absolute measure of the force refers to its causal origin located at the center towards which the force of gravitation is directed. The motive force is defined as the force a composite body experiences. The accelerative force is the force experienced by each of the body's parts. Moreover, Newton asserts that the motive force is related to the accelerative force as the momentum is related to velocity, i.e., $f_m = mf_a$. The motive force f_m is the product of the mass of the body and the accelerative force f_a operating on each part.

Stein argues that Newton's notion of accelerative force functions as an acceleration field. This acceleration field describes the disposition of any body to accelerate according to the inverse-square law, and it is clear that Newton ascribes this disposition to the place a body occupies rather than to the body itself. Newton also describes how these dispositions are distributed from the center of the attracting body to the surrounding places, and so that the "accelerative quantity of force" describes the efficacy of the gravitational force at these places.

An important inductive step, according to Stein, is Newton's hypothesis that the acceleration field exists:

> Newton's inductive conclusion is that the accelerations toward the sun are *everywhere* – i.e., even where there are no planets – determined by the position relative to the sun . . . that argument cannot be made without the notion of a field.
>
> (Stein 1970b, p. 268)

Stein's account suggests that the disposition of the gravitational force to generate accelerations, where that disposition is attributed to particular *places* rather than particular existing bodies, enables Newton to generalize from the particular cases observed to a universal rule. Thus according to Stein it is the notion of a field, describing a set of dispositions spread out throughout space, which provides the gravitational force its universal validity, including its validity in counterfactual cases. Only if we assume that the attracting body generates an acceleration field, can we say that a body *would* experience a gravitational force had it been placed at a certain distance from the attracting body.

However, contrary to Stein's assertion, it seems as if the notion of acceleration field is not a necessary conceptual tool for making the generalization. If we take Newton's laws of motion to hold in counterfactual cases, then every time the *Area Law* and the *Harmonic Rule* apply, the body's acceleration would be proportional to $1/R^2$, independently of its mass or material constitution. If Newton is able to justify the claim that all gravitating bodies are likely to obey the *Area Law* and the *Harmonic Rule*, then a body would accelerate in proportion to the "intensity" of the gravitational force at that particular position. Thus, the notion of field is not *necessary* for taking the inductive step. If we can justify extending the premises of Arguments I and II to counterfactual cases, the acceleration field would be the *result* of Newton's inductive conclusion, not an aid in reaching it.

In fact, there is an alternative explanation for Newton's inductive step. It seems as if the inductive leap occurs when Newton takes the empirical claim (i.e., that gravitational acceleration does not depend on the mass of the body), and reconceptualizes it as a structural assumption. In the concluding remarks in Proposition 6 Newton states as follows:

> But further, the weights [or gravities] of the individual parts of each planet toward any other planet are to one another as the matter in the individual parts. For if some parts gravitated more, and others less, than in proportion to their quantity of matter, the whole planet, according to the kind of parts in which it most abounded, would gravitate more or gravitate less than in proportion to the quantity of matter of the whole.
>
> (Newton 1999, p. 808)

The theoretical fact which best accounts for the empirical fact is the assertion that the gravitational (motive) force operating on a composite body is the sum of the gravitational (accelerative) forces operating on the parts. Moreover, we know that the accelerative force f_a operating on each part is independent of the nature of that part. It does not matter whether a body is made of gold or of coal, each of its parts will experience the same gravitational acceleration. Because Newton was able to formulate a structural assumption governing the gravitational force, and because the force does not take into account any property restricted to a particular kind of body, Newton reaches the conclusion that this structural assumption applies to all bodies in all circumstances. He thus concludes that the motive force is related to the accelerative force via $f_m = mf_a$. This argument can be summarized as follows:

Argument III-1		
Premise	Gravitational acceleration is independent of mass	(*Conclusion II*)
Conclusion III-1	The motive force f_m is the sum of the accelerative forces f_a operating on the parts; $f_m = mf_a$	

Argument III-1 is not a deductively valid argument. Rather, it is an argument where an empirical claim is reconceptualized and elevated to the status of a structural assumption. Thus, the inductive step that Newton takes does not rely on the notion of a field. Rather, it relies on the criterion for universalizing properties implicit in Rule 3. This criterion is not unique for gravitation and is employed for all universal properties such as extension, impenetrability, and inertia. Since we can divide the gravitational force operating on the composite body to the forces operating on the ultimate parts, we cannot separate this gravitational force from the ultimate parts of matter. Thus, the gravitational force operates on all ultimate parts of matter in the same way.

The second part of the third step of the derivation concludes with the universal nature of the force of gravitation. As we have seen from step 2, we know that the gravitational acceleration is proportional to $1/R^2$ independently of the mass of the body, so that:

$$f_a \propto \frac{1}{R^2}. \tag{6.6}$$

We also know that the motive force is proportional to the mass, since we take the acceleration of the composite to arise from the accelerations of the parts, so that:

$$f_m \propto \frac{m}{R^2}. \tag{6.7}$$

Newton then uses the structural assumption governing the gravitational force together with the third law of motion to conclude the universal nature of the force of gravitation:

> Since all the parts of any planet A are heavy [or gravitate] toward any planet B, and since the gravity of each part is to the gravity of the whole as the matter of that part to the matter of the whole, and since to every action (by the third law of motion) there is an equal reaction, it follows that the planet B will gravitate in turn toward the whole of the planet as the matter of that part to the matter of the whole.
>
> (Newton 1999, p. 810)

The reasoning here may be described as follows. If body *A* gravitates toward body *B*, then the motive force operating on *A* is proportional to the mass of *A* over the distance squared, so that:

$$f_A = k_A \frac{m_A}{R_{AB}^2} \tag{6.8}$$

where k_A is some constant. But according to Newton's third law of motion, the gravitational force operating on *A* is equal in magnitude and is opposite in direction to the force operating on *B*. This force is gravitational in nature, so it too is the composite of the forces operating on *B*'s parts. Thus, the force operating on *B* is:

$$f_B = k_B \frac{m_B}{R_{AB}^2}. \tag{6.9}$$

From the third law of motion it therefore follows that $f_A = -f_B$, which implies that the gravitational force is proportional to the product of the bodies' masses:

$$f_G = G \frac{m_A m_B}{R_{AB}^2}. \tag{6.10}$$

We can summarize the third step with the help of the following two arguments:

Argument III-2

Premise 1	$f_m = mf_a$	(*Structural assumption – Conclusion III-1*)
Premise 2	$f_a \propto \frac{1}{R^2}$	(*Conclusion II*)
Premise 3	Momentum conservation	(*Structural assumption*)
3.1	Newton's third law of motion	
Conclusion III-2	$f = G\frac{m_1 m_2}{R^2}$	

We can see that arguments I, II, and III-2 all follow the DI method. These arguments use phenomenal laws and background assumptions as premises and deductive reasoning to conclude the nature of the force generating the phenomena. However, it is important to note that the background assumptions used in these arguments are of a very particular nature. Other than mathematical propositions they are all structural assumptions. A necessary inductive step is the claim that a structural property is a universal property.

It seems tempting to think of these two structural assumptions (i.e., the conservation of momentum and the rule of composition governing the force of gravitation), as de facto hypotheses. In a heuristic sense, yes. These are theoretical propositions that are not deductively entailed by empirical claims.

It is conceivable that these structural assumptions will be replaced in the future with new, more adequate assumptions. The universal validity attributed to these assumptions is not metaphysical in nature, since it may be that a more adequate structural assumption will be introduced in a future theory.

However, these structural assumptions are *not* hypotheses in the sense used by philosophers of science. First, they are not arbitrary conjectures governed solely by the imagination and luck of individual scientists. It may have required the imagination and courage of Newton to conceive of these structural assumptions and to take them as applying universally, but these assumptions are certainly not arbitrary, as we have spelled out what singles out these assumptions over others. Second, it is clear why these assumptions acquire the universal validity that is attributed to them. Unlike enumerative induction, where we articulate universal propositions linking composite properties, structural assumptions enable us to argue that we have stumbled on the properties of the ultimate parts over which it seems safer to generalize. If structural assumptions are valid, they are valid universally in both actual and counterfactual cases, if we assume that ultimate parts of matter are all alike. Finally, structural assumptions are not hypotheses because they are closely related to the results of experiments and observations (i.e., only after carefully assembling all the evidence, can structural assumptions be introduced into the theory).

6.4 Conclusion

For Newton, the conservation of momentum and the compositional nature of the gravitational force are more than just hypotheses; they are structural assumptions. Newton was well justified in perceiving himself as deducing his universal law of gravitation from the phenomena. He was not employing in his inductive method hypotheses that function as inspired guesses. He introduced structural assumptions based on a careful procedure. First, an empirical fact universally confirmed by all observations is singled out. Then, this empirical fact is reconceptualized as a structural assumption, in which no property restricted to a particular kind of material is utilized. Finally, the structural assumption is recognized as universally valid due to its role in making composite physical systems intelligible.

It is clear that this procedure does not depend on the sociological role of background assumptions as standardizing rules for solving scientific problems. Newton was creating a new paradigm through the introduction of structural assumptions. Once Newton started using the conservation of momentum as a universally valid rule, it gave the impetus to the generations that followed to emulate him. But it is not their currency in the eyes of his peers that gave Newton the confidence to apply these scientific propositions universally; it is their nature as structural assumptions. The universal validity of structural assumptions are derived from Newton's belief that the properties of composite systems must be constructed from properties of the ultimate parts of matter.

Our analysis of Newton's argument also indicates that Popper is right in that there is no *universal* 'rule of induction' that applies to all inductive arguments. The reason why Newton takes his structural assumptions to hold universally is that it seems unlikely that composite physical processes follow different rules of composition depending on the context. However, it may be that the particular theoretical and experimental context determines which structural assumption is "suggested" by the evidence. It is very possible that new structural assumptions will end up replacing older ones. Thus, these structural assumptions should be treated as local rules of inductive inference. The generation of a structural assumption follows a regulated procedure, but we should not think that this procedure is mechanical nor is it incorrigible. Nature dictates the nature of structural assumptions, but nature does not show how to read them directly from the phenomena. There is an element of stipulation in formulating these structural assumptions, however, this stipulation is not arbitrary and is not a convention.

The conclusions of DI arguments are able to introduce theoretical terms not present in the phenomena, carry a natural necessity extending the regularities present in the phenomena, and hold the status of being more exact than phenomenal laws. The DI arguments are only able to do so because of the Newtonian procedure of elevating approximate empirical laws into exact structual assumptions. The success of the DI method crucially depends on anchoring some of its premises in structural properties.

Our overall conclusion is that Newton followed the method of deduction from the phenomena. He does take the phenomena to dictate the conclusions of his scientific inferences. It is now clear that Newton had a valid method for distinguishing between hypotheses and propositions that are derived from the phenomena. However, we have to be clear that the derivation does not rely exclusively on deductive rules. The DI method gets part of the story right, but not the whole story. There is an important inductive step that Newton utilizes to secure the background assumptions, which serve as premises in his DI arguments. The strength of these DI arguments crucially depends on the strength of these background assumptions.

7

Newton, Huygens, and Euler

empirical support for laws of motion

WILLIAM HARPER

7.1 Basic empirical support for laws of motion

7.1.1 Huygens's rules for the motion of bodies arising from mutual impact

Huygens, along with Christopher Wren and John Wallis, responded to the Royal Society's invitation to produce their meditations and discoveries on the laws of motion.[1] Huygens's basic rule is a geometrical method of using line segments to give velocities after impact from velocities at impact, together with the center of gravity. Wren, independently, gave the same geometrical method for determining velocities after impact for perfectly elastic collisions in which the total velocity is conserved. Huygens's gives conservation of momentum and conservation of what we call kinetic energy, for such perfectly elastic collisions. The following remarks are evidence that Huygens had, already by 1669, achieved for collisions a quite extraordinary understanding of some important fundamental concepts and results that Newton would present in his *Principia.*[2]

> In all these cases the Author considers bodies of the same material, or would have us estimate the mass [*moles*] from the weight.
>
> He adds, moreover, that he has observed a certain wonderful law of Nature, which he affirms he can demonstrate in spherical bodies, whatever velocity v be given, and in all others whether hard or soft, and whether impacting directly or obliquely, namely: The common center of gravity of two, three, or any number of bodies, always advances uniformly in the same direction and in a straight line, both before and after impact.

1 See Murray, Harper and Wilson (2011) for translations by Curtis Wilson together with comments by Harper, Wilson and Murray.

2 The translation of this passage is by Curtis Wilson.

The first sentence suggests that Huygens may have anticipated Newton's distinction between mass and weight.[3] The second paragraph shows that, for collisions, Huygens did anticipate the basic idea of corollary 4 of Newton's laws of motion.[4]

7.1.2 Newton on projectiles and pendulums

Newton characterizes his laws of motion as "accepted by mathematicians and confirmed by experiments of many kinds" (Newton 1999, p. 424). He recounts the confirmation afforded to the first two laws and the first two corollaries by their role in accounting for the – by then familiar – idealized patterns that dominate and make intelligible the ubiquitous phenomena of free fall and projectile motion. He then cites the confirmation afforded to those first two laws and corollaries from their application to pendulums and clocks. He continues with a reference to the papers in which Wren, Wallis, and Huygens reported finding the rules of the collisions and reflections of bodies and the following remark about pendulum experiments by Wren:

> But Wren additionally proved the truth of these rules before the Royal Society by means of an experiment with pendulums, which the eminent Mariotte soon after thought worthy to be made into the subject of a whole book.
>
> (Newton 1999, pp. 424–425)

For Newton, these pendulum experiments provided measurements of the equality of actions and reactions for applications of Law 3 to collisions.[5] On the – by then well-known – idealized theory of uninterrupted pendulum motion in a vacuum, the velocity of a pendulum in its lowest point is as the chord of the

3 Howard Stein (1990, p. 25 and note 31) was the first I know of to argue that Huygens had anticipated Newton's distinction between mass and weight. Stein (1990, pp. 21–22) gives a very informative account of impressive theoretical developments and arguments from Huygens's theoretical paper on the motion of colliding bodies. This paper of Huygens was first published in 1703, eight years after his death. See Blackwell (1977, note 1).

4 **Corollary 4:**

> The common center of gravity of two or more bodies does not change its state whether of motion or of rest as a result of the actions of the bodies upon one another; and therefore the common center of gravity of all bodies acting upon one another (excluding external actions and impediments) either is at rest or moves uniformly straight forward.
>
> (Newton 1999, p. 421)

5 **Law 3:**

> *To any action there is always an opposite and equal reaction; in other words, the actions of two bodies upon each other are always equal and always opposite in direction.*
>
> (Newton 1999, p. 417)

arc that it has described in falling and as the chord of the equal arc it traverses from that lowest point to its highest point on the other side. In the application to collisions, these chords of the arcs before and after reflection represent the velocities before and after impact.

Newton remarks that these earlier experiments did not take into account air resistance or the elastic force of the colliding bodies.[6] He gives his own improved versions of these experiments to take into account air resistance. He also extends such experiments to collisions between bodies of imperfect elasticity, where there is loss of velocity in the collision. He reported trials which afforded measurements of the relative elasticities of balls of tightly wound wool, of steel, of cork, and of glass. This information was then applied in further experiments which measured the equality of action and reaction when the loss of velocity appropriate to the elasticities of the colliding bodies was taken into account. These experiments afford theory-mediated measurements of velocities that are fit by the equality of action and reaction calculated in accordance with the reductions appropriate to the differing elasticities.

These experiments suggest that the third law of motion applies to collisions quite generally. What are to be counted as the action and reaction are to be appropriately reduced to take into account losses due to imperfect elasticity, as well as any losses due to damage of the bodies in the collision.

These extensions of the classic pendulum experiments to take into account air resistance and differing elasticities are examples of Newton's often-cited achievement of going beyond idealized cases to take into account details of interfering factors. Like Huygens's collision rules, however, Newton's treatment of collisions is limited to relations between the states of motion of the bodies before the collision and the states of motion of those bodies after the collision. Newton makes no attempt to give any detailed account of the collision process itself.

7.1.3 Newton on evidence from machines

In his discussion of corollary 2 of the laws of motion (Newton 1999, pp. 418–420), Newton gives an example of resolving the forces to move a wheel of weights hung from unequal spokes going out from the center.[7] This example illustrates pulleys, stretched strings, and the law of the lever.[8] It is then further

6 Murray *et al.* (2011) gives a detailed account of Newton's versions of these experiments.

7 This topic will be treated in somewhat more detail in chapter 3 section II.3 of Harper (forthcoming).

8 Chandrasekhar (1995, pp. 24–25) provides a very nice explication of Newton's resolution of these forces.

elaborated to show how to resolve forces corresponding to the actions of a wedge, a hammer, and a screw. He concludes with the following remarks.

> Therefore, this corollary can be used very extensively, and the variety of its applications clearly shows its truth, since the whole of mechanics – demonstrated in different ways by those who have written on the subject – depends on what has just now been said. For from this are derived the forces of machines, which are generally composed of wheels, drums, pulleys, levers, stretched strings, and weights ascending directly or obliquely, and the other mechanical powers, as well as the forces of tendons to move the bones of animals.
>
> (Newton 1999, pp. 419–420)

Newton is not just saying that the forces of machines and the forces of tendons to move the bones of animals can be derived from this corollary. He is further saying that all these extensive varieties of applications of mechanics depend on it. These applications of mechanics would not be available if this corollary did not hold to sufficiently good approximation.

The last three paragraphs of his Scholium to the laws are devoted to a recounting of the enormous empirical support afforded to Law 3 from these applications to machines and devices (Newton 1999, pp. 428–430). He is at some pains to point out that these applications of Law 3 are not restricted to idealized cases in which resistance from sources such as friction can be ignored. Newton gave a detailed account of the applications of Law 3 to weights on a balance and how such calculations can be extended to account for weights interfered with by oblique planes or other obstacles, as well as weights raised by ropes over pulleys and combinations of pulleys. This was followed by instructions for such calculations for engaged gears in clocks, as well as for the force of a hand turning the handle of a screw-driving machine to the force of the screw to press a body. It concluded with the action–reaction calculation of the forces by which a wedge presses the two parts of the wood it splits to the force impressed upon it by the hammer. He suggests that if the empirical phenomena corresponding to these applications were to change so as to appreciably violate these applications of Law 3, we would soon know about it.

7.1.4 Euler on space, time and empirical support for the laws of motion

Euler's 1748 paper, 'Reflections sur l'espace et le tem[p]s' appeals to empirical support for what he characterizes as two principles of mechanics to defend Newton's appeal to absolute space.[9] Here, from the first of the twenty-one

9 Kant cites this paper of Euler's in helping motivate his 1768 appeal to incongruent counterparts to defend Newton's absolute space from objections based on commitments to a relational theory of space and time. See Harper (1991) for an account of Kant's interesting and powerful appeals to incongruent counterparts.

numbered paragraphs which constitute this short paper, is his characterization of the support which makes the principles of mechanics so well established that it would be wrong to doubt their validity.

> 1. The principles of mechanics have already been established on such a sound basis that one would greatly err if he wished to encourage any doubt as to their validity. Even if one were not in position to demonstrate them by the use of general principles of metaphysics, the excellent agreement of all the conclusions which one draws from them by means of the calculus, with all the movements of bodies both solids and liquids, on the earth, and likewise with the movements of the heavenly bodies, would be sufficient to place the truth of the principles of mechanics beyond doubt.
>
> (Koslow 1967, p. 116)

Euler is not claiming that these principles are established by measurement from phenomena. He, nevertheless, claims that this wonderful conformity of consequences drawn from them with motions of bodies should be sufficient for making the truth of these principles beyond doubt.

This first numbered paragraph goes on to give what he takes to be the principles so established. Euler defends Newton's provision for distinguishing absolute rest. He emphasizes this by distinguishing two principles,

> a body being once at rest will remain continually at rest unless it be disturbed in its state of rest by some external force.
>
> (Koslow 1967, p. 116)

and

> a body, being once set in motion, will continually move with the same speed in the same direction provided it does not meet with obstacles contrary to the preservation of that state.
>
> (Koslow 1967, p. 116)

These correspond to the two states of motion in Newton's more elegantly formulated first law of motion.[10] Like Newton's, Euler's provision for distinguishing absolute rest goes beyond what is afforded according to Newton's corollary 5 of his laws of motion.[11]

10 **Law 1**

> *Every body preserves its state of being at rest or of moving uniformly straight forward except in so far as it is compelled to change its state by forces impressed.*
>
> (Newton 1999, p. 416)

11 **Corollary 5**

> *When bodies are enclosed in a given space, their motions in relation to one another are the same whether that space is at rest or whether it is moving uniformly straight forward without circular motion.*
>
> (Newton 1999, p. 423)

The second numbered paragraph endorses the theme that these laws of motion ought to inform the study of the nature and the properties of bodies.

> 2. Since these two truths are so certainly verified, it follows with absolute necessity that they depend on the nature of bodies: and since it is the purpose of Metaphysics to study the nature and the properties of bodies, the knowledge of these truths of mechanics is capable of serving as a guide in these intricate researches (of metaphysics). For one would be right in rejecting in this science (of metaphysics) all the reasons and all the ideas, however well founded they may otherwise be, which lead to conclusions contrary to these truths (of mechanics); and one would be warranted in not admitting any such principles which cannot agree with these same truths. The first ideas which we form for ourselves of things, which are found outside of ourselves, are ordinarily so obscure and so indefinite that it is extremely unsafe to draw from them conclusions of which one can be certain. Thus it is always a great step in advance when one already knows some conclusions from some other source, at which the first principles of metaphysics ought to finally arrive: and it will be by these conclusions, that the principal ideas of metaphysics will be necessarily regulated and determined.
>
> (Koslow 1967, pp. 116–117)

It is interesting to see Euler strongly endorsing this important theme of using the principles of mechanics to inform the metaphysics of bodies.

After arguing against the capacity of relational accounts of motion to characterize the distinction between rest and motion in his first principle, Euler goes on to argue that such relational accounts cannot recover his second principle.

> 17. The reality of space will be found again established through the other principle of mechanics, which involves the preservation of uniform motion in the same direction. For if space and place are not the reference of co-existing bodies, what is meant by the same direction? One would be very much confused in expressing an idea of the mere relation of mutually co-existing bodies, without making use of that immovable space. For in some manner, while bodies move and change the positions of their situation, that does not prevent one from preserving a distinctly clear idea of one fixed direction which bodies tend to follow in their movement, in spite of all the alterations which other bodies experience. From this it is evident, that the identity of direction, which is a distinctly essential factor in the general principles of motion, cannot be absolutely explained by the relation, or the order of co-existing bodies. Thus again there must necessarily be some other real existence, outside the bodies, to which the idea of one same direction corresponds; and there is no doubt that this would be the space, the reality of which we have established.
>
> (Koslow 1967, pp. 122–123)

This argument by Euler is similar to the sort of argument against relational accounts of motion given by Newton in his unpublished *De Gravitatione* manuscript.[12]

7.2 Law 3 for attractions?

7.2.1 Euler on Newton's argument for attractive forces proportional to masses

Euler expresses his negative assessment of the empirical evidence supporting the proposition that gravitational attraction toward heavenly bodies is proportional to their masses in a letter to Mayer dated 25 December 1751.

> I consider the objection that the attraction must not necessarily be proportional to the masses, to be of no great importance, as it is still not decided by any single phenomenon that the attractive forces of heavenly bodies are proportional to their masses. On the contrary, Newton tried to determine the masses on this basis since there is no other way of specifying them. As soon as one now places the statement that the attractive forces are proportional to the masses (which is founded on a crude hypothesis) in doubt, this objection against my idea is completely eliminated.
>
> (Forbes 1971, p. 44)

Unlike Newton, who regards the proportionality of attraction to mass as a requirement that any adequate proposal for a cause of gravity must meet,[13] Euler regards it as not sufficiently established.

7.2.2 Cotes' query about Newton's application of Law 3

Euler's characterization of the claim that gravitational attraction is proportional to the mass of the attracting body as a crude hypothesis, though less tactful, is in line with an objection by Cotes, the editor of the second edition of the *Principia*. Cotes objects to Newton's application of his Law 3 in corollary 1, proposition 5, book 3.

> And since, by the third law of motion, every attraction is mutual, Jupiter will gravitate toward all its satellites, Saturn toward its satellites, and the earth will gravitate toward the moon, and the sun toward all the primary planets.
>
> (Newton 1999, p. 806)

The following passage is from Cotes's letter to Newton dated March 18, 1713.

12 Newton (2004, pp. 12–39).

13 See the first part of paragraph 8 of the General Scholium (Newton 1999, p. 943).

> But in the first Corollary of the 5th I meet with a difficulty, it lyes in these words *Et cum Attractio omnis mutua sit.* I am persuaded that they are then true when the Attraction may properly be so call'd, otherwise they may be false. You will understand my meaning by an Example. Suppose two Globes *A* & *B* placed at a distance from each other upon a Table, & that whilst *A* remains at rest *B* is moved towards it by an invisible Hand. A by-stander who observes this motion but not the cause of it, will say that B does certainly tend to the centre of *A*, & thereupon he may call the force of the invisible Hand the Centripetal force of *B*, or the Attraction of *A* since ye effect appears the same as if it did truly proceed from a proper & real Attraction of *A*. But then I think he cannot by virtue of the Axiom [*Attractio omnis mutua est*] conclude contrary to his Sense & Observation, that the Globe *A* does also move towards the Globe *B* & will meet it at the common center of Gravity of both Bodies.
>
> (Newton 1959–1977, vol. V, p. 392)

He goes on to indicate that this stops him from giving a popular account of the reasoning in Newton's argument for the 7th proposition in book 3.[14]

14 Newton's proposition 7 is the culmination of his basic argument for universal gravity. Here is the proposition and its proof.

> **Proposition 7, Theorem 7**
>
> *Gravity exists in all bodies universally and is proportional to the quantity of matter in each.*
>
> We have already proved that all planets are heavy [or gravitate] toward one another and also that the gravity toward any one planet, taken by itself, is inversely as the square of the distance of places from the center of the planet. And it follows (by book 1, prop. 69 and its corollaries) that the gravity toward all the planets is proportional to the matter in them.
>
> Further, since all the parts of any planet A are heavy [or gravitate] toward any planet B, and since the gravity of each part is to the gravity of the whole as the matter of the part is to the matter of the whole, and since to every action (by the third law of motion) there is an equal reaction, it follows that planet B will gravitate toward all the parts of planet A, and its gravity toward any one part will be to its gravity toward the whole of the planet as the matter of that part to the matter of the whole. Q.E.D.
>
> (Newton 1999, pp. 810–811)

The two paragraphs correspond to two parts of the proof. The first paragraph argues that the gravity toward each planet is proportional to the quantity of matter of that planet. This argument appeals to book 1 proposition 69 and its corollaries. Book 1 proposition 69 applies the 3rd law of motion to argue that if bodies attract each other by inverse-square accelerative forces then the attraction toward each will be as its quantity of matter or mass.

The second paragraph argues to extend this result to all the parts of planets. This second argument appeals directly to the 3rd law of motion. In both arguments the forces are interpreted as interactions between the bodies in such a way as to allow Law 3 to apply, by counting the attraction of each toward the other as action and equal and opposite reaction.

> This is what stops me in the train of reasoning by which as I said I would make out in a popular way the 7th Prop. Lib. III. I shall be glad to have your resolution of the difficulty, for such I take it to be. If it appeares so to You also; I think it should be obviated in the last sheet of Your Book which is not yet printed off, or by an Addendum to be printed with ye Errata Table. For 'till this Objection be cleared I would not undertake to answer anyone who should assert You do *Hypothesim fingere* I think You seem tacitly to make this Supposition that the Attractive force resides in the Central Body.
>
> (Newton 1959–1977, vol. V, p. 392)

Cotes points out that what seems to be Newton's tacit assumption that the attractive force resides in the central body appears to count as an assumed hypothesis on which the argument is based, rather than as a conclusion supported by the evidence adduced.[15]

7.2.3 Newton's initial response to Cotes's query about Attraction

In his initial response, Newton instructs Cotes to follow up the famous *hypotheses non fingo* passage, in the General Scholium being added to book 3, with what became the following remarks.[16]

> For whatever is not deduced from the phenomena must be called a *hypothesis*; and hypotheses, whether metaphysical or physical, or based on occult qualities, or mechanical, have no place in *experimental philosophy*. In this experimental philosophy, propositions are deduced from the phenomena and are made general by induction. The impenetrability, mobility, and impetus of bodies, and the laws of motion and the law of gravity have been

15 This objection, originally made by Cotes, has been revivified and extended by Howard Stein. See his (1991, p. 217) as well as his earlier (1967, pp. 179–180). My discussion here develops from my more extensive discussion in (Harper 2002b) and in (Harper forthcoming).

16 Here is Newton's instruction with the Latin of his proposed revision.

> And for preventing exceptions against the use of the word Hypothesis I desire you to conclude the next paragraph in this manner
>
> Quicquid enim ex phaenomenis non deducitor Hypothesis vocanda est, et ejusmodi Hypotheses seu Metaphysicae seu Physicae use Qualitatum occultarum sue Mechanicae in Philosophia experimentali locum non habent. In hac Philosophia Propositions deducunter ex phaenomenis & reddunter generales per Inductionem. Sic impenetrabilitas mobilitas & impetus corporum & leges motuum & gravitis innotuere. Et satis est quod Gravitas corporum revera existat & agat secundum leges a nobis expositas & ad corporum caelestium et maris nostri motis omnes sufficiat.
>
> (Newton 1959–1977, vol. V, p. 397)

The printed Latin replaces [, et ejusmodi] in line 2 with [; &]. It also italicizes *hypothesis* in line 2 and *philosophia experimentali* in line 4 (Newton 1972, p. 764).

> found by this method. And it is enough that gravity really exists and acts according to the laws that we have set forth and is sufficient to explain all the motions of the heavenly bodies and of our sea.
>
> (Newton 1999, p. 943)

Key additions to what was already there[17] include Newton's specific negative characterization of his *experimental philosophy* as one in which hypotheses have no place, together with his positive characterization of it as one in which propositions are deduced from the phenomena and made general by induction.[18]

These, together with his already specified characterization of *hypotheses* as whatever is not deduced from phenomena, make it clear that what Newton counts as deductions from the phenomena have to be construed widely enough to include propositions made general by induction.[19]

The other significant addition included in this passage sent to Cotes is Newton's specific claim that

> The impenetrability, mobility, and impetus of bodies, and the laws of motion and the law of gravity have been found by this method.

Consider Newton's claim that the laws of motion and the law of gravity have been found by this same method. We have seen that Newton's pendulum experiments afford theory-mediated measurements supporting his application of Law 3 to collisions. The key theoretical background assumption is the proportionality of the velocity of a pendulum in its lowest point to the chords of

17 Here is a translation of the corresponding original passage Newton had earlier sent to Cotes:

> For whatever is not deduced from phenomena is to be called a hypothesis; and I do not follow *hypotheses*, whether metaphysical or physical, whether of occult qualities or mechanical. It is enough that gravity should really exist and act according to the laws expounded by us, and should suffice for all the motions of the celestial bodies and of our sea.
>
> (Newton 1999, p. 276 [reader's guide])

18 It appears that this very significant positive characterization of Newton's method was a direct response to this challenge from Cotes.

19 As Stein (1991, p. 219) has pointed out, on Newton's usage any appropriately warranted conclusion inferred from phenomena as available evidence will count as a deduction from the phenomena. Newton's identification of "hypotheses" with "whatever is not deduced from the phenomena" makes counting something as a mere hypothesis equivalent to counting it as not appropriately warranted on the basis of available evidence.

The *Oxford Classical Dictionary* gives a relevant usage of the Latin verb "deduco" as "derive" in a sense general enough to include the origin of a word (*OCD*, p. 527). The *Oxford English Dictionary* gives as an example for "deduction" as "The process of deducing or drawing a conclusion from a principle already known or assumed" as late as 1860, "the process of deriving facts from laws and effects from causes"(*OED*, p. 358). These usages are clearly not restricted to logically valid or mathematically demonstrated inferences.

the arcs of its downward and upward swings. For Newton, this assumption is an application of his first two laws of motion to an idealized theory of pendulum motion in a vacuum under a uniform acceleration of gravity. Similarly, the laws of motion, together with the theorems of books 1 and 2, are accepted as background assumptions that can be appealed to in supporting the inferences from phenomena in Newton's argument for universal gravity in book 3. The central role of these theorems is to support subjunctive conditionals expressing the systematic dependencies that make Newton's classic inferences into *theory-mediated* measurements of features of forces from the cited phenomena.[20] We

20 For example, Newton infers the centripetal direction of the force maintaining Jupiter in its orbit about the Sun from Kepler's area rule. A planet satisfies Kepler's area rule just in case it moves in a plane intersected by the center of the Sun and the rate at which it sweeps out areas by radii from that center of the Sun is constant.

Newton's inference is backed up by systematic dependencies following from the laws of motion. The first two propositions of *Principia* book 1 (together with their corollaries) yield an equivalence between the area law phenomenon for a body orbiting with respect to an inertial center and the centripetal direction of the total force deflecting that body from its own inertial motion. They also yield that having the area rate be increasing would require that the total force be directed off center forward, while having the area rate be decreasing would require that the total force be directed off center backward. These systematic dependencies make the area law phenomenon count as measuring the centripetal direction of the force.

We can think of the second derivative of area being swept out as a phenomenal magnitude, which takes the value zero when the areal rate is constant, is positive when the rate is increasing and is negative when the rate is decreasing. Propositions 1 and 2 together make zero value of this phenomenal magnitude equivalent to having the total force directed toward the center. The additional dependencies (in corollary 1 of prop. 2) make alternatives to this zero value carry information about alternative directions of the force. Positive values carry the information that the force is off-center in a forward direction while negative values carry the information that the force is off center in the opposite direction.

These systematic dependencies support subjunctive conditionals. The suggestion that making do with truth functional material conditionals avoids the metaphysically suspect commitments of subjunctive conditionals is not driven by problems internal to the practice of science. The claim that subjunctive conditionals are suspect is, rather, a slogan for a philosophical project of revising the commitments of scientific practice to what can be accommodated by extensional language alone.

The problematic aspect of this sort of reductive analysis can be illustrated by comparing Newton's classic inferences from orbital phenomena with Clark Glymour's proposal to interpret such inferences as examples of his conception of bootstrap confirmation. (See Harper 1998 and Harper forthcoming, chapter 3.IV.4–5). Glymour's bootstrap confirmation was an attempt to explicate such theory-mediated inferences using material conditionals. (See Glymour 1980, pp. 127–133.) The initially enthusiastic response to Glymour's proposal was followed by later work raising problems for it. The most influential of these problems have been counterexamples proposed by David Christensen based on constructing "unnatural" material conditionals entailed by theory to use as background assumptions (Christensen 1983, 1990). These eventually led to the demise of bootstrap confirmation as a serious candidate for explicating scientific inference.

have seen that Newton shows how to adjust the application of this basic proportionality of velocities to chords of arcs to take into account air resistance. We shall see that Newton also shows how to correct the application of his basic theorems about measuring features of forces from orbits, which are one-body idealizations, to take into account gravitation toward other bodies.

Newton clearly takes this role of the laws of motion and theorems as accepted background theory for his inferences from phenomena to features of gravity as compatible with his counting them as empirical. One way this is clear is that if the empirical phenomena corresponding to machines were to change so as to appreciably violate the applications of Law 3, we would soon know about it. Less disruptive of our form of life and more interesting for illustrating the depth of Newton's philosophy, I claim that the transition from Newton's theory to Einstein's is in accordance with Newton's own methodology.[21]

Let us consider the first part of the paragraph from the General Scholium to the *Principia* in which Newton's *hypotheses non-fingo* passage and the revision he sent Cotes occurs. The properties of gravity cited by Newton in this passage are ones he counts as having been established by the end of his *Principia*. They are counted as properties that any adequate account of the cause of gravity would have to recover. His arguments in support of them can inform our understanding of what he counts as a deduction from the phenomena.

The first sentence makes the familiar but important claim that Newton's failure to yet find a cause of gravity does not undercut his achievement in explaining the phenomena of the heavens and of our sea by the force of gravity.

> Thus far I have explained the phenomena of the heavens and of our sea by the force of gravity, but I have not yet assigned a cause to gravity.
>
> (Newton 1999, p. 943)

Newton's explanation of phenomena of our sea includes his explanation of the basic phenomenon of two high tides, one corresponding to the Moon above and one corresponding to the Moon on the other side of the Earth. As

The counterfactual supporting nature of the dependencies backing up Newton's inferences make them immune to counterexamples based on constructing "unnatural" material conditionals.

This failure of Glymour's proposal illustrates the inadequacy of material conditionals for characterizing the systematic dependencies underwriting theory-mediated measurements in science.

21 Harper (2007) argues that Newton's 4th rule would endorse the transition from Newton's theory of gravity to Einstein's theory of General Relativity. (See also Harper 2009, and George Smith forthcoming b).

I think it is clear that the empirical measurements supporting the constancy of the speed of light with respect to all inertial frames together with Einstein's argument for the relativity of simultaneity would make Rule 4 endorse the transition from Newton's laws of motion to Special Relativity.

Cotes points out in his preface to the second edition, Newton's explanation makes such phenomena confirm gravitational attraction of the Earth toward the Moon. Here is the relevant remark from Cotes's preface:

> Further, by mutual action, the earth in turn gravitates toward the moon, a fact which is abundantly confirmed in this philosophy, when we deal with the tide of the sea and the precession of the equinoxes, both of which arise from the action of the moon and the sun upon the earth.
>
> (Newton 1999, p. 389)

The fact that Newton's detailed account of the tides, like his detailed account of the precession, required considerable revision does not undercut the fact that it was legitimate for Cotes to regard such phenomena as confirming differential action in accordance with inverse square gravitational attraction toward the Moon on parts of the Earth and seas at different distances from the center of the Moon.

Newton goes on to outline some features of gravity that can be inferred from its explanation of these phenomena.

> Indeed, this force arises from some cause that penetrates as far as the centers of the sun and planets without any diminution of its power to act, and that acts not in proportion to the quantity of the surfaces of the particles on which it acts (as mechanical causes are wont to do) but in proportion to the quantity of solid matter.
>
> (Newton 1999, p. 943)

The phenomena cited in Newton's argument for proposition 6, book 3, afford agreeing measurements, for any given distance from the center of any one planet, of the equality of the ratio of the weight of a body toward that planet to the quantity of matter contained in that body for all bodies at that distance.[22] For example, these measurements afford strong evidence that Jupiter's weight toward the Sun is proportional to its total mass, to the total quantity of matter of this three-dimensional (i.e. solid) body, not to the quantities of its surface areas exposed to impact from different directions. Janiak has argued that this very well established proportionality of weight to mass rules out the sort of contact action on the surfaces of bodies that Leibniz assumed to be the only physical alternative to action at a distance.[23]

The passage we have been quoting goes on to characterize additional features required by a cause from which the force of gravity is to arise:

22 See Harper (1999, pp. 91–93) for an account, together with a table specifying bounds for phenomena cited by Newton, afforded by data available to Newton, as well as bounds from later data available today from corresponding tests of the weak equivalence principle. This will be treated in more detail in Chapter 7 of Harper (forthcoming).

23 See Janiak (2007, pp. 142–144) and Janiak (2008, pp. 74–80).

> and whose action is extended everywhere to immense distances, always decreasing as the squares of the distances. Gravity toward the sun is compounded of the gravities toward the individual particles of the sun, and at increasing distances from the sun decreases exactly as the squares of the distances as far as the orbit of Saturn, as is manifest from the fact that the aphelia of the planets are at rest, and even as far as the farthest aphelia of the comets, provided that those aphelia are at rest.
>
> (Newton 1999, p. 943)

Newton's precession theorem and its corollaries afford systematic dependencies that make precession of an orbit measure the power law of a centripetal force maintaining a body in such an orbit.

As Newton (1999, p. 802) points out, even a quite small departure from the inverse-square would result in a precession that would easily show up after many revolutions. The famous 43 seconds a century precession of Mercury, unaccounted for by Newtonian perturbations (but a second-order phenomenon made possible by the application of the developed Newtonian theory),[24] would measure the -2.00000016 power of distance for gravitation toward the Sun.[25] This example also illustrates the fact that the systematic dependencies can be applied to planets subject to perturbations. For any planet for which all orbital precession can be accounted for by perturbations, the zero left over precession measures the inverse-square variation of gravitation toward the Sun. Newton's proposition 45 of book 1 and its corollaries are proved for orbits that are very nearly circular. The results can be extended to orbits of arbitrarily great eccentricity.[26] Indeed, it turns out that the larger the eccentricity, the more sensitive absence of unaccounted for precession is as a measure of inverse-square variation of forces maintaining bodies in orbit.[27]

The passage we have just quoted from the first part of the paragraph, in which Newton goes on to discuss *hypotheses* and his *experimental philosophy*, highlights Newton's applications of his precession theorem to infer inverse-square variation with distance from the Sun of the force of gravity that maintains planets in their solar orbits. These inferences from the stability of these orbits are impressive examples of deductions from the phenomena that count as theory-mediated measurements from the phenomena. Newton also highlights that gravity toward the Sun is proportional to its quantity of matter and is compounded of the gravities toward the individual particles of the Sun. The

24 See George Smith (forthcoming b).

25 According to *ESAA* (Seidelmann 1992, p. 704) the period of Mercury is 0.24084445 Julian years. This gives $(1/0.24084445)(100) = 415.205748$ revolutions per Julian century and $43/415.205748 = 0.10356$ sec/rev or $0.10356/602 = 0.0000288$ degrees per revolution. On Newton's formula (corol. 1 Prop. 45 bk1) the corresponding power law for the centripetal force is as the $(360/360.0000288)^2 - 3 = -2.00000016$ power of distance.

26 See Valluri *et al.* (1997) for a modern proof. George Smith (manuscript) has shown that Newton's own proof can be so extended.

27 See Valluri *et al.* (1997).

phenomena cited in Newton's argument for proposition 6 do not directly afford measurements of the proportionality of gravitational attraction to the mass of the attracting body.

In his argument for proposition 7, Newton appeals to the challenged application of the third law of motion to infer that contested proportionality.[28] If counting the Sun's gravitational attraction toward Jupiter as the equal and opposite reaction of Jupiter's gravitational attraction toward the Sun can be adequately supported by reasoning from phenomena, then Newton's inference can be counted as a deduction from the phenomena. If the inference cannot be adequately supported enough to count as a deduction from the phenomena, then the outcome of Newton's argument in proposition 7 is threatened to count as a mere hypothesis.

7.2.4 Newton's arguments to extend Law 3 to attractions[29]

In his initial letter to Cotes, Newton supplements his characterizations of *hypotheses* and his *experimental philosophy* with an appeal to arguments for extending Law 3 to attractions in his Scholium to the laws. These arguments begin with a thought experiment in which an obstacle is interposed between two bodies that attract each other so as to prevent their coming together. Unless these oppositely directed attractive forces were equal, the obstacle would not be in equilibrium. If they were in empty space, the system of the two bodies and the obstacle would accelerate in the direction of the push on the obstacle exerted by the more strongly attracted body. To the extent that the system consisting of the two bodies and the obstacle can be treated as a body,[30] this would violate the first law of motion. This argument appeals to the first two Laws of Motion together with the application of Law 3 to contact pushes to extend the application of Law 3 to some attractions. The immediate applications of Law 3 to pushes between body A and the obstacle and body B and the obstacle make the equilibrium of the system support the further application of Law 3 to the attraction between A and B themselves.

Newton goes on to outline an actual experiment in which there is such an attraction between two bodies.

> I have tested this with a lodestone and iron. If these are placed in separate vessels that touch each other and float side by side in still water, neither one will drive the other forward, but because of the equality of the attraction

28 See note 14 above.

29 See Harper (2002b) for a more extensive discussion of these *Scholium* arguments. This will be further expanded upon in Chapter 9 of Harper (forthcoming).

30 Here we see an example of Newton arguing to extend the concept of body to include what can be counted as closed systems of interacting bodies. Corollary 4 of the laws insures that the center of mass of such a system will not be accelerated by the interactions among themselves of such bodies.

> in both directions they will sustain their mutual endeavors toward each other, and at last, having attained equilibrium, they will be at rest.
>
> (Newton 1999, p. 428)

The conclusion of the thought experiment is empirically shown to hold for an actual case of attraction found in nature. Magnetic attraction exhibits general regularities – phenomena – that make it count as an interaction. You can make either lodestone or iron move towards the other by holding the other still. You can feel the pull on the lodestone towards the iron, just as you can feel the pull on the iron towards the lodestone. Moreover, the directions of these pulls towards one another are independent of orientation with respect to the still water on which the vessels containing the lodestone and iron float.

As Newton points out, magnetic attraction does not exhibit the proportionality of attraction to mass that characterizes gravitation. This does not, however, give grounds to dismiss these experiments as irrelevant to Newton's controversial application of Law 3 to gravity. Newton appeals to his experiments with magnetic attraction as measurements affording empirical support to back up the application of Law 3 to attractions in general argued for in his thought experiment. Given that Law 3 applies to attractions in general, a defense of Newton's controversial application of Law 3 to gravity depends on how well one can make a case for counting gravity as an attraction.

Newton follows up his discussion of the experiment with lodestone and iron with an equilibrium argument for the claim that gravity is such a mutual attraction between the Earth and its outer parts. Suppose the Earth is cut by a plane to carve off an outer part. The weights toward the center of the Earth of all its pieces will make this outer part press on the rest of the Earth. In his earlier *System of the World*, Newton argued that unless the rest of the Earth were attracted toward that outer part with an equal and oppositely directed force, the whole Earth would be accelerated by this outer part pressing upon it.[31] A refined version of this argument in which another outer piece of equal weight to the first is cut off by a parallel plane on the other side of the Earth generates the equal and oppositely directed pressings from the assumption that weights of outer parts of the Earth are directed towards its center of gravity and distributed about it so as to be in equilibrium. This refined version of the equilibrium argument was added in the second edition. It shows that the third law of motion applies to gravity between the Earth and its outer parts. As any body lying on the Earth can count as an outer part, this argument shows that all of the large class of cases of gravity as weight toward the center of the Earth count as attractions between the Earth and terrestrial bodies.

Let us examine the extent to which such arguments, as the foregoing gravitational equilibrium argument or Newton's appeal to Law 1, can be applied to extend Law 3 to count gravitation of Jupiter toward the Sun as an attraction

31 See (Newton 1934, p. 570).

between them. The orbits of the moons of Jupiter afford agreeing measurements of the strength of an inverse-square acceleration field directed toward Jupiter. The orbits of the primary planets afford agreeing measurements of the strength of an inverse-square acceleration field directed toward the Sun. Given the contested application of the third law of motion, these two acceleration fields can be combined to preserve the ratios of the measured centripetal accelerations toward the Sun and Jupiter by having each orbit their common center of mass. It turns out, however, that as long as the ratio of their distances from the center about which they orbit is inversely as the measured strengths of their respective centripetal acceleration fields, their motions with respect to one another will be the same, even if the center is not their center of mass.[32] We can continue this construction, adding as many bodies as we want, consistently with our information about the relative strengths of their acceleration fields. This shows that, so long as they do not result in collisions, the motion phenomena resulting from gravitational interactions among these bodies will not put any bounds on the ratios among their masses if Law 3 does not apply to these interactions.

The orbital phenomena corresponding to gravitational interactions do not directly measure the equality of the relevant oppositely directed weights. They do, however, provide additional empirical support for the conclusion that gravity satisfies the crucial criterion – that these bodies maintain forces towards one another as they move about – which distinguishes what Newton counts as attraction from Cotes's example. The oppositely directed weights towards one another resulting from combining these acceleration fields is maintained as the bodies move about. Perhaps, this would support accepting that gravity counts as an attraction. This might make the equalities of the oppositely directed forces in attractions between bodies cited in Newton's arguments to extend Law 3 to attractions support his application of Law 3 to construe gravity as a universal force of attraction between bodies.

7.3 Rule 4

7.3.1 Newton's Rule 4 applied

Newton's Fourth Rule for doing natural philosophy was developed as an additional response to Cotes's challenge.[33]

32 The center of mass will move uniformly on a circle about the center of the orbits. Even if the center of mass is construed as a body this is not obviously a violation of Law 1 (Harper 2002b, p. 92).

33 Newton's initial letter to Cotes ends with the remark

> I have not time to finish this Letter but intend to write to you again on Tuesday.
>
> (Newton 1959–1977, vol. V, p. 397)

His Tuesday letter, dated 31 March 1713, opens with the following remarks.

> Rule 4. In experimental philosophy, propositions gathered from phenomena by induction should be considered either exactly or very nearly true notwithstanding any contrary hypotheses, until yet other phenomena make such propositions either more exact or liable to exceptions.

Newton comments that the point of this rule is to defend arguments based on induction from being undercut by hypotheses. The rule tells us to consider propositions gathered from phenomena by induction as "either exactly or very nearly true" and tells us to maintain this in the face of "any contrary hypotheses." The defense against undercutting by hypotheses focuses the engine of revision on empirical phenomena.

We can clarify the difference between what are to count as propositions gathered from phenomena by induction and what are to count as mere hypotheses

> Sr
> On saturday last I wrote to you representing that Experimental philosophy proceeds only upon Phenomena & deduces general Propositions from them only by Induction. And such is the proof of mutual attraction. And the arguments for ye impenetrability, mobility & force of all bodies & for the laws of motion are no better. And he that in experimental philosophy would except against any of these must draw his objection from some experiment or phaenomenon & not from a mere Hypothesis, if the Induction be of any force.
>
> (Newton 1959–1977, vol. V, p. 400)

The last sentence, which rejects objections from mere hypotheses while endorsing objections drawn from phenomena, is a clear anticipation of the method advocated in the Fourth Rule for doing natural philosophy that was first printed in the third edition of 1726.

Newton had included a somewhat longer statement in an un-sent draft of this letter (Newton 1959–1977, vol. V, p. 401).

> Sr
> On Saturday last I wrote to you representing that Experimental philosophy proce[e]ds only upon Phenomena & makes Propositions general by Induction from them. In this Philosophy neither Explications nor Objections are to be heard unless taken from phaenomena. Nor are Propositions here made general by arguments a priori by [*read* but] only by Induction without exception. And upon such an Induction the mutuall and mutually equal Attraction is founded. One may suppose that there may be bodies penetrable or immoveable or destitute of force, or with attraction mutually unequal, but such suppositions without any instance in Phaenomena are mere hypotheses & have no place in experi[ment]al Philosophy: & to introduce them into it would be to overthrow the Arguments from Induction upon wch all the general Propositions in this Philosophy are built.

This somewhat more extended discussion anticipates Newton's comment on Rule 4,

> This rule should be followed so that arguments based on induction may not be nullified by hypotheses.
>
> (Newton 1999, p. 796)

as well as the method advocated in that Rule.

by considering Newton's explicit appeal to Rule 4 to back up his inference to extend gravity to all the planets in his Scholium to proposition 5, book 3.[34] What would it take for an alternative proposal to succeed in undermining this generalization of gravity to planets without moons to measure centripetal forces toward them? Consider the skeptical challenge that the argument has not ruled out the claim that there is a better alternative theory in which these planets do not have gravity. Rule 4 will count the claim of such a skeptical challenge as a mere contrary hypothesis to be dismissed, unless such an alternative is given with details that actually deliver on measurement support sufficient to make it a serious rival, or provide other phenomena making the proposition inferred by Newton liable to exceptions.

On vortex theories it would be the changes in motion of invisible vortical particles resulting from their pushing the planets into orbital motion – rather than the gravitation of the Sun towards the planet – that counted as the equal and opposite reaction to the weight of a planet towards the Sun. On Newton's Rule 4, the agreement among the measurements of relative inertial masses among solar system bodies provided by orbital phenomena counts as evidence supporting Newton's application of Law 3 to count gravitation among solar system bodies as pair-wise interactions between them. To avoid counting as a mere hypothesis to be dismissed, a theory construing gravitation toward the Sun as an interaction between planets and vortical particles that did not recover Law 3 between the Sun and planet would have to develop some comparably accurate way of measuring these relative masses, provide some alternative realization of agreeing measurements sufficient to offset them, or provide other phenomena making the propositions inferred by Newton liable to exceptions.

7.3.2 Huygens's proposal for a cause of gravity

Some years ago I gave a paper at which I dismissed vortex theories as merely predictive hypotheses that completely failed to realize Newton's stronger ideal of empirical success. Mike Mahoney and Simon Schaffer pointed out that Huygens had been able to use his measurement of the strength of surface gravity to calculate the speed with which the spherical shells of vortical matter

34 **Scholium** Hitherto we have called "centripetal" that force by which celestial bodies are kept in their orbits. It is now established that this force is gravity, and therefore we shall call it gravity from now on. For the cause by which the moon is kept in its orbit ought to be extended to all the planets, by rules 1, 2, and 4.

Here are

Rule 1. *No more causes of natural things should be admitted than are both true and sufficient to explain their phenomena.*

and

Rule 2. *Therefore, the causes assigned to natural effects of the same kind must be, so far as possible, the same.*

surrounding the Earth would have to rotate to account for surface gravity in his theory. This is a theory-mediated measurement of a theoretical parameter by a phenomenon it purports to explain.

For Huygens, as for other proponents of the Mechanical Philosophy – the dominant approach to Natural Philosophy at this time – applications of principles of mechanics in natural philosophy are restricted to making natural phenomena intelligible by showing how they could be caused by the different magnitudes, figures, and motions of bodies.[35] Huygens makes his hypothesis recover the phenomenon of weight toward a center by having many different very small layers or shells of vortical particles swirling in all different directions.[36] The centripetal tendencies imparted by the actions of these very tiny layers of vortical fluid matter on the parts of a body will add together to produce its weight, while the transverse tendencies imparted by the actions of these very tiny layers of vortical particles will cancel out.[37]

Huygens appeals to collision experiments to establish that weights of bodies are proportional to the quantities of matter that compose them.[38]

> To this end, I will point out what occurs during the impact of two bodies when they meet in horizontal motion. It is certain that the resistance that causes bodies to be moved horizontally, as a ball of marble or lead placed on a very level table would be, is not caused by their weight toward the Earth, since the lateral motion does not draw them away from the Earth, and so is not at all contrary to the action of gravity that pushes them down.
>
> There is nothing then in the quantity of matter attached together contained in each body that produces this resistance. So, if two bodies each contain as much matter as the other, they will reflect equally, or both will remain completely motionless, depending on whether they are hard or soft. But experience shows that every time two bodies reflect equally in this way or stop one another, having come to meet with equal velocities, these bodies are of equal gravity. It follows then from this that those bodies that are composed of equal quantities of matter are also of equal gravity.
>
> (Huygens 1690b, p. 140)

These experiments measure the equal weights of bodies which collisions have shown to have equal quantities of matter by their equal reactions when having come together horizontally with equal velocities. The measured equality of ratios of weight to quantity of matter are taken to be empirically demonstrated as a requirement for Huygens's hypothesis about the cause of gravity.

35 Huygens's version of this Cartesian Mechanical Philosophy is revisionist in that he rejected Descartes's identification of matter with extension to allow for void spaces and corpuscles (see Stein, 2002).

36 Huygens (1690b, p. 135). 37 Huygens (1690b, p. 137).

38 If this passage was written before Huygens read Newton's *Principia* then we have another striking anticipation of an important result of Newton's.

Huygens provides some details which suggest that such a version of his hypothesis is not obviously impossible.[39] The parts of his hypothesized fluid matter need to be small enough to penetrate through bodies to interact with the particles which compose them. He construes his account so that the weight of a body is proportional to the total volume of the impenetrable particles which make it up.

> This matter then passes easily through the interstices of the particles that compose the bodies, but not through the particles themselves; and this causes the various gravities (weights), for example, of rocks, metal, etc. This is because the heavier of these bodies contain more of such particles, not in number but in volume: For only in their place (only in the places unoccupied by the particles) is the fluid matter able to rise.
>
> (Huygens 1690b, p. 139)

If the particles are all equally solid (have the same density) then the weights of bodies will correspond to the quantities of *solid* matter that comprise them. Such a version of Huygens's proposed local acting cause of gravity would recover weight proportional to mass.[40]

If we considered Huygens's theory vs. Newton's theory just up to proposition 6, then even if we used Newton's Rule 4, informed by his richer notion of empirical success, Huygens might have had a good case. Any version of his proposal that recovered the inverse-square centripetal attraction toward planets with weight proportional to mass would recover all of the measurements cited up to that point in Newton's argument. Like Newton, Huygens considers all the features established by these agreeing measurements as features that any adequate account of the cause of gravity must recover. Also like Newton, Huygens has not given details of a causal account of gravity that recovers the inverse-square or the proportionality of weight to mass.

Unlike Newton, Huygens would count his having given a causal account of centripetal forces, that could be consistently extended to include these additional features, as a positive achievement unmatched by Newton's argument to extend proposition 6 to parts of bodies. The velocities of his spinning shells, a key causal parameter in his account, are measured by phenomena measuring the acceleration of gravity at the surface of the Earth. Huygens was able to exploit Newton's measurements of surface gravities of the Sun, Earth, Jupiter, and Mars to measure the velocities of the corresponding spherical shells. The speed for the Sun is so great,

> 49 times greater than what we have found near the Earth, which was already 17 times greater than the velocity of a point at the equator.
>
> (Huygens 1690b, p. 168)

39 See Huygens (1690b, pp. 137–140) and Harper (forthcoming, chapter 5 section I.4) for more details.

40 Contrary to Janiak (2007, p. 145) and Janiak (2008, p. 78).

that it suggests a speculation about the cause of the brilliant light of the Sun. Here we have a second phenomenon, the brilliant light of the Sun, which this same causal parameter might also be purported to explain. Huygens does not, however, have systematic dependencies that would turn the brightness of the light of the Sun into an agreeing measurement of this awesome speed required to have his rotating shells explain the surface gravity of the Sun.

7.3.3 Accumulating empirical successes affording increased support

The role of accumulating empirical successes in affording increased support is another important aspect of the methodology of Rule 4. In corollary 2 of proposition 8, book 3, Newton exploited his application of Law 3 to make orbital phenomena measure the relative masses of the Sun, Earth, Jupiter and Saturn from the orbits about them. The agreeing measurements of the mass of the Sun afforded from the orbits of its six known planets, of the mass of Jupiter from the orbits of its four known moons, of the mass of Saturn from its five known satellites, and of the mass of the Earth from the orbit of its moon and from pendulums measuring its surface gravity, are all realizations of Newton's ideal of success as convergent accurate measurements by phenomena. None of these is available to Huygens's alternative proposal.

Hypothetico-deductive methodology that construes accurate prediction as confirmation of assumed hypotheses shares the basic idea that accumulating empirical success affords increased support. On such hypothetico-deductive models, empirical success is limited to accurate prediction of data. The above examples illustrate the superior power of Newton's richer ideal of empirical success to afford empirical discrimination among proposed alternative theories. An important role of accepting the laws of motion and theorems derived from them is to afford subjunctive conditionals – the systematic dependencies – that make the phenomena count as theory-mediated measurements of the inferred features of forces.

Huygens's alternative can further reinforce this lesson. Huygens resisted Newton's arguments to combine the separately argued for inverse-square centripetal acceleration fields into a single system. He did not accept Newton's center of mass resolution of the two chief worlds systems problem on which Kepler's orbits are good starting approximations from which to generate more accurate accounts by correcting for perturbations due to gravitation toward other planets. Huygens was convinced by Newton's arguments to inverse-square attraction of gravity toward the Sun that Kepler's elliptical orbits with force toward the Sun at a focus were exact descriptions of the motions of the planets. This makes empirical establishment of perturbation corrections of the basic Keplerian elliptical orbits clear empirical counterexamples to Huygens's alternative.

It turns out that it was a surprisingly long time before applications of Newton's theory led to significant, empirically established, corrections to take into account perturbations. Curtis Wilson offers the following remarks on the astronomical tables of Cassini II, published 1740, and those of Halley, published posthumously in 1749:

> These tables, the most highly respected at the time of their publication, and still in use in the 1780s, were purely Keplerian in principle, except for Halley's inclusion in his tables of an anomalous acceleration of the mean motion of Jupiter and an anomalous deceleration in the mean motion of Saturn.
>
> (Wilson 1985, p. 16)

Halley, of course, was aware of and in support of Newton's contention that on his theory of universal gravity one would expect Jupiter and Saturn to mutually perturb one another's orbital motions. The problem was how to give a detailed account of the empirical correction needed as well as of the Newtonian perturbation explaining it.

Wilson plots the dominant Jupiter–Saturn perturbation, the great inequality, along with the next two largest perturbational inequalities of each.[41] This shows the great inequality to have a period of about 900 years, with Jupiter speeding up and Saturn slowing down for half the cycle and Jupiter slowing and Saturn speeding up for the other half.[42] It also suggests that the pattern reverses in the mid 1700s, so that Halley's correction becomes increasingly inaccurate after the 1760s.[43] In 1773, Lambert showed that Halley's supposition was empirically untenable.[44]

Finally, on 23 November 1785, Laplace announced to the Paris Academy that the anomalies in the mean motions of Jupiter and Saturn could be accounted for on the assumption of universal gravitation. Wilson goes on to describe the enormous import of Laplace's achievement:

> in the wake of Laplace's 'Théorie de Jupiter et de Saturne' and primarily as a result of it, the practice of predictive astronomy had been transformed. On the basis of the new procedures introduced by Laplace, the way appeared open to a marked reduction in the gap between tables and observations, and a new period of advance, both theoretical and observational, was entered upon.
>
> (Wilson 1985, p. 23)

This new period of advance was the extraordinarily successful research program that, from the work of Laplace at the turn of the nineteenth century and up through the work of Simon Newcomb at the turn of the twentieth century, led to increasingly accurate perturbation-corrected orbits fitting increasingly

41 Wilson (1985, p. 35). 42 *Ibid.* 43 *Ibid.* 44 Wilson (1985, p. 20).

precise data and affording increasingly accurate measurements of the masses of solar system bodies.[45]

The solar tables of Lacaille of 1758 were the first to include perturbations of a planet. They included perturbations of the motion of the Earth, due to the Moon, Venus, and Jupiter. They also were the first tables to take account of the aberration of light and the nutation of the Earth's axis, two effects that had prevented advances in telescopes and clocks from achieving precision far exceeding the best naked eye observations. Bradley announced his discovery of aberration of light in 1729 and of the nutation of the Earth's axis in 1748. The successful theoretical treatment of nutation as a Newtonian perturbation was by d'Alembert and Euler. D'Alembert took his treatment of nutation as affording striking confirmation of attraction of the Earth toward the Moon. Wilson quotes the following passage from d'Alembert's memoir of 1749.

> The nutation of the terrestrial axis, confirmed by both the observations and the theory, furnishes, it seems to me, the most complete demonstration of the gravitation of the Earth toward the Moon, and consequently of the principal planets toward their satellites. Previously this tendency had not appeared manifest except in the ocean tides, a phenomenon perhaps too complicated and too little susceptible to a rigorous calculation to silence the adversaries of reciprocal gravitation.
>
> (Wilson 1995, p. 48)

Euler refined d'Alembert's treatment and was inspired by it to develop the first treatment of mechanics of rigid bodies.[46]

7.3.4 *Newton vs. Euler on action at a distance*

Euler's objection to Newton's inference to gravity proportional to the mass of an attracting body was in a letter to Mayer, who was extending the work of Euler and Clairaut on the lunar precession to develop the first really accurate lunar tables. Euler's reaction to Clairaut's solution to the lunar precession problem is expressed in the following quotation from a letter to Clairaut of 29 June 1751:[47]

> the more I consider this happy discovery, the more important it seems to me . . . For it is very certain that it is only since this discovery that one can regard the law of attraction reciprocally proportional to the squares of the distances as solidly established; and on this depends the entire theory of astronomy.
>
> (Waff 1995, p. 46)

45 See George Smith (forthcoming b).

46 See Wilson (1995, p. 53).

47 See Waff (1995, pp. 35–46), for a brief account of the lunar precession problem and Clairaut's eventual solution to it. Waff (1976) is a wonderfully detailed account.

This allowed the inverse-square to be assumed confidently in the research project of finding perturbation-corrected orbits. Euler's letter to Mayer of 25 December shows that removing his doubts about the inverse-square did not remove his doubts about Newton's controversial application of Law 3.

Euler wanted to avoid action at a distance. He continued to look for a cause of gravity that would avoid action at a distance and he continued to look for phenomena that would afford evidence of an aether that would require modification of Newton's theory, even as he developed fundamental contributions to the analytic treatment of perturbations within Newton's theory.

Proposition 69 of book 1 is where Newton shows that applying Law 3 to a system of mutually attracting bodies makes the absolute measure (the strength) of the centripetal force toward each body proportional to the mass of that attracting body. In the Scholium to that proposition, Newton tells us:

> I use the word "attraction" here in a general sense for any endeavor whatever of bodies to approach one another, whether that endeavor occurs as a result of the action of the bodies either drawn toward one another or acting on one another by means of spirits emitted or whether it arises from the action of aether or of air or of any medium whatsoever – whether corporeal or incorporeal – in any way impelling toward one another the bodies floating therein.
>
> (Newton 1999, p. 588)

His counting gravity as an attraction is compatible with having it arise "from the action of aether or of air or of any medium whatsoever – whether corporeal or incorporeal – in any way impelling toward one another the bodies floating therein".

Newton is suggesting that his application of Law 3 does not rule out local causation. An analogy suggests that causes that would support the application of Law 3 without action at a distance are indeed conceivable. Suppose the separate causes pushing the bodies together acted as though they were produced by the two jaws of a pair of tweezers.[48] Even though the direct applications of Law 3 would be between the bodies and the separate causes driving them towards one another, being able to count on having these impulses so coordinated would support the additional application of Law 3 to the motive forces of the two bodies. Such causes would extend to separated bodies the indirect applications of Law 3 of Newton's experiment with magnetic attraction and his argument that terrestrial gravity is an attraction between the Earth and its outer parts.

Newton's separation of his philosophical commitment to avoid action at a distance from his methodological commitment to make theory-mediated measurements afford empirical answers to questions about the force of gravity that any adequate cause of gravity must account for is sharper than Euler's.

48 This analogy was suggested to me in conversation by Howard Stein.

For Newton, the apparent commitment to action at a distance generated by the proposed application of Law 3 carries no weight to offset the convergent agreeing measurements to which it leads.[49]

7.4 A concluding remark

I want to conclude by arguing against turning Newton's ideal of empirical success as theory-mediated measurement into a necessary criterion for counting a proposition as gathered from phenomena by induction. We have seen that in his initial response to Cotes, Newton points out that he does not intend axioms to count as hypotheses. Deductions from the phenomena should include his characterization of the confirmation afforded to the first two laws and the first two corollaries by their role in accounting for the patterns that dominate the ubiquitous phenomena of free fall and projectile motion, as well as the confirmations afforded by applications of the laws of motion to account for the effectiveness and usefulness of machines. This widening of Newton's construal of deductions from the phenomena in their application to laws of motion suggests that such a widening of what can count as deductions from the phenomena may be part of Newton's response to Cotes's challenge.

It has been pointed out that the first direct measurement establishing the proportionality of gravitational attraction to the mass of the attracting body was not achieved until Cavendish's laboratory measurement of the gravitational constant in 1798.[50] Until Cavendish there was no direct measurement supporting the application of Law 3 to count gravitation as a pair-wise attraction between separated bodies. This might suggest that Newton's application of Law 3 to count gravity as a pair-wise attraction holding between solar system bodies should not have been counted as a proposition gathered from phenomena by induction before Cavendish completed his experiment.

I have suggested that, on Newton's methodology as explicated in Rule 4, Newton's convergent agreeing measurements of relative masses of solar system bodies in corollary 2 of proposition 8 of book 3 may well put his theory sufficiently far ahead to count Huygens's alternative as a mere hypothesis. Newton's own responses to Cotes suggest that he would appeal to the whole of the *Principia*.[51] Cotes, in his preface, focuses on the tides and precession of the equinoxes as affording phenomena that testify to attraction of our sea and Earth

49 It is interesting that by Kant's day the empirical support for gravity as a universal force of mutual interaction between bodies was so great that it was taken to empirically settle the causal question in favor of action at a distance.

50 George Smith pointed this out in his vivid masterful presentation in Leiden. As I suspect he intended, it was a surprise to many that direct measurement of the mutuality of gravitation between separated bodies was so late in coming.

51 This is also Howard Stein's position. See Stein (1991).

toward the Moon. Even if one were to reject Newton's own treatments of the tides and precession of the equinoxes as insufficient to count the attraction of the Earth by the Moon's gravity as acceptable as an approximation, d'Alembert's and Euler's treatment of nutation and Lacaille's tables of 1758 with perturbations of the Earth by the Moon would each afford clear empirical support for attraction of the Earth toward the Moon. According to the methodology of Newton's Rule 4, by the time of Laplace's solution to the great inequality of the Jupiter–Saturn mutual perturbation the indirect support afforded by convergent agreeing measurements of parameters that had accumulated to back up universal gravity was sufficiently great that it is no surprise that few, if any, regarded Cavendish's measurement as removing a serious obstacle to accepting Newton's theory.

There is a long history of philosophers turning powerful sufficient conditions for some apparently clear cases of knowledge into necessary conditions, which then lead to skeptical arguments designed to undercut other commonly accepted cases of knowledge. Descartes's example of knowledge by perception limited to subjective contents, that not even a *malin génie* could deceive one about, got turned into a strict subjectivist empiricism according to which knowledge of external bodies is problematic. I suggest that turning the Newtonian ideal of convergent accurate measurement of parameters from diverse phenomena into a necessary condition for acceptance of theoretical propositions in natural philosophy would be an example of this practice that promotes unwarranted skepticism.

8

What did Newton mean by 'Absolute Motion'?

NICK HUGGETT

Newton's *Scholium* on time, space, and motion is familiar material, but its relation to some important recent work in philosophy makes it well worth revisiting. In particular, I wish to discuss Newton's views on the foundations of mechanics, as they appear in his published works of natural philosophy: what he wished to make public to his intellectual community, what he took to be crucial to his programme in mechanics, and what he took to be scientifically defensible in the same way as, say, universal gravitation. Recent work in the foundations of spacetime theories has developed a new (or perhaps rediscovered a disregarded) understanding of the nature of spacetime in mechanics; thus expanding the space of known logical possibilities expands the space of interpretational possibilities, and we can fruitfully ask whether Newton proposed anything like the new/rediscovered account. Indeed, one of the main purposes of this chapter is to discuss critically such a reading.

First, a brief sketch of the relevant foundational issue. As we'll discuss, canonical mechanical theories require space and time to have a geometric structure that cannot be defined in terms of the relative positions and motions of bodies. The foundational question is what to make of that structure; should we just think fairly literally of spacetime as a manifold of points endowed with a geometry, much as Newton's absolute space is usually conceived? There is another tradition, starting at least with James Thomson (1884; brother of Lord Kelvin),[1] which views the structure as a feature of the laws, not of 'spacetime itself'. This idea can be cashed out by starting with frames (smooth assignments of co-ordinates to the points of space and time) and physical laws (Newton's or Maxwell's, for example). Then take those laws to say that the

Thanks to Ori Belkind, Gordon Belot, Andrew Janiak, Eric Schliesser, Howard Stein, and Daniel Sutherland for criticisms and comments. In particular, I have received some illuminating correspondence from Stein on an early draft, for which I am very grateful. Of course, my thanks carry no implication of endorsement of any kind. Parts of this paper were presented at the *Issues in Modern Philosophy Conference* at NYU, 10–11 November 2006.

1 See DiSalle (1991) and Barbour (1989, chapter 12) for further discussion.

laws hold in *some* frames: by definition, the 'inertial' frames. If these frames are related by Galilean transformations then they correspond to a geometry of 'Galilean spacetime' (discussed below), and if by Lorentz transformations then to the Minkowski geometry of special relativity. That is, the geometry is not ascribed to spacetime directly, instead the symmetries of the laws are projected onto spacetime as a geometry with the same symmetries. That is, spacetime is interpreted 'dynamically', as fundamentally a feature of the laws of physics (without any implication that the laws couple geometry to matter so that it evolves). Such a view may or may not be 'relational' as that is usually understood: while the laws can be thought of as *governing* the relative motions of bodies, whether the laws themselves can be '*reduced*' to relations is a further issue.

In a recent book that has attracted considerable attention (and the 2006 Lakatos Award), Brown (2005) offers an account along these lines. DiSalle (1994) develops a similar position (as does Huggett 1999, 2006). The clarification of these ideas makes it possible to look at earlier thinkers in a new light to see whether their work can be understood in such terms: perhaps they have been erroneously ascribed views because we were unclear on the conceptual possibilities. In a terrific new book, DiSalle (2006) has described how the understanding of the nature of spacetime structure was developed through the history of mechanics, from Newton to Einstein. The philosophical question of the book is how scientists understood the 'a priori' assumptions of theories, particularly spacetime structure, especially as these were discovered and rejected experimentally. DiSalle's discussion is extremely insightful, and reveals a great deal about the relation of philosophy to physics, but here I focus on his argument – following, he claims, Stein (1967) – that Newton developed a version of the dynamical interpretation in his *Scholium* on time, space and motion. More specifically, I want to demonstrate that DiSalle's readings of both Newton and Stein are implausible. Although Stein's paper appeared over 40 years ago, and is commonly cited by philosophers of physics, it bears further scrutiny for a number of reasons: for its relation to the important new work of DiSalle, because it is often not well understood despite its significance, and because it is not as widely known to historians of philosophy as it should be.

The plan of this paper is to lay out DiSalle's position (Section 8.2) briefly, then explain Stein's views and some of the aspects of the *Scholium* that they fail to encompass (Section 8.3–8.4). In particular, I will point out that Newton does not introduce 'true motion' and 'absolute motion' as synonyms; instead the former is the sense of motion implicit in the laws, while the latter is defined as motion in absolute space – Newton argues that these two conceptions refer to the same thing. Then, (Section 8.5) I will be able to return to the question of how DiSalle is at odds with Newton and Stein. But first there is an important question hanging over the whole discussion.

8.1 Newton's definitions

The following discussion of how Newton defined these terms may seem paradoxical, because he himself thought the concept of motion to be too obvious to stand in need of definition; he starts the *Scholium* (to the definitions) by declaring that he does 'not define time, space, place and motion, as being well known to all'.[2] But on the other hand, the *Scholium* clearly tells us what Newton thinks time, space and motion are, which amounts to telling us what he takes the corresponding terms to mean – itself tantamount to defining them!

But we can avoid any tension here, if we understand what role Newton takes to be played by definitions – i.e., those explicitly headed as such in the *Principia*. The definitions that he does give under that designation are all of new or contentious technical terms whose precise senses are appealed to in the various mathematical proofs of the *Principia*. Concepts that he takes to be familiar, such as volume and velocity, he appeals to in making his definitions without explication; the common meanings of space, time and motion he takes to be similarly sufficiently precise for the proofs he offers – thus they need no definitions either. But that is not to say that his work does not raise further questions about their meanings, and it is those issues that the *Scholium* addresses. And thus in the sense that it explicates what the terms mean, he does offer 'definitions' – but since they are not meanings directly appealed to in the proofs, they are not 'definitions' of the formal system of mechanics developed in the *Principia*.

To illustrate the point, consider the treatment of time. In formal proofs, time is of course treated in the 'well-known' way, as the parameter with respect to which positions change, according to the laws of motion: as an 'absolute', rather than with reference to any physical 'clock'. But in *Book III* the experimental meaning of time is important, because, for instance, the *Phenomena* report the observed motions of the 'planets' – the observed variations in their positions with respect to measured, 'relative' time. So, while the parameter of change is sufficiently 'well known to all' for formal proofs, its connection to time measurements needs to be discussed to make clear how Newton's formal results are to be brought to bear on observations. And that is exactly what Newton does in the *Scholium*, explaining how the solar day must be corrected using the 'equation of common time' to take account of its variation as the Earth moves in its orbit (p. 410; all references to Newton are from 1999,

2 At least he does, according to Cohen and Koyré, in the manuscript and first two editions of the *Principia* (Newton 1972, p. 46), and hence in older translations, such as Motte's (Newton 1729, p. 9), from which the phrase is quoted. For reasons I do not understand – though plausibly related to the topics of this essay – the phrase is omitted from the third edition, and hence from Cohen and Whitman's translation (Newton 1999, p. 408), from which I generally quote in this essay. Stein's work avoids ascribing such definitions to Newton; DiSalle's does not.

unless otherwise indicated).[3] Thus there is no formal definition of 'time', but there is discussion of its meaning in physics. We can sensibly also expect discussions of the meanings of 'space, place and motion'. Such discussions may similarly clarify the connection between absolute and experimental quantities, but it would be hasty to conclude on the basis of the example of time that Newton will restrict himself to such clarifications. We should consider the possibility of further, non-experimental elucidations of absolute space and motion.

With all of the preceding in mind, we can say straight off that in a loose sense – not in the sense of a definition in the formal system of the *Principia*, but in the sense of telling us what he means by the term – Newton *does* define 'absolute motion'. Namely, it 'is the change of position from one absolute place to another' (p. 409). We shall have more to say about what is involved here, but as usually understood, Newton intends absolute motion to be taken with respect to some enduring Euclidean frame;[4] thus bodies have well-defined, 'absolute' positions, velocities, accelerations and rotations.

8.2 *Understanding Space-Time*

In (2002) and (2006) – *Understanding Space-Time* – DiSalle proposes a reading of Newton's views on space and time in the *Scholium* which is considerably at odds with that usually offered. There is of course a considerable literature on this topic, but in most recent treatments there has been considerable agreement that Newton's arguments are aimed at establishing that space is ontologically robust; debate has focused on how Newton argued (see, for instance, Rynasiewicz 1995b). DiSalle instead starts with a new understanding of what space(time) was for Newton.

According to DiSalle, Newton's arguments from 'properties, causes and effects' amount to a *definition* by conceptual analysis of absolute motion (pp. 16–17, §2.5; all references to DiSalle are from 2006, unless otherwise indicated). The kind of definition DiSalle has in mind here is more like that given by a linguist reporting on linguistic practice than by a logician introducing a new term into a formal language: Newton explicates how natural philosophers use 'motion' in their applications of mechanics, rather than giving an independent meaning in terms of which laws can then be formulated.

3 In modern terms, Newton is referring to the correction of 'apparent' solar time to 'mean' solar time. (Thanks to Bill Harper for some discussion of Newton here.) Note further that advances in physics do not change the underlying point that no real clocks allow perfect accuracy: even atomic clocks are affected by variations in the ambient electromagnetic field.

4 Arthur (1994) points out some ways in which Newton's 'spaces' are not reference frames in the contemporary sense, but those differences are not important for our purposes.

Newton's subsequent argument is that Descartes's definitions are incompatible with those uses, including his own.

According to DiSalle, the 'properties' to which Newton appeals are pre-scientific and unempirical, and thus fail to give satisfactory definitions: for instance, 'bodies truly at rest are at rest in relation to one another' (p. 411). However, Newton's attempts to give a definition of absolute motion in terms of its causes and effects (i.e., forces and the inertial effects in rotational motion) are more successful, since they are measurable (p. 31). Specifically, the results of analysing how the concepts of motion are used in the mechanics of impelled and rotating bodies are that:

(a) 'Newton defines true motion as that which cannot change without the action of a force, and which must change when a force is applied' (p. 31).
(b) Similarly, 'true rotations are by definition those that give rise to centrifugal effects' (p. 35).

What is important to emphasize here is that the specific definitions identified by DiSalle are supposed to be 'empirical': absolute quantities are defined in terms of (more) directly observable quantities.

At stake are both Newton's metaphysical views regarding absolute space in the *Scholium* and the overall logic of the text. Canonical views hold that Newton postulates absolute space as a *subsistent* background frame relative to which 'true' motion should be understood. Interpretations differ on the role of the following discussion: does Newton fallaciously claim that the convex surface of the water in his bucket can only be correlated with motion relative to absolute space (as positivistic critics such as Mach 1893, pp. 279–296, allege)? Or is absolute space part of the best explanation of the surface (as substantivalist proponents' arguments urge: e.g. Earman 1989, p. 125)?

According to DiSalle, when Newton 'distinguishes' absolute motion from relative he is not defending some prior understanding of motion at all; certainly not that of motion with respect to a subsistent space. Instead he explicates the meaning of 'motion' in the laws of mechanics: in the 'true, philosophical and mathematical' (p. 408) sense. That is, *Understanding Space-Time* proposes that the idea of a 'dynamical' interpretation of mechanics goes right back to its origin. The proposed logic of the *Scholium* is completely different from the other accounts – and in particular Newton eschews metaphysical claims about space altogether.[5]

DiSalle claims (17) that his reading is in agreement with that of Stein's (1967) celebrated paper, 'Newtonian Space-Time'. I want to demonstrate that his interpretation, as just sketched, is both incompatible with Newton, and

5 In a review of DiSalle (Huggett 2009) I question whether such a dynamical interpretation of spacetime actually transcends substantival-relationist issues; for one can still ask what fundamental spatiotemporal properties the laws govern.

overstates Stein's in important ways. To that end, and to clarify Stein's views and comment on their adequacy as an interpretation of Newton, we will next turn to his paper; later we shall return to our discussion of DiSalle's Newton.

8.3 'Newtonian Space-Time'

Until 1967, the prevailing, positivist-influenced reading of Newton's *Scholium* on space, time and motion held it to be a gratuitous metaphysical – therefore unscientific – intrusion into a scientific treatise. A main goal of Stein's 'Newtonian Space-Time' (all references to Stein are from 1967, unless otherwise indicated) is to show that, far from being unscientific, the *Scholium* serves an important role in the overall argument of the *Principia*. First there is the very important point that Newton is responding to Descartes's two conceptions of motion (from *Book II* of *The Principles of Philosophy*); at least his view that motion is to be taken with respect to arbitrary bodies was a very serious scientific position at the time. Stein argues that Newton shows the Cartesian conceptions to be inadequate for the new physics, but that 'absolute motion' can do the work demanded of it. His analysis (in part *I* of the paper) of the essential role of absolute motion in the *Principia*, especially in the demonstration of the absolute motions of the solar system is, by any standard, a classic, masterful explication of a scientific argument – a revelation.

Here I want to concentrate on what 'Newtonian Space-Time' says specifically about the *Scholium*.[6] According to Stein, and in direct reply to the positivists, Newton shows that

> the principles of dynamics, already discovered by earlier investigators and applied successfully to many phenomena, distinctly require a view of motion and therefore place and space that cannot be explicated in terms simply of the geometrical relations among bodies. [Thus he] adopts that conception of space and motion on which alone dynamics can be based – which implies [that] considerations of *force*... must be brought to bear in order to determine the true state of motion or rest of bodies.
>
> Viewed so, Newton's analysis is, but for the one shortcoming [discussed just below], a classic case of the analysis of the empirical content of a set of theoretical notions.
>
> (p. 197)

In this section and the next my purpose is to evaluate this as a description of the *Scholium*. The two main tasks facing us are to elucidate the concepts that 'Newtonian Space-Time' attributes to Newton, and to determine, with reference to the text, whether those are his concepts and whether it is best read

6 I refer largely to Part *II* and especially to the remarks on pp. 190–198.

as an empirical analysis. I want to show that, as an account of the *Scholium*, this description leaves out some important points.

An important caveat about this approach: Stein's paper is as much about how Newton should be evaluated – not according to some narrow positivistic conception – as about giving an interpretation of Newton. So in the first place, it aims to lay out the *Scholium* in a way that reveals its positive contributions and, in contrast, its shortcomings; but Stein does not say explicitly that he intends the above as an exhaustive description of the *Scholium*.[7] Thus it might seem that I am setting up a straw man, but that would only be the case if the following were a criticism of 'Newtonian Space-Time' – and it is not (nor of Newton). Instead, my goal is to see how comprehensive this description is to see what *else* we can learn about Newton.

Now to elucidate Stein's description. First, Newton's conception of absolute space, and the 'shortcoming' identified by Stein: Newton's conception corresponds to that of full 'Newtonian spacetime'. Very briefly and informally, in Newtonian spacetime the relations of temporal and spatial distance are defined between any pair of points; thus it makes sense to ask how far apart in space and time two points along a worldline are – and so to ask the (average) 'absolute' speed with respect to space, the distance divided by the interval. The problem with this conception is that Galilean relativity means that absolute velocity cannot be measured, as Newton demonstrates in *Corollary V* of the laws: 'When bodies are enclosed in a given space, their motions in relation to one another are the same whether the space is at rest or whether it is moving uniformly straight forward without circular motion' (p. 423).

However, the mathematically weaker conception of motion in 'Galilean spacetime' is sufficient for Newtonian mechanics, and avoids this problem.[8] In Galilean spacetime, while any two points have a temporal separation, they only have a spatial separation if they are simultaneous; thus no absolute velocity can be assigned to worldlines. However, a relation of co-linearity is defined over all triples of points; worldlines composed of mutually co-linear points are those of bodies in uniform linear motion, those of *zero acceleration*. (In Newtonian spacetime, worldlines of arbitrary, *constant absolute velocity* are straight, so also composed of co-linear points.) More generally, the relations of Galilean spacetime suffice to make acceleration well-defined, as the laws clearly require, without making an unobservable absolute velocity well-defined.

7 Indeed, because it is not relevant to my discussion, I have put to one side Stein on the *Scholium*'s discussion of the meaning and methods of 'philosophy'.

8 Stein refers to this spacetime as 'Newtonian' (p. 175), but the usage I have followed has become fairly standard in philosophy of physics since. I apologize for reviewing material that will be familiar to many, but I was surprised to discover during an earlier presentation of this work that these ideas are not universally known to historians of early modern philosophy, despite their significance. See Stein for a more thorough and rigorous presentation.

So, Stein explains, by introducing absolute velocities, Newton presupposed more than his mechanics needed.[9] However, he is not inclined to fault Newton too harshly, because a proper understanding took another 300 years of mathematical development, and because Newton understood and acknowledged the problem. (On the other hand, Stein does take Newton to have failed here in comparison with his contemporary, Huygens, who, he claims, saw the possibility of well-defined rotation without a well-defined, unique velocity.)

To summarize, according to Stein, Newton's postulation of absolute space is the postulation of the structure of Newtonian spacetime for the 'spatio-temporal framework of events' (p. 182). (He also emphasizes that this structure is 'independent' in the sense that it cannot be defined in terms of the relative positions and motions of bodies.) Many commentators are tempted thus to understand Newton as a 'manifold substantivalist' (to use Earman's phrase, 1989, pp. 125–126) – to take the 'spatio-temporal framework' as literally a differentiable manifold, an object of primary predication. It's easy to see why, for the alternative seems to be that Newton thought that mechanical processes merely evolved *as if* there were a manifold with the appropriate geometry, while in fact there was none; but Newton does not seem to offer any such instrumentalist gloss. But 'Newtonian Space-Time' takes another view, denying that the question of whether or not there is 'really' a manifold is addressed in the *Scholium* at all. All that Newton wishes to establish is that the structure of Newtonian spacetime (or at least that part that it shares with Galilean spacetime) is 'in some sense really exhibited by the world of events' (p. 193).[10]

9 Note that motion in Galilean spacetime is, as Newton would have recognized, not fully observable according to his laws of motion. Just as absolute velocity is unobservable according to *Corollary V*, so a common acceleration is shown to be unobservable by *Corollary VI*: 'If bodies are moving in any way whatsoever with respect to one another and are urged with equal accelerative forces along parallel lines, they will all continue to move with respect to one another in the same way as they would if they were not acted on by those forces' (p. 423). (That is, if at each instant every body in a system accelerates in the same direction at the same rate, then the system will be indistinguishable from an identical one in which there is no common acceleration.) For instance, a system in free fall in a constant gravitational field behaves just like one not in a gravitational field – indeed what makes *Corollary VI* important is the existence of a force like gravity in which all matter has the same acceleration. However, such unobservable differences in motion correspond to distinct motions in Galilean spacetime, since they differ in the rate of acceleration.

If in addition to Newton's laws it is also postulated that there are no source-free forces then *Corollary VI* will only apply to subsystems, whose behaviour will be observably different with respect to some more encompassing system. The point is that motion in Galilean spacetime is not fully observable according to Newton's laws of motion, only according to some stronger set of laws, such as the laws of motion *plus* Newton's law of gravity.

10 In fact, it is not entirely clear that Stein intends to make exactly this claim – hence the earlier caveat about my approach. Perhaps what he means is that when one evaluates

The question of just *what* sense brings us to Newton's conception of absolute motion. One of the important themes of 'Newtonian Space-Time' is that 'absolute motion' is a theoretical term whose meaning is given in part by Newton in terms of the theory of mechanics (e.g., p. 190). That is, absolute motion just is, 'by definition', that sense of motion appearing in the laws, in their canonical formulation. If correct, this understanding sheds new light on Newton's bucket argument. Assuming suitable forces (i.e, gravity and inter-atomic forces), Newtonian mechanics deductively entails the rotation of the water, which, on the proposed understanding of what Newton means by 'absolute motion', *is absolute rotation*: absolute rotation is simply a consequence of the laws (given the physics of the system). Since absolute motion is 'by definition' motion in the sense of the laws, then this result has no immediate consequences at all for the question of manifold substantivalism – consistent with the view that the *Scholium* does not take a stand on the issue. Instead, such examples show that the structure of absolute space is 'exhibited by events': mechanical processes are governed by laws in which 'motion' cannot be taken relatively, but only with respect to the structure of Galilean spacetime. Put another way, the example shows that absolute space is exhibited by events in the sense that it shows that absolute motion has empirical content – the term appears in meaningful, true (supposing Newtonian mechanics) sentences concerning phenomena.

To make these ideas clearer for further discussion, I want to introduce a more precise understanding of how terms get meaning from theories in which they appear. A reasonable answer, which will be of great help to us, is provided by the Carnap–Lewis account of the definition of theoretical terms (e.g., Lewis 1970). There are a number of nice features of this account, not least that it explains how theories can be interpreted even if they contain terms with no antecedent meanings, and which have no plausible empirical significance. Very briefly, suppose the vocabulary of a new theory, $T[t; U]$ contains some antecedently understood terms U – for instance, because they have rather direct empirical meaning, or because they are understood theoretical terms of existing theories – plus a novel theoretical term t. Then t is implicitly defined to be *the* x such that $T[x; U]$ (the Ramsey sentence for T with all occurrences of t replaced by x). In the case of absolute motion, we proceed, rather schematically, as follows: let N[motion] be Newton's laws, so N[x] is the laws with every occurrence of the term 'motion' replaced by the free variable x; then 'absolute motion' is defined to be *the* x such that N[x].

Of course Newton does not put things this way, and neither does Stein. But 'Newtonian Space-Time' proposes that absolute motion is a theoretical term

Newton the further questions can be logically separated from the empirical analysis of absolute concepts and postponed. As I indicated earlier, it matters little here, for we seek to learn more about Newton by seeing what is left out of 'Newtonian Space-Time''s description of Newton's positive achievements.

of mechanics, and this formulation certainly gives a clear statement of just what content it obtains from the theory (though, as we shall discuss below, Newton means more than this). By comparing this more explicit formulation against the *Scholium* we can understand more precisely the sense, if any, in which Newton introduced absolute motion and space as theoretical terms in the *Scholium*. But before we pursue that issue, a remark is in order to avoid confusion.

It is no part of Lewis' account that such a definition is an 'empirical' one, in terms of some privileged vocabulary; rather the definitions are given in terms of whatever terms are already understood at the time that the new theory is formulated. Thus the formulation in no way amounts to a definition of absolute rotation in terms of what can be observed in, say, a bucket of water – in no way does it take Newton's empirical analysis of the concepts as itself a definition. One might still ask whether all definitions terminate in empirical ones: is there a chain of Carnap–Lewis definitions that eventually leaves only empirical terms undefined? But Lewis' account is simply not aimed at offering a general theory of meaning, only at showing how new terms can obtain meaning from old terms and new theory. All that is important in the present case, is that somewhere in the chain are terms with empirical meaning; for that makes the analysis of the empirical content of absolute quantities a non-trivial enterprise – i.e., they have some! So in particular, the definition is not itself intended as an empirical analysis, but an account of what the term means, a precondition for performing such an analysis. I stress this point because Stein explicitly rejects the analysis of empirical content by translation to an observation language (p. 190) – while, on the contrary, we saw earlier that DiSalle reads Newton's analyses as empirical definitions.[11]

The spacetime framework allows a precise statement of the formal properties of motion thus defined: it possesses all the properties of motion in Galilean spacetime (and no more, *modulo* footnote 9). And so we have the precise sense in which events exhibit the structure of absolute space: the laws governing mechanical processes implicitly define a conception of motion which has exactly the properties of motion in Galilean spacetime. Conversely Galilean spacetime has structures strong enough to make the motive quantities necessary for the laws well-defined.

Now, Newton of course intended absolute motion to be a richer conception of motion, one in which rest is well-defined: that is, motion in Newtonian

11 In addition, the proposed schema is called a 'definition', while Newton denied that he defined motion at all. But we saw that Newton meant something quite specific by 'definition', and did not mean to deny that he was saying what he took 'time', 'space', 'place' or 'motion' to mean. So the analysis is offered in that sense: if the *Scholium*'s absolute concepts are only as wide as mechanics requires, then the proposed schema captures them. As long as we bear in mind the distinct role of explicit definitions in the *Scholium*, it matters little whether we call the formulation a definition or not, though I will generally refrain.

spacetime. So the formulation proposed only captures part of his notion, that part which can properly be considered 'theoretical'. It is for the additional 'hypothetical' content that Stein faults him (pp. 182–183). Equivalently, only part of the structure of Newtonian spacetime is 'exhibited by events' – that part which it has in common with Galilean spacetime.

The theoretical definition also clarifies the empirical analysis of the concept. The clearest example of such an analysis in 'Newtonian Space-Time' is that centrifugal forces are 'associated universally with [absolute rotation]' (p. 195).[12] That fact of course does follow if absolute motion has the proposed meaning, for it is a consequence of the laws appearing in the definition. And so the formulation helps clarify how the term gets empirical content, and how the analysis works – how the structure of 'absolute space' is exhibited in particular phenomena, like the rotating bucket.

To summarize, 'Newtonian Space-Time' proposes that in the *Scholium* Newton argues *(i)* that existing, relative conceptions of motion are inadequate for mechanics, and *(ii)* that the laws implicitly (partially) define a new conception of motion, which he logically strengthens to arrive at the idea of 'absolute motion' with the properties of motion in a Newtonian spacetime; in addition, *(iii)* Newton avoids addressing questions of 'the ontology of spacetime' beyond introducing this geometrical structure – which he calls 'absolute space' – and *(iv)* beyond an explication of the empirical content of his concepts. Stein's paper certainly shows that this interpretation *of Newtonian mechanics* is adequate (substituting Galilean for Newtonian spacetime); the claim that concerns us though is whether it presents an accurate account of the *Scholium* – whether it is an adequate interpretation of *of Newton*. In the next section (with my earlier caveat in mind) we will discuss the parts of Stein's interpretation in turn.

8.4 The Scholium

(i) First, Newton certainly does give incontrovertible arguments showing the inadequacy of Descartes's conceptions, even with respect to his own mechanical pronouncements.[13] (However, the arguments of the *Scholium*, unlike those of *De Gravitatione* (2004, II), do not mention Descartes specifically; it's reasonable to think they were intended to have force against relational definitions of motion

12 In discussing this proposition Stein does not refer to an argument of Newton's at all, but rather analyses Foucault's pendulum experiment and compares it with Cavendish's experiment. This strategy demonstrates that absolute rotation is indeed a theoretical term with empirical content, but it offers little support for reading the *Scholium* as making a theoretical 'definition' and analysing its empirical content – after all, these experiments occurred 164 and 110 years, respectively, after its publication. Presumably the bucket and globes examples are supposed to do similar work for Newton.

13 The interpretation I defend here is derived from Rynasiewicz (1995a), and makes – specifically in relation to Stein – a number of similar points.

in general.) Consider how Newton distinguishes absolute and relative motions by their 'properties, causes and effects'.

One 'property' of motion is that 'parts which keep given position in relation to wholes participate in the motions of such wholes' (p. 411), for instance, because in mechanics all parts of a rotating body recede from the axis of rotation. Indeed, Descartes is in explicit agreement: for example in stating that the planets move with the ambient 'heavenly fluid', in which they are locally at rest (*Principles of Philosophy* III.26 and 140; all references to Descartes are to 1991). But an enclosed part is at rest in the sense of Descartes's motion 'properly understood': with respect to 'those bodies immediately contiguous to it' (II.25; see II.28–31 for the 'clarification' of the concept). Hence that definition of 'motion' is at odds with Descartes's use of the term in mechanics.[14]

Or again, the 'causes' of motion – forces, percussive or perhaps otherwise – can be applied to a body when there is no acceleration relative to a given reference body (if it also experiences forces), and vice versa (if forces are applied only to the reference body). So mechanical principles, including Descartes's Rules of Motion (II.45–52), are also in conflict with the Cartesian definition of motion 'ordinarily understood': with respect to arbitrary reference bodies (II.24).[15]

And most famously, there are the 'effects' of motion, namely the inertial forces associated with rotation, as illustrated by 'Newtons's bucket'. Describing a spinning bucket of water hanging from a cord, Newton writes:

> when the relative motion of the water [with respect to the bucket] decreased, its rise up the sides of the vessel revealed its endeavor to recede from the axis, and this endeavor showed the true circular motion of the water . . . becoming greatest when the water was relatively at rest in the vessel. [*(1)*] Therefore, that endeavor does not depend on the change of position of the water with respect to the surrounding bodies, and thus true circular motion cannot be determined by means of such changes of position. [*(2)*] The true circular motion of each revolving body is unique, corresponding to a unique endeavor as its proper and sufficient effect, while relative motions are innumerable in accordance with their varied relations to external bodies.[16]
>
> (pp. 412–413)

14 See Garber (1992, chapter 8) for an argument that 'proper' motion is the sense used in Descartes's mechanics.

15 Two cautions: according to *Rule 4* (II.49) the percussive force of a smaller body cannot put a resting body in motion. And 'ordinary' motion is not entirely arbitrary but depends on a pre-scientific attribution of 'action' to moving bodies.

16 The passage continues, 'and, like relations, are completely lacking in true effects except insofar as they participate in that true and unique motion'. It is unclear whether the claim that relative motion is causally inert is a metaphysical premise to a further argument, or a further conclusion from what has just been shown. I tend to the latter reading, so I will not discuss it further; but some may read it as another example of Newton's metaphysical assumptions.

I see two arguments (disagreeing slightly with Rynaseiwicz): *(1)*, that inertial effects are anticorrelated with the motion of the water with respect to the sides of the bucket – i.e., with Cartesian 'proper' motion. *(2)*, the surface of the water always rises to a unique height, corresponding to a unique rate of rotation, while its 'relative motions are innumerable in accordance with their varied relations to external bodies'. Newton means that while the water has many Cartesian 'ordinary motions', the rotation revealed by the tendency of the water to recede is unique. So, once again, Newton demonstrates how Descartes's conceptions fail to work in mechanics, even his own. (In Book III of the *Principles*, Descartes regularly appeals to the tendency of bodies in the heavenly vortex to recede from the center of rotation.) The error is as if Euclid's definition of 'circle' were incompatible with his use in the theorems.

Pointing out the character of Newton's arguments is one of the incredibly illuminating insights of 'Newtonian Space-Time', and really changed how we look at the *Scholium*. I want to flag two points to which we shall return. First, as Rynasiewicz emphasizes, the whole series of arguments from 'properties, causes and effects' are clearly intended to have the common end of establishing the inadequacy of the Cartesian conceptions (and implicitly the adequacy of the absolute conception in the situations discussed). So, their rhetorical purpose is the same. Second, DiSalle rejected the arguments from properties as unscientific, but we have just seen one at least is premised on the mechanical conception of motion, specifically on centrifugal effects. Later we will return to the others.

For now, consider point *(ii)*, and hence the question in the title of this paper. From other texts we know more about what Newton thought about space and motion than is revealed in the *Scholium*: but the question here is *what sense he intended to convey to the reader of this text*, not what he might have believed in general. As usual, Newton is trying to avoid superfluous 'hypotheses', in part to establish a foundation of agreement with his contemporary readers: views that can be reliably 'gathered from phenomena' (1999, p. 796). So when Newton talks about 'true and absolute' motion does he mean – did he intend to be taken as conveying – nothing but that sense of motion implicit in the laws of mechanics, strengthened to make rest uniquely defined? There is something right about this idea, but there is rather more to say.

Consider the conjunction: of course Newton took 'true' quantities to be 'absolute' (and 'mathematical'), but did he intend the terms to be synonymous? In the discussion of space and motion, the *Scholium* first contrasts absolute and relative space and motion (in the paragraphs numbered 2 and 4) without use of the modifier 'true': relative space is 'any movable measure' of absolute space, and absolute (relative) motion is with respect to absolute (relative) space. Only when he turns to illustrate the difference between absolute and relative motion with the example of a ship on the moving Earth does Newton speak of 'true rest' as rest with respect to 'unmoving space'. So read carefully, Newton introduces 'absolute space' as that unmoving space, and absolute motion as motion with respect to it.

Moreover, in the subsequent arguments distinguishing 'absolute and relative rest and motion', Newton is very consistent in attributing the 'properties, causes and effects' to 'true' *not* 'absolute' motion. In the first place I refer readers to the text (pp. 411–413) to satisfy themselves of this fact. To avoid a long and tedious discussion here, in support of the claim I will simply report the relevant statistics: if we look just at the arguments explicitly involving those 'properties, causes and effects' that follow from the laws of mechanics, then of 19 distinct references to true or absolute motion or rest, 18 are to 'true' motion and only 4 to 'absolute'. (If we look at the entire series of arguments, the numbers are 25, 20 and 8, respectively.) Newton's use of the modifiers in these passages is very natural if we take him as accepting a special, 'privileged' sense of motion, which he labels 'true motion', picked out by the 'properties, causes and effects' – the concept appearing in mechanics.

Thus, in the natural reading of the *Scholium*, 'absolute' and 'true' are not synonyms, rather 'true motion' is implicitly (partially) defined by the laws of mechanics but 'absolute motion' is not. This distinction makes a difference to the logic of Newton's arguments. For example, we saw above that if absolute rotation were, by definition, that rotation implicit in the laws, then it would follow, from the laws and the forces present, that by definition the water with a concave surface is rotating absolutely. But on the reading I have just given of the *Scholium* Newton argues that from what we know of true motion – especially from mechanics – it cannot be identified with either Cartesian conception, but can be understood as the *independently defined* notion of absolute motion. The implicit form of argument is that of a disjunctive syllogism: Descartes gives two possible accounts of motion – ordinary and proper – and Newton a third – absolute motion – the only one compatible with the properties, causes and effects of true motion. Logically, such an argument is open to the criticism (essentially made by Mach) that the disjunction is not exhaustive, but it is certainly valid.

Of course it is the identification of the two kinds of motion that leads Newton sometimes to speak of 'true and absolute motion': not because the terms are synonymous but because they refer to the same thing.

Now, even though this (conceptual) distinction changes the logic of Newton's argument, there is a way of rereading Stein's *(ii)* to accommodate it. That is, consider the possibility that while Newton defined absolute motion by reference to a fixed frame, he defined true motion theoretically; then he demonstrated that, of the possible conceptions available, only absolute motion is compatible with those of true motion. On this view Newton has a purely theoretical conception of motion – the concept of true motion is just the concept of the x such that $N[x]$. The additional hypothetical features – absolute rest – arise because he identifies motion thus conceived with the distinct, stronger conception of motion relative to absolute space. On a familiar view of identity, that absolute velocity is a feature of true motion is a synthetic but necessary fact. If this were how Newton thought, it would make sense of how in *(ii)* he opted for a stronger conception of motion than he needed.

However, things are not so straight forward, because the concept of true motion is itself logically stronger than the theoretical conception. To see this we shall return to an earlier point: that not all 'properties' of motion are those required by mechanics. The first argument that I postponed follows from the premise that 'bodies truly at rest are at rest in relation to one another' (p. 411). This is contrary to the Cartesian conception of proper motion, since bodies in relative motion may be at rest in their own immediate surroundings; also to ordinary motion if we take the motions of different bodies relative to different reference bodies. Thus Descartes's conception is inadequate given Newton's assumption. But the assumption does not follow from mechanics: since mechanics does not make a notion of rest well-defined at all, it cannot imply the premise. What can be said is that Newtonian mechanics is naturally formulated in 'rigid Euclidean' frames, in which points at fixed distances always have the same co-ordinate distances (i.e., $\sqrt{\Delta x^2 + \Delta y^2 + \Delta z^2}$ is fixed). In such frames, Newton's premise is true, so it's plausible that he thought here that one of the frames must be the one (up to Euclidean symmetries) relative to which true rest is defined. But that is to assume more of true motion than follows from mechanics; it is of course to assume one of the features of Newtonian spacetime.[17]

So, true motion is not just the theoretical concept for Newton, contrary to the suggestion I mooted above: although the concepts of true and absolute motion are distinct, they both allow the idea of rest. That is, the concept is only *partially* defined by the theory. Moreover, Newton's argument is that Descartes's conceptions do not match that of 'true motion', and here we have a case where it is non-mechanical properties of true motion that are relevant. Hence here Newton's argument is not just that Descartes's conception is inadequate for mechanics, and *(i)* is not the full story about the *Scholium.*

One might – like DiSalle – think that all of the arguments from properties also assume more than follows from mechanics. But that would be hasty: we saw above that one appeals to centrifugal forces, and Belkind (2007, §5) argues convincingly that premises concerning the motions of parts of moving wholes are connected by Newton to the defined theoretical concept of 'quantity of motion' (p. 404), or momentum. If so, they follow the pattern of the arguments from causes and effects more closely than previously recognized. So

17 That bodies with the same *motions* are at relative rest follows from the structure of Galilean spacetime; so it is the additional reference to rest that takes Newton beyond mechanics.

it is primarily in the first argument, just discussed, that Newton goes beyond mechanics and a theoretical definition.[18]

Now to *(iii)*. What is encompassed by Newton's concept of absolute space? Compared to other writings, the *Scholium* is much more cautious about any metaphysical speculations about space: for instance, claims that space is 'as it were' God's sensorium (*Optiks*, 2004, Query 28) or an 'emanative effect of God' that 'approaches more nearly to the nature of substance' than accident (*De Gravitatione*, 2004, pp. 21–22) are absent from the *Principia*. Instead, famously we have only that 'Absolute space, of its own nature without reference to anything external, always remains homogeneous and immovable' (p. 408). The phrase 'of its own nature' avoids a detailed account of just what that nature is: finite spirit? infinite spirit? matter? accident? substance? nothing? none of the above? Proposing such a theory – and we know from *De Gravitatione* that he held the latter view – would make no difference to the role of absolute space in mechanics, and moreover would likely cause distracting discord with his readers. Hence he left his detailed views unstated here; what is required is that space is not determined by any 'external' relations to other bodies, and that qualitatively indiscernible places persist in fixed relative positions, so space is 'homogeneous and immovable'.[19] With his implicit assumption of Euclidean geometry (of course made without comment in his mathematical proofs), absolute space grounds the structure of Newtonian spacetime.[20]

But that is not quite all that Newton says about absolute space; in arguing for its immobility he does say more about its ontology, something hypothetical or metaphysical. He says that 'It is of the essence of spaces to be places, and for primary places to move is absurd' (p. 410). The absurdity follows because (primary) places are ('as it were') the places of all things including themselves; so if they moved – i.e., changed places – they would move 'from themselves', which is nonsense. It is the idea that parts of space have essences which is relevant to us; for something cannot be said to have an essence unless it is in some sense a thing. So Newton is apparently saying more than that the world

18 The final argument (pp. 411–412) is curious for it does not refer to 'true motion' like the others, but consistently to 'absolute motion'. But it follows in virtue of the meaning of this term that such motions are referred to unmoving places, so it is not clear what the argument is supposed to achieve.

19 Of course the contrast here is with Descartes. First, there are the arguments we've been discussing against the Cartesian accounts of place as defined by external relations. Second there is Descartes's picture of a mobile, inhomogeneous Universe, with matter (i.e., space) of different grades of 'solidity' at different locations, in constant motion.

20 Note, however, that Newton describes space similarly as having 'its own manner of existing' in the more metaphysical *De Gravitatione* (p. 21); his point there is to say that space does not fall in traditional metaphysical categories, not that he won't discuss its nature.

in some way has the structure of Newtonian spacetime, he is saying that it has it as the structure *of some thing* – be it accident, substance, or something else. To make the point stronger, consider a way that the world might possess the structure other than as the structure of some thing, and how the argument for immobility then breaks down.

We saw that the structure of Newtonian spacetime is not completely exhibited by mechanical processes, for these do not depend on an invariant state of rest. But the world could have the structure in the following way: take the absolute motive quantities[21] to be metaphysically primitive, not possessed in virtue of motion with respect to anything. Then the world would have the structure of Newtonian spacetime in the sense that all the well-defined quantities of Newtonian spacetime are indeed well-defined. And one could introduce 'space' in a sense, as a rigid Euclidean frame such that the motion of each body in the frame agrees with its absolute quantities of motion: so that bodies at true rest are at rest in the frame, and so on. But this frame is just a logical construct from the absolute quantities, a way of expressing the correlations between them: e.g., that bodies truly at rest are at relative rest. So it is no 'thing', except in a loose sense.

For Newton, places are (occupied) volumes of space (p. 409), and indeed regions of the frames just constructed are at rest with respect to the frame. But that follows from the Newtonian structure of the primitive, absolute quantities, not from considerations to do with the 'natures' of primary places. It would twist Newton's words a good deal to see the argument for immobility as depending on the absolute quantities of bodies themselves in this way. Rather, as Nerlich (2006, p. 121) argues, Newton has in mind an ontologically robust entity (not a construct) with metaphysically essential features.[22]

Moreover, such a construction goes against the spirit of the claim that absolute space is 'without reference to anything external', for the frame would be with reference to bodies and their primitive absolute quantities. Now perhaps that statement is supposed to rule out the construction of absolute space from the *relations* of bodies, but if Newton had any intention of stretching the conception of space to a frame in this way then his qualification is remarkably

21 I.e., rest, velocity, acceleration, and rotation. Stein (1967, pp. 183–184) explains how these quantities are represented in a spacetime theory, and how they are expressed in particular frames.

22 Note further that *De Gravitatione* (2004, p. 25) offers the same argument for the immobility of absolute space, together with one based on identity criteria for points (discussed by Stein 1967, p. 194). As I show in Huggett (2008), that argument does not appear in the *Principia*, which is just as well since it entails that all motion is the relative motion of bodies. DiSalle (2002) uses the argument to try to show that Newton would deny the possibility of 'Leibniz shifting' all matter in space, but that can hardly be attributed to Newton in the *Scholium* since the argument is not used there – and in general, for the reason just given, it can hardly be one of Newton's core views.

careless. Additionally, a letter to Maizeaux, co-authored by Newton and Clarke, concerning the Leibniz–Clarke correspondence, corrects Clarke's claim that space is a property, but does not dispute Clarke's acceptance that the material Universe could have been 'shifted' elsewhere in space (see Koyré and Cohen, 1962, pp. 94–103). This silence is evidence that Newton also accepted such shifts; but they are only meaningful if the Universe has its place relative to space as an entity, and not according to the construction, which involves no absolute position. None of this is to say that if the idea of space as a construct had occurred to Newton then he would have rejected it in the *Scholium*; the point is that as a matter of fact what he does say is incompatible with that concept of 'absolute space'.

The bottom line is that the 'hypothetical' elements of the *Scholium* are not exhausted by the additional structure of Newtonian spacetime over Galilean. Especially, when he defines absolute motion it is relative to a robust thing, absolute space; he is not simply neutral on the nature of absolute motion, open to the possibility that it is primitive for instance. However, Newton makes no specific proposal concerning the nature of space, and so traditional, positivistic complaints that he detoured into metaphysics here are grossly over-stated.

It is unremarkable that Newton would think that true motion was change of position with respect to something since almost everyone who considered the issue, from Aristotle until the twentieth century, had that conception. In *Physics IV.4* Aristotle argues that motion is with respect to 'innermost surroundings', (or perhaps innermost *stationary* surroundings) – while the *Categories 7* contains a concept of arbitrary relative motion. These relational conceptions are of course echoed by Descartes's 'proper' and 'ordinary' conceptions.[23] This observation also makes sense of why Newton introduced a stronger conception of motion than that required by his mechanics; the question of whether a body is at rest has a well-defined answer if motion is change of position with respect to something. That is, the theoretical definition is better understood as saying that true motion is 'the *motion* x such that N[x]', where Newton, quite naturally in historical context, understood motion to be velocity with respect to something, in this case 'unmoving space'.

Finally, *(iv)*: Newton gives an analysis of the empirical content of 'absolute motion'. To address this idea we need to distinguish two parts of the *Scholium*: there are the arguments that we have discussed, from 'properties, causes and effects', which end after the bucket argument. But there is also the distinct

23 Exceptions to the view that motion is with respect to something, are Leibniz and (perhaps) Huygens. The former held that relative motion was ideal, and that true motion was really force. The latter, in his unpublished writings, arguably developed the idea of an inertial frame. (I am very grateful to Marius Stan for sharing and discussing his draft translation of Huygens's *Oeuvre* with me.)

subsequent discussion of the rotating spheres (p. 414, after Newton's comments on the meaning of 'motion' in scripture).

In the former, Newton is arguing that true motion is absolute motion, and in general not referring to directly observable states of affairs; so it is misleading to describe it as an analysis of empirical content. It is true that the bucket argument relies on the observable properties of rotation, but we should read it as respecting Newton's intention that the arguments all be taken as a piece – after all, that is how they are introduced. Thus the argument turns on the fact that *true* rotations are correlated with centrifugal effects, while motions in Descartes's sense are not. Again, it seeks to show that the conception of motion relative to absolute space is the only adequate notion on the table. Moreover, the idea that motion in the sense of mechanics – true motion – was correlated with centrifugal forces, was not something that required any demonstration: just what contemporary thinkers, for instance Descartes, Huygens, and Leibniz meant by the term. So it is somewhat misleading to describe even this argument as an empirical analysis of absolute motion.

On the other hand, the latter section clearly is concerned to explain how absolute motions can be determined empirically: Newton has shown them to be true motions, and hence how they can be inferred from mechanical processes. For instance, even without observing any relative motions, the rate and sense of rotation of the sphere system can be determined from the tension in the connecting rod, and the change in that tension when forces are applied to the sides of the spheres. So it is correct to say that here Newton explicates how the truth or falsity of statements about absolute motions can be determined experimentally – certainly an empirical analysis.

The passage concerning the rotating spheres is a key one for 'Newtonian Space-Time', for it expresses clearly the idea that absolute motions can be determined by phenomena: 'the purpose for which [Newton] composed the [*Principia*]' (p. 414). Just as the motion of the spheres can be determined from phenomena, so can that of the solar system (of course, things are more complicated in the latter case, for the forces themselves also have to be determined). So it is wrong to see the spheres as another argument for absolute motion; Newton has completed that part of the discussion, and is explaining how to find such motions. But to see the bucket argument as part of this discussion is to make the converse mistake; it is not intended to show how to determine absolute motions, but that the motion in question is absolute.

The upshot of the last few pages is that while the spacetime formalism and the idea of empirical analysis of concepts provide a framework that sheds considerable light on the logic of the *Scholium* as Stein demonstrated, there are a number of important aspects that this account leaves out. I have attempted to lay out various points made in the text that Newton would have expected his contemporaries to understand, which should be counted in his views, but which are not captured by 'Newtonian Space-Time'. I make clear once more that despite what I have shown, Newton indeed makes most of the positive contributions attributed to him by Stein – contributions that were almost entirely misunderstood by previous commentators.

8.5 DiSalle's Newton

Let's now turn to DiSalle's reading, and see why it does not do full justice to Newton's words, or the view developed by Stein (though clearly DiSalle is close in many ways). Recall, according to *Understanding Space-Time*, when Newton distinguishes absolute motion from relative (by their properties, causes and effects), in addition to demonstrating the failure of Descartes's conceptions, he is in fact offering a definition of the term; the specific definitions identified by DiSalle were labelled (a) and (b) in Section 8.2.

It should be clear from our earlier discussion how this description is at odds with Stein's.[24] In that account, in addition to demolishing the Cartesian conceptions, Newton offers an analysis of the empirical content of the concept of absolute motion. But such an analysis is not a *definition*. If it were, then Newton would not be giving a theoretical definition as I described above, but an empirical one: the meaning of absolute rotation would be nothing but the empirical criteria for its assertion. Stein explicitly denies that he attributes such a view to Newton (and indeed questions the general cogency of empirical definitions). What Stein does say is that Newton's conception of motion is (partially) defined to be that sense found in the laws of mechanics, and that via the laws of mechanics that theoretical quantity is correlated with empirical quantities – thus it can be meaningfully used in true sentences about the empirical world. But that is not the same as thinking those correlations are the meaning, for that ignores the theoretical aspect of the conception.

It should also be clear that I have considerable sympathies with the general idea that Newton's conception of motion – specifically *true* motion – is grounded in the laws. What I have argued is that his conception of absolute space cannot be understood to be nothing more than the corresponding spacetime structure, manifested in any way. Further, Newton introduces the conception of absolute motion differently from true motion, and not theoretically. Many of the comments made previously apply to *Understanding Space-Time*, but the idea that the *Scholium* significantly amounts to a definition faces additional problems.

First, there is the question of what the purported definitions (a)–(b) actually say and whether they are sensible ones for Newton to assert. Definition (a) might sound like a definition of (true) motion through the second law, but it

24 I want to acknowledge here that DiSalle's account of Stein added enormously to my understanding; the remarks I have made would not have been possible without DiSalle's roadmap.

is not, because Newton's discussion of the causes of motion does not involve *quantifying* it; there is no statement of proportionality to the forces. Only the discussion of effects describes quantifying motion, in terms of the magnitude of inertial effects, but in DiSalle's reading, (a) comes rather from the discussion of the causes of motion. Thus if Newton's discussion of causes was intended as a definition then it was remarkably sloppy, for absolute motion is quantitative.

But the discussion of effects, from which (b) comes, only quantifies rotational motions – by the tendency to recede from an axis of rotation – not linear accelerations. Moreover, that inertial effects do not necessarily quantify linear accelerations is part of the content of *Corollary VI* to the *Laws* (423, discussed in footnote 9), according to which bodies 'urged by equal accelerative forces along parallel lines' exhibit no inertial effects as a result. Thus (b) does not fully quantify true motions either. Thus overall, if (a)–(b) were intended to define true motion, insofar as it can be defined empirically, then they fail to do so, and in a way that Newton would have seen and could have fixed. It makes more sense to think that he is offering arguments not definitions.

Second, DiSalle reveals a problem with his own conception, when he claims that Newton can't be wrong that the water in the bucket is in absolute rotation because a tendency to recede is definitional of absolute rotation. But it is hard to believe that Newton thought that such things were true by definition: even given the laws, the concave surface of the water is correlated with absolute motion *conditional* on the presence of the expected (intermolecular and gravitational) forces. It is compatible with Newtonian mechanics that bucket and water not rotate, and yet the the surface be concave – if appropriate forces are present. Newton would surely accept such possibilities, and would not wish to define them away, so would not define (true) rotation directly in terms of a 'tendency to recede'. Instead it is defined as the concept implicit in the laws, which *under the appropriate circumstances* is correlated with centrifugal effects. So when Newton discusses causes, and indeed relies directly on the correlation, it makes best sense to think he is talking about the consequence of his theoretical conception, not defining the conception.

Next there is the fact that DiSalle (pp. 30–31) downplays Newton's discussion of the properties of absolute motion, whereas as we have already noted, there is nothing in the text to suggest that Newton viewed such arguments as playing any different or lesser role than the arguments from causes and effects. DiSalle sees these three arguments as unsuccessful attempts to define absolute motion, because they rely on metaphysical rather than mechanical features; thus they do not constitute an empirical definition. (Although we have also seen or mentioned ways in which some of them do appeal to mechanical features.) Thus this reading of the text artificially divides the arguments, while taking them as disjunctive syllogisms from premises shared with his interlocutors (as Rynosiewicz proposes) renders them a whole.

The final problem facing the view that absolute motion is defined by its properties, causes and effects, follows from the distinction between absolute and true motion, pointed out earlier. For (as DiSalle, p. 25, acknowledges) Newton defines absolute motion as change of absolute place *before* the purported empirical definition, so *Understanding Space-Time* has Newton implausibly defining absolute motion twice over. Perhaps DiSalle means that Newton claims that motion in a spacetime with the geometry of absolute space and time *represents* motion as defined by causes and effects. However, such a reading does not sit well with Newton's words: absolute motion '*is*' change of absolute place, not merely 'as if' it were motion in absolute space. Taken at face value, Newton is not making a claim merely about isomorphism, but being.

In addition to diverging from 'Newtonian Space-Time' in regards to the definition of absolute motion, *Understanding Space-Time* also goes further in attributing metaphysical views about space to Newton. Stein's position is that the *Scholium* takes no stand on the metaphysics of absolute space; it simply has 'its own nature'. Newton leaves open the options for his contemporary readers: matter, finite or infinite spirit, accident, and so on. However, according to DiSalle (pp. 37–38) Newton is not merely neutral, but is in fact a deflationist: he rejects all views that hold spacetime structure to inhere in any 'substantial' manifold, thereby rejecting these metaphysical options.

It's a small step from thinking that the *Scholium* proposes no view of the nature of spacetime, to thinking it proposes a 'no-view' view, but it is a step. DiSalle supports it by the idea that absolute motion is defined empirically; Newton just rejects any further non-empirical claims about absolute space (absolute rest aside). That position was criticized above. DiSalle further supports his claim by reference to *De Gravitatione*, and Newton's views on the identity criteria of the parts of space, invoked in an argument for their immobility (see footnote 22). There is a great deal to say about this argument, but ultimately I don't believe Newton's views there are relevant to the *Scholium*, because they entail that all motion is the relative motion of bodies, which he clearly denies (see Huggett 2008). (At any rate, claiming that Newton's metaphysical views in the *Scholium* are of a piece with those elsewhere is contrary to the ideas of 'Newtonian Space-Time'.) So I do not see any support for DiSalle's account of Newton's metaphysics of absolute space.

8.6 Conclusions

The argumentative parts of this essay have focused on revealing limitations in 'Newtonian Space-Time''s reading of the *Scholium* and criticizing *Understanding Space-Time*'s account of Newton's position. In conclusion I want to highlight briefly the central positive points of the discussion.

There is something quite accurate and very illuminating to Stein and DiSalle's view that Newton has a conception of motion which gets its meaning from the role of the concept in the contemporary science of mechanics (drawing not just on his insights, but on the contributions of Descartes, Huygens, Wren, and others). But the role the concept plays is subtle. First, 'absolute' and 'true' are not synonyms: especially, 'absolute motion' connotes change of absolute place, while 'true motion' connotes a special privileged sense of motion. More specifically, the latter concept gets meaning from the laws of mechanics, it is the concept of motion implicit in the laws. So while Stein's point is important, it does not do full justice to the *Scholium* because it does not illuminate the meaning of absolute motion, only true. And without this conceptual distinction it is impossible to make sense of Newton's arguments, since they are to the effect that the two concepts denote the same thing.

And even with respect to true motion the theoretical meaning is only partial. True motion's role in mechanics is central among its definitional characteristics, but Newton's arguments place non-mechanical features on a par. As Rynasiewicz emphasized, we should take the arguments from properties, causes and effects as a whole, working from similar premises; those 'properties' of true motion that do not follow from mechanics are just as constitutive as the mechanical ones.

In other words, in 'true motion' especially, Newton consciously held an extremely sophisticated conception of motion. The theoretical part of the concept is (absolute rest aside) indeed that of contemporary 'dynamical' interpretations, which also hold motion to be that which the laws refer to – motion in the frames in which the laws hold. But for the reasons just explained (and because of his conceptions of absolute space and motion) Newton cannot be said to have advocated a purely dynamical view in the *Scholium*, but rather the view that motion with respect to absolute space satisfied the dynamical concept.

9

From velocities to fluxions

MARCO PANZA

Newton reached the main results that would later constitute his theory of fluxions between the end of 1663 and the Fall of 1666.[1] Many notes dating back to this period have been conserved, and D. T. Whiteside has published them in the first volume of Newton's *Mathematical Papers* (Newton 1967–1981, I). They can be used to reconstruct the evolution of Newton's ideas at the very beginning of his mathematical researches and his progressive achievements.[2]

In none of these notes does the term 'fluxions' appear. Newton used it for the first time in the *De Methodis*, which he probably composed in the winter of 1670–1671 (Newton 1967–1981, III, pp. 3–372) but never published during his life.[3] The role that this term plays in this treatise and in the later presentations of Newton's theory is, *mutatis mutandis*, played in his first notes by several other terms like 'motion', 'determination of motion', and 'velocity'.

Though the *De Methodis* results, for its essential structure and content, from a re-elaboration of a previous unfinished treatise composed in the Fall of 1666 – now known, after Whiteside, as *The October 1666 tract on fluxions* (Newton 1967–1981, I, pp. 400–448) –, the introduction of the term 'fluxion' goes together with an important conceptual change concerned with Newton's understanding of his own achievements. I shall argue that this change marks a crucial step in the origins of analysis, conceived as an autonomous mathematical theory.

In Section 9.1, I shall distinguish three different senses in which the term 'analysis' can be used in historical contexts concerned with classic and early modern mathematics. This will allow me to clarify what I mean by speaking of the origins of analysis conceived as an autonomous mathematical theory. This

1 I thank Annalisa Coliva, Mary Domski, Massimo Galuzzi, Michael Friedman, Christian Houzel, Andrew Janiak, Vincent Jullien, Sébastien Maronne, Eric Schliesser, George Smith, André Warnstel, and Josu Zabaleta for valuable comments and/or suggestions.

2 This is what I have done in Panza (2005). The present paper develops some points I have made in that book.

3 The *De Methodis* first appeared, in an English translation by J. Colson, in 1736: cf. Newton (1736).

is what I suggest we call 'Eulerian analysis', a term I shall clarify by contrasting it with 'Aristotelian analysis' and 'Vietian analysis'.

In Section 9.2, I shall compare, in the light of the distinctions introduced in Section 9.1, the senses in which Newton speaks of analysis in the *De Analysis* (presumably written in 1669) and in the *De Methodis*, and argue that what he calls, in the latter, 'field of analysis' is much more extended than the domain of application of the analytical techniques described in the former.

In Section 9.5, I shall argue for the main thesis of the chapter, namely that Newton's field of analysis is, in fact, the original kernel of Eulerian analysis. My main point will be that fluxions were conceived by Newton as abstract quantities related to other abstract quantities called 'fluents', whereas that which he called 'motion', 'determination of motion' or 'velocity' in his previous notes were understood as (scalar components of) punctual speeds of motions generating particular geometric magnitudes, typically segments.

In order to clarify this point, I shall reconstruct, in Sections 9.3 and 9.4, some of Newton's arguments and achievements concerning motion dating back to the years 1664–1666. This will allow us to appreciate the evolution of his ideas on this matter up to the *October 1666 tract*, and also make a comparison with the new approach of the *De Methodis* possible.

Namely, in Section 9.3, I shall consider Newton's proof of a theorem showing an intrinsic link between the problems of tangents and normals and the problem of areas for curves referred to a system of Cartesian co-ordinates. This proof manifests a crucial idea that Newton will henceforth never abandon, that of considering related geometric magnitudes as generated by motions whose punctual speeds are mutually dependent on each other. But this theorem is also relevant in connection with a claim made in another note, according to which – when these motions are rectilinear and the generated segments are related by a polynomial equation and are taken as Cartesian co-ordinates of a curve – the problem of determining the ratio of (the scalar components of) their speeds is equivalent to the problems of tangents and normals for this curve. It follows that, for curves like these, these last problems and the problem of areas are connected with appropriate problems concerned with motion.

In Section 9.4, I shall show how Newton tackles and responds to the question of knowing whether this link holds also in general for any sort of curve. The (positive) response will come through his researches into Roberval's method of tangents. Newton succeeded in unifying this method in a unique, quite general proposition (proposition 6 of the *October 1666 tract*) concerned with the trajectory of the intersection point of two rigid curves that move separately from each other. This is a modality of composition of motion to which any other modality involved in Roberval's method can be reduced. Hence, Newton's theorem provides a recursive rule that can be applied to find tangents for any curve described by a composed motion. In the light of this proposition, the connection between the problems of tangents, normals and areas and appropriate problems concerned with motions – which Newton had shown to hold for curves expressed, with respect to a system of Cartesian co-ordinates, by a polynomial equation – appears to be a particular case of a more fundamental and general connection. This is the base of Newton's theory of fluxions. This theory appeared as such, when Newton, in the *De Methodis*, replaced the motions of lines with the variation of fluents, conceived, as noted, as abstract quantities.

Finally, in Section 9.6, I shall address some conclusions by discussing, in quite general terms, the links of this theory with Newton's natural philosophy.

9.1 Analysis

The term 'analysis' is highly polysemic. In order to understand the point I would like to make, it is necessary to distinguish three different senses in which it is habitually used by historians of mathematics. These senses reflect three different ways in which this term and its translated forms and cognates have been used by mathematicians up to the eighteenth century. They do not of course exhaust the spectrum of significations that it has taken and continues to take in mathematics and related fields.

In the first of these senses, 'analysis' refers to a pattern of argumentation largely used in Greek, Arabic and early modern mathematics – especially geometry –, often (but not always) in the context of the application of a twofold method, called 'the method of analysis and synthesis'. In order to avoid misunderstandings, call this pattern of argumentation 'Aristotelian analysis'. This appellation is justified, since Aristotle used 'ἀνάλυσις' and its cognates in this sense on different occasions.[4]

Aristotelian analysis is the common pattern of any argument which is based on the consideration of something that is not actually available as if it were available. Aristotle's clearest example (*Nicomachean Ethics*, III, 5) is deliberation: this is an argument through which one comes back from an imaginary situation that one aims to obtain to the actual one, so as to suggest a way for obtaining the former by operating on the latter.

Pappus' classical description of the method of analysis and synthesis and the corresponding distinction between theorematic and problematic analysis (*Mathematical Collection*, VII, 1–2) clearly refer to Aristotelian analysis.

According to Pappus, a theorematic analysis applies when a certain proposition has to be proved. It consists in deducing from it an accepted principle, a proved theorem, or their negations.

4 For example in: *Posterior Analytics*, 78*a* 6–8, 84*a* 8, 88*b* 15–20; *Sophistical Refutations*, 175*a* 26–28; *Metaphysics*, 1063*b* 15–19; *Nicomachean Ethics*, pp. 1112*b* 20–24. For a discussion of these passages and a reconstruction of Aristotle's views on analysis, see Panza (1997, pp. 370–383 and 395).

A problematic analysis applies, instead, when a geometrical problem asking for the construction of a geometric object satisfying certain spatial conditions relative to other given objects, is advanced.[5] One begins by supposing that this problem is solved and representing its solution through a diagram involving both the given and the sought after objects. Then, by reasoning about this diagram, and possibly by extending it through licensed constructions, one isolates a configuration of given objects and known data concerned with them, based on which the sought after objects can be constructed and thus the problem solved.[6]

Both Pappus' theorematic and problematic analyses are reductions. A theorematic analysis in Pappus' sense can provide *ipso facto* a proof (by *reductio ad absurdum*) if that which is deduced is the negation of an accepted principle or a proved theorem. This case apart, both a theorematic and a problematic analysis, as described by Pappus, are preliminary arguments suggesting another and conclusive argument, generally called 'synthesis': a theorematic analysis suggests a valid proof; a problematic one suggests an admissible construction.

There is no doubt that Pappus' theorematic and problematic analyses are both forms of Aristotelian analysis. Still, they are not the only possible forms that Aristotelian analysis has actually taken in classical, medieval, and early modern mathematics. Another relevant form of Aristotelian analysis occurring in classical, medieval, and early modern mathematics applies when a certain geometrical problem, asking for the construction of a geometric object satisfying certain purely quantitative conditions, is advanced.[7] In this case, the

5 An example is the following: suppose that two straight lines, two points on them and a third point outside them are given (in position); find a straight line from this last point that intersects the given lines so as to cut on them – together with the given points on them – two segments that stand to each other in a given ratio. This is the problem considered in Apollonius' *Cutting-off of a Ratio.*

6 To be a little bit more precise, consider the relevant problem as a configuration $C_{g,x}$ constituted by a system O_g of geometric objects which are taken as given (in the example mentioned in footnote 5, the two given straight lines and the three given points), an amount D of data (in the example, the given ratio), and a characterization O_x of some objects to be constructed based on O_g and D (in the example, the sought after straight line, or better, the points at which it has to intersect the given ones). The analysis begins by supposing that the problem is solved. This is the same as supposing that some objects satisfying O_x are given. The configuration $C_{g,x}$ can thus be represented by a diagram representing both the objects included in O_g and the objects satisfying O_x. Insofar as these last objects are not taken as given and the solution is only supposed, such a diagram cannot be completely obtained by applying the licensed constructive clauses, but is partially freely traced, so as to represent the relevant spatial relations between the relevant objects and to reflect the data. By reasoning about it and possibly by extending it through licensed constructions, one isolates a sub-configuration C_g on the basis of which the objects satisfying O_x can be constructed.

7 An example is the classical problem of finding two mean-proportional segments between two other given ones.

analysis aims to transform this condition into another equivalent but different one capable of suggesting a way for constructing the sought after objects.[8] Also in this case, the analysis is a reduction. But it is now the reduction of a given problem to a new and equivalent, yet still distinct, one.[9]

This last form of Aristotelian analysis may also apply if the relevant problems are not stated using the symbolic language introduced by Viète and Descartes and the related formalism. The possibility of appealing to some crucial theorems included in the *Elements* (especially in books II, V and VI) is enough for allowing the required transformations.[10] Still, this form of Aristotelian analysis naturally applies to the solution of problems stated by means of equations using this formalism. In this case, it consists in appropriate transformations of these equations according to the rules of such a formalism. Viète's *Zeteticorum Libri* (1591)[11] contains many examples of this form of Aristotelian analysis. This is because, after Viète, it became quite usual to employ the term 'analysis' to refer – not to a pattern of argumentation – but to the formalism or family of techniques that these transformations depend on. In order to avoid misunderstandings, call this formalism 'Vietian analysis'.

Under this meaning, the term 'analysis' is often used in early modern mathematics as a synonym for 'algebra', another highly polysemic term. For the sake of simplicity, I shall not use this last term in the present paper, and I shall use the adjective 'algebraic' in a modern sense, as opposed to 'transcendent'.

8 To be a little bit more precise, consider the relevant problem as a configuration $C_{g,x}$ constituted by a system Q_g of given quantities (in the example mentioned in footnote 7, the two given segments), and a characterization Q_x of some other quantities to be determined (that is, calculated or constructed) based on Q_g (in the example, the sought after mean-proportional segments). In this case, the analysis needs no diagram and, rather than isolating a sub-configuration C_g of $C_{g,x}$, transforms the latter into a new configuration $C'_{g,x}$ constituted by a system Q'_g of given quantities that can be determined based on the quantities included in Q_g, and a new characterization Q'_x of the same quantity characterized by Q_x (in the example, the condition $a{:}x = x{:}y = y = b$ is possibly transformed into the system of proportions $a{:}x = x{:}y$ and $x{:}y = y{:}b$ providing the symptomata of two parabolas an intersection of which determines the sought after segments).

9 For a more comprehensive description of these two forms of Aristotelian analysis applied to mathematical problems, cf. Panza (2008).

10 A nice example of this possibility is found in Thābit ibn Qurra's treatise on the 'restoring of the problems of algebra through geometrical demonstrations' (cf. Luckey (1941)); a French translation of Thābit's treatise is provided by the conjunction of the three quotations inserted in al Khwārizmī (2007, pp. 33–34, 37–38 and 41). The first of the three second-order equations of al Khwārizmī is here understood as the problem of looking for a segment x such that $S(x) + R(a, x) = S(b)$, where a and b are two given segments, $S(x)$ and $S(b)$ are the squares constructed on them, and $R(a, x)$ is the rectangle constructed on a and x. The appeal to proposition II.6 of the *Elements* is enough to allow Thābit to transform this problem into that of looking for the segment x such that $S(b) + S(\frac{a}{2}) = S(x + \frac{a}{2})$, which can be easily solved using the Pythagorean theorem.

11 A recent very comprehensive study of Viète's treatise is Freguglia (2008).

Newton's theory of functions and Leibniz's differential calculus are largely dependent on Vietian analysis, which occurs in them under the form that it takes in Descartes's *Geometry* (1637). They can even be viewed as appropriate extensions of it. The development of these theories went together with other, and partially independent, extensions of Vietian analysis, for example those connected with power series expansions. From this process, the crucial notion of function emerged and acquired a quite central role in mathematics. In his *Introductio in analysin infinitorum* (1748a), Euler launched a foundational programme aimed at a reformulation of any mathematical theory within the general frame of a theory of functions, defined as appropriate expressions expressing abstract quantities.[12] The following part of the chapter will be devoted to a partial clarification of this notion of abstract quantity through the reconstruction of the intellectual path that led Newton to the connected notion of fluxion. For the time being, it is enough to say that in the first half of the eighteenth century, the term 'analysis' and its cognates began to be used to refer to a general theory of functions conceived as abstract quantities, and to some of its features and connected developments. In order to avoid misunderstandings, call this theory 'Eulerian analysis'. It is to this form of analysis that I refer when I claim that the conceptual change that goes together with Newton's introduction of the term 'fluxion' in the *De Methodis* is a crucial step in the origins of analysis, conceived as an autonomous mathematical theory.

9.2 From the *De Analysis* to the *De Methodis*

On 20 June 1669, Isaac Barrow, at that time Lucasian Professor of Mathematics at Cambridge, replied to Collins, who had sent him a copy of Mercator's *Logarithmotechnia* (1668), with these words (Newton 1959–1977, I, p. 13; cf. also Newton 1967–1981, II, p. 166 (footnote 11), and Westfall 1980, p. 243):[13]

> A friend of mine here, that hath a very excellent genius to those things, brought me the other day some papers wherein he hath sett downe methods of calculating the dimension of magnitudes like that of Mr. Mercator concerning the hyperbola, but very generall; as also of resolving æquations; which I suppose will please you.

Ten days later, Barrow sent to Collins an example of this genius: a short treatise that is today known as the *De Analysis per Æquationes Numero Terminorum Infinitas* (Newton 1967–1981, II, pp. 206–247). Collins made a copy of it and circulated it. As a result, the young Newton and some of his early results became known in the English scientific community, though he did not allow

12 For a clarification of Euler's notion of function as I understand it, and some related bibliographical references, cf. Panza (2007).

13 For the factual pieces of information contained in this section, cf. Westfall (1980) and the critical apparatus of Newton (1967–1981, vols. II and III).

the publication of his treatise before 1711, when it appeared, in fact, as a piece of history (Newton 1711), in order to support the thesis of Newton's priority in the famous *querelle* with Leibniz.

Because of its circulation among the members of Collins's circle, the *De Analysis* is often considered as the first public presentation of Newton's theory of fluxions. This is not properly correct, however. It is rather a sort of instant book, which Newton wrote to expound only some of his results: those equivalent or similar to Mercator's.

After presenting two rules (Newton 1967–1981, II, pp. 206–210) for squaring curves expressed by equations of the form

$$y = ax^{\lambda} + bx^{\mu} + cx^{\nu} + \&c. \tag{9.1}$$

where $\lambda, \mu, \nu, \ldots$ are rational exponents and $\&c.$ means that the right-side member is either a finite or an infinite sum, he devotes the main part of his treatise to the detailed exemplification of a third rule (1967–1981, II, pp. 211–213; for the examples, cf. 1967–1981, pp. 212–242):[14]

> If the value of y or of some of its terms is more composed than the previous ones, it should be reduced to simpler terms by operating on letters in the same way as the arithmeticians get decimal numbers by division, extract roots and solve equations.

To say it more explicitly, Newton supposes that curves be expressed by algebraic equations $F(x, y) = 0$ of different forms, and shows how to operate on these equations so as to transform all of them into equations of the form (9.1), by applying to literal expressions procedures derived, by generalization or infinitary extension, from the arithmetic rules used for calculating with numbers.

Finally, he considers some mechanical curves, like the cycloid, and shows how to express also these curves by means of (infinitary) equations of the form (9.1), through the application of some appropriate yet peculiar tricks.

The term 'analysis' and its cognates occur quite seldom in Newton's treatise (1967–1981, II, pp. 206, 222, 240, 242), and always to refer to, or to speak of, Vietian analysis.[15] One could say, however, that the *De Analysis* includes several examples of Aristotelian analysis performed through Vietian analysis.

14 I quote Whiteside's translation. Here is Newton's original (1967–1981, II, pp. 210–212): 'Sin valor ipsius *y* vel aliquis ejus terminus sit præcedentibus magis compositus, in terminos sempliciores reducendus est, operando in literis ad eundem modum quo Arithmetici in numeris decimalibus dividunt, radices extrahunt, vel Æquationes solvunt'.

15 Cf., for example, the following quotation (1967–1981, II, pp. 241 and 240): 'And whatever common analysis performs by equations made up of a finite number of terms (whenever it may be possible), this method [the method of quadrature previously expounded] may always perform by infinite equations: in consequence, I have never hesitated to bestow on it also the name of analysis.' ['Et quicquid Vulgaris Analysis per æquationes ex finito terminorum numero constantes (quando id sit possibile) perficit, hæc per æquationes infinitas semper perficiat: Ut nil dubitaverim nomen Analysis etiam huic tribuere.']

They allow for the expression of different families of algebraic curves and some transcendent curves by equations of the form (9.1). This makes it possible to apply to these curves the following rules of quadrature:

$$y = ax^{\lambda} \Rightarrow \mathcal{A}(y) = \frac{a}{\lambda + 1} x^{\lambda+1}; \mathcal{A}(y+z) = \mathcal{A}(y) + \mathcal{A}(z)$$

(where '$\mathcal{A}(w)$' denotes the area of the trapezoid delimited by the curve of Cartesian orthogonal ordinate w, which, in modern terms, corresponds to $\int_0^x w(t)dt$, supposing that $w(0) = 0$).

This is merely a small fragment of the huge amount of mathematical results that Newton had obtained between 1663 and 1666. Despite that, on 29 October 1669, based on the samples of his competence offered in the *De Analysis* and in some other short notes that he had probably showed to Barrow, Newton was appointed as Lucasian Professor of Mathematics at the University of Cambridge, to replace Barrow himself.

Hence, though attracted by other topics, like natural philosophy, spectral colours and alchemy, he could not refuse Barrow and Collins' invitation to prepare some additions to be annexed to the Latin edition of Kinckhuysen's *Algebra* (1661), which Mercator had just translated from Dutch. On 11 July 1670 Newton was convinced that he had finished his work and sent it to Collins. But Collins had the bad idea of sending it back to Newton with the request of some further clarifications about the roots of binomials. He never received back either of these clarifications or the old version of Newton's additions.

On 27 September 1669, Newton had informed Collins that he had decided to replace his additions with a new treatise, which he wrote, in fact, but did not finish before 1683 (1967–1981, V, 54–532). Then he kept silent until 20 July 1670 when he sent a letter to Collins including the following passage (1959–1977, I, p. 66; cf. also: 1967–1981, II, pp. 288, and III, 5 pp. and 32 (footnote 1), and Westfall 1980, p. 268):

> The last winter... partly upon Dr Barrow's instigation I began to new methodiz the discourse of infinite series, designing to illustrate it with such problems as may (some of them perhaps) be more acceptable then the invention it selfe of working by such series. But... I have not yet had leisure to returne to those thoughts, & I feare I shall not before winter. But since you informe me there needs no hast, I hope I may get into the hummour of completing them before the impression of the introduction, because if I must helpe to fill up its title page, I had rather annex something which I may call my owne & which may bee acceptable to Artist as well as the other to Tyros.

This 'something acceptable to Artist' that Newton was planning to annex to Kinckhuysen's *Algebra* was just the *De Methodis*: a treatise quite different from the *De analyisis* in which he aimed to expound his own new theory and all of its extensions.

Here is the how the treatise begins (1967–1981, III, p. 33):[16]

> Observing that the majority of geometers, with an almost neglect of the ancient's synthetic methods, now for the most part apply themselves to the cultivation of analysis and with its aid have overcome so many formidable difficulties that they seem to have exhausted virtually everything apart from the squaring of curves and certain topics of like nature not yet elucidated: I found it not amiss, for the satisfaction of learners, to draw up the following short tract in which I might at once widen the boundaries of the field of analysis and advance the doctrine of curves.

Though he seems reluctant to admit that the 'neglect of the ancients' synthetic methods' is a symptom of progress, Newton clearly inscribes his own results into the 'field of analysis'. But it seems to me that he is no longer speaking of Vietian analysis, as in the *De Analysis*: his aim is no longer to show how the problem of quadratures can be solved by series for any algebraic curve and for some mechanical ones, based on a preliminary transformation. As a matter of fact, the methods expounded in the *De Analysis* are also expounded in the *De Methodis*, and a new quite powerful one of the same kind – the so-called method of Newton's parallelogram – is added to them. But these methods are now conceived as nothing but preliminary material only concerned with some 'modis computandi' (1967–1981, III, p. 70).

After having expounded them, Newton writes (1967–1981, III, p. 71):[17]

> It now remains, in illustration of this analytical art, to deliver some typical problems and such especially as the nature of curves will present.

It seems thus that, for him, the 'analytic art' does not merely consist in some appropriate techniques to be used in preparing the solution of some problems, but is rather concerned with these problems as such, and thus also with their solutions. It has taken on a peculiar form, and at the least is no longer Aristotelian or Vietian analysis.

This extension of the 'field of analysis' is not independent of an appropriate reduction of these problems to other ones. But this reduction is no longer the mere reduction of a certain configuration of given and ungiven quantities to a

16 I quote Whiteside's translation. Here is Newton's original (1967–1981, III, p. 32): 'Animadvertenti plerosque Geometras, posthabitâ fere Veterum syntheticâ methodo, Analyticæ excolendæ plurimum incumbere, et ejus ope tot tantasque difficultates superasse ut pene omnia extra curvarum quadraturas et similia quædam nondum penitùs enodata videantur exhausisse: placuit sequentia quibus campi analytici terminos expandere juxta ac curvarum doctrinam promovere possem in gratiam discentium breviter compingere.'

17 I quote Whiteside's translation. Here is Newton's original (1967–1981, III, p. 70): 'Jam restat ut in illustrationuis hujus Artis Analyticæ tradam aliquot Problematum specimina qualia præsertim natura curvarum ministrabit.'

new and more suitable one. It is rather a transformation of the very nature of these problems.

This is in fact a double reduction. Firstly, problems concerned with curves are reduced to problems concerned with motion. Secondly, problems concerned with motions are reduced to problems concerned with fluxions. The former reduction was already at work in the *October 1666 tract*. The theory expounded in that treatise is, indeed, a theory of motions and speeds to be used to solve geometrical problems: the aim of this treatise is of showing how to solve geometrical problems by motion. The latter reduction, however, is new and constitutes the essential novelty of the *De Methodis*, which I would like to emphasize.

Let us clarify this matter.

Here is how Newton describes the former reduction (1967–1981, III, p. 71):[18]

> But first of all I would observe that difficulties of this sort may all be reduced to these two problems alone, which I may be permitted to propose with regard to the space traversed by any local motion however accelerated or retarded:
>
> 1. Given the length of the space continuously (that is, at every [instant of] time), to find the speed of motion at any time proposed.
>
> 2. Given the speed of motion continuously, to find the length of the space described at any time proposed.

As a matter of fact, the language used by Newton to state these problems is more general than that used in the *October 1666 tract*. Here is, indeed, how these problems are stated in propositions 7 and 8 of this treatise (1967–1981, I, pp. 402–403):

> 7. Haveing an Equation expressing the relation twixt two or more lines x, y, z &c: described in the same time by two or more moveing bodys A, B, C, &c: the relation of their velocities p, q, r, &c may bee thus found, viz: . . . [19]

18 I quote Whiteside's translation. Here is Newton's original (1967–1981, III, p. 70): 'Sed imprimis observandum venit quod hujusmodi difficultates possunt omnes ad hæc duo tantùm problemata reduci quæ circa spatium motu locali utcunque accelerato vel retardato descriptum proponere licebit. 1. Spatij longitudine continuò (sive ad omne tempus) data, celeritatem motûs ad tempus propositum invenire. 2. Celeritate motûs continuò datâ longitudinem descripti spatij ad tempus propositum invenire.'

19 Newton supposes that the equation expressing the relation between x, y, z, *&c.* is polynomial; the suspension points stand, thus, for the description (in fact for three equivalent but different descriptions) of the well-known algorithm that, in the simplest case of two variables, leads from

$$\sum_{i=0}^{n}\sum_{j=0}^{i} A_{i-j,j}x^{i-j}j^{j} = 0$$

> 8. If two Bodys *A* & *B*, by their velocitys *p* & *q* describe the line *x* & *y*. & an Equation bee given expressing the relation twixt one of the lines *x*, & the ratio $\frac{p}{q}$ of their motion *p* & *q*; To find the other line *y*.

The difference between these statements and those of the *De Methodis* seems to be quite relevant: by avoiding the supposition that the spaces described are to each other in a relation expressed by a polynomial equation,[20] Newton seems to transform two problems concerned with the transformation of polynomial equations – that is, two algorithmic problems belonging to Vietian analysis – into two genuinely geometrico-mechanical problems. The comparative consideration of propositions 1–6 of the *October 1666 tract* (1967–1981, I, pp. 400–402) and the second reduction that Newton performs in the *De Methodis* suggests, however, a quite different picture. Before considering this second reduction, and in order to understand its real meaning, it is thus necessary to consider these propositions more carefully.

They aim to provide a quite general theory of composition of motions that is completely independent of the possibility of expressing the relation of the spaces described by means of algebraic equations. When looked at in light of this theory, the algorithms involved in the subsequent propositions 7 and 8 thus appear as local tools to be used in this theory in some particular situations for determining appropriate ratios or relations. The purpose of the next two sections is to reconstruct the essential aspects of this theory and the evolution in thought that led Newton to it.

9.3 Motions and geometry

Newton's first appeal to motions and their properties for proving geometrical theorems and solving geometrical problems occurs in a note composed in the Summer of 1664 (1967–1981, I, pp. 219–233; for the dating of this note, cf. Panza 2005, pp. 183–184), after his reading of the second Latin edition of Descartes's *Géométrie* (1659–1661).

In this edition, Descartes's treatise is supplemented by a large number of commentaries, other treatises on connected topics, and notes. Among this material, there is a letter of H. van Heuraet (Descartes 1659–1661, I, pp. 517–520), containing an important theorem about quadratures and rectifications: If curves AML (Figure 9.1) and END are such that, for every point P taken on

to

$$\sum_{i=0}^{n}\sum_{j=0}^{i}(i-j)A_{i-j,j}x^{i-j-1}j^{j}p+\sum_{i=0}^{n}\sum_{j=0}^{i}jA_{i-j,j}x^{i-j}j^{j-1}q=0.$$

20 Cf. footnote 19, above.

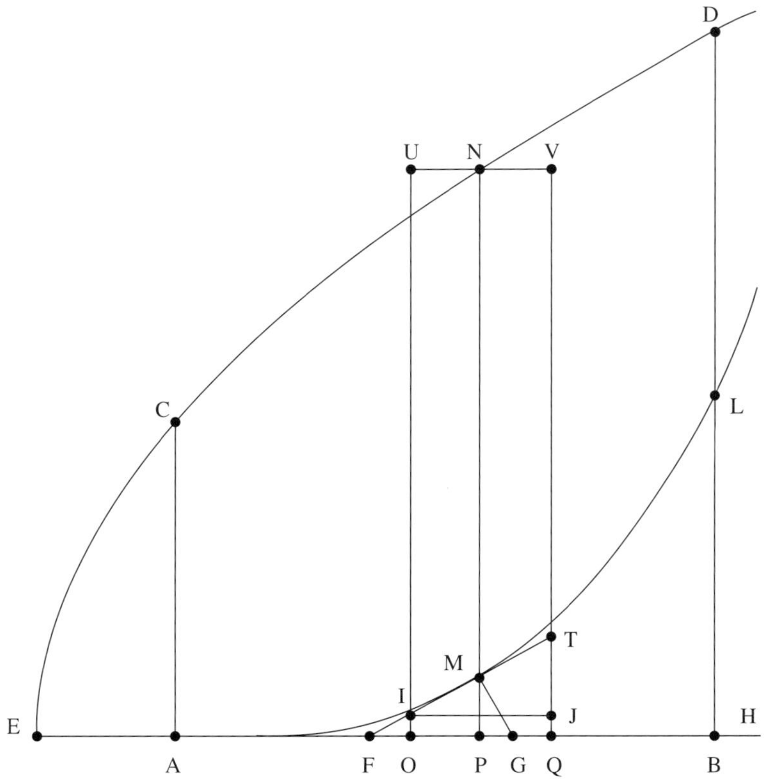

Figure 9.1 Diagram relating to van Heuraet's theorem about quadratures and rectifications

their common axis EH, their ordinates PM and PN comply with the following proportion

$$\text{PM} : \text{MG} = K : \text{PN}, \tag{9.2}$$

where MG is the normal to AML in M, and K any constant segment, then the trapezoid ABDNC is equal to the rectangle constructed on K and another segment equal to the arc AML.

In his note, Newton applies a slightly modified version of this theorem: if PM and PN are such that

$$\text{PM} : \text{PG} = K : \text{PN}, \tag{9.3}$$

where PG is the sub-normal to AML in M, then the trapezoid ABDNC is equal to a rectangle constructed on K and BL.

Van Heuraet's theorem allows one to rectify some curves provided that the areas of some other appropriate curve is known. Newton's modified version allows one to square some curves, provided that the normal or tangent of some other appropriate curve is known. More generally, it provides an intrinsic link between the problem of tangents and the problem of quadratures.

The two theorems can be proved in the same way, by a simple application of the method of indivisibles. Suppose that OQ = IJ is an indivisible portion of the base AB and remark that

$$\text{PM} : \text{MG} = \text{IJ} : \text{IT} \text{ and } \text{PM} : \text{PG} = \text{IJ} : \text{JT}.$$

Then, compare these proportions with the proportions (9.2) and (9.3) respectively, so as to derive that

$$R(\text{IJ}, \text{PN}) = \text{QVUO} = R(\text{IT}, K) \text{ and } R(\text{IJ}, \text{PN}) = \text{QVUO} = R(\text{JT}, K),$$

where, for any pair of segments α and β, $R(\alpha, \beta)$ is the rectangle constructed on these same segments. Finally, sum up all the rectangles such as QVUO, $R(\text{IT}, K)$, and $R(\text{JT}, K)$, and get the theorems. It is essentially in this way that van Heuraet proves his theorem.

Newton's argument for proving the second theorem (1967–1981, I, pp. 222–229) is quite different. He refers to another figure (Figure 9.2), where the segment K is identified with the constant base DB = QA of the rectangle DBCE and the ordinates PM and PN of the curves YV and ZW are, as before, such that the proportion (9.3) holds, supposing that PG is the sub-normal to YV relative to M. Then he remarks (1967–1981, I, pp. 228–229):[21]

> . . . supposeing the line PN always moves over the same superficies in the same time, it will increase in motion from QL in the same proportion that it decreaseth in length and the line DB will move uniformly from EC, soe that the space ECBD = NPQL.[22]

The statement 'it will increase in motion from QL in the same proportion that it decreaseth in length' makes manifest that Newton is here understanding motions as scalar quantities, that is, as (scalar components of) punctual speeds. He seems to take for granted that which is the main object of the previous argument through indivisibles, namely, the equality of the elements of the rectangle ECBD and the trapezoid NPQL. Then, he appeals to motions for proving that which is taken for granted in this argument, namely, that the equalities of elements entails the equality of the whole figures. Instead of appealing to an infinite sum of indivisibles or infinitely small elements, he considers figures

21 For reasons of uniformity, I change the letters used by Newton to refer to the points in the diagram.

22 Note that ECBD is the rectangle constructed on QA = K and the difference of the ordinates QE and PM depends on the choice of point B on the axis HA. It follows that the equality ECBD = NPQL expresses, with respect to the curves represented in Figure 9.2, the same result that, with respect to the curves represented in Figure 9.1, is expressed by the claim that the trapezoid ABDNC is equal to a rectangle constructed on K and BL.

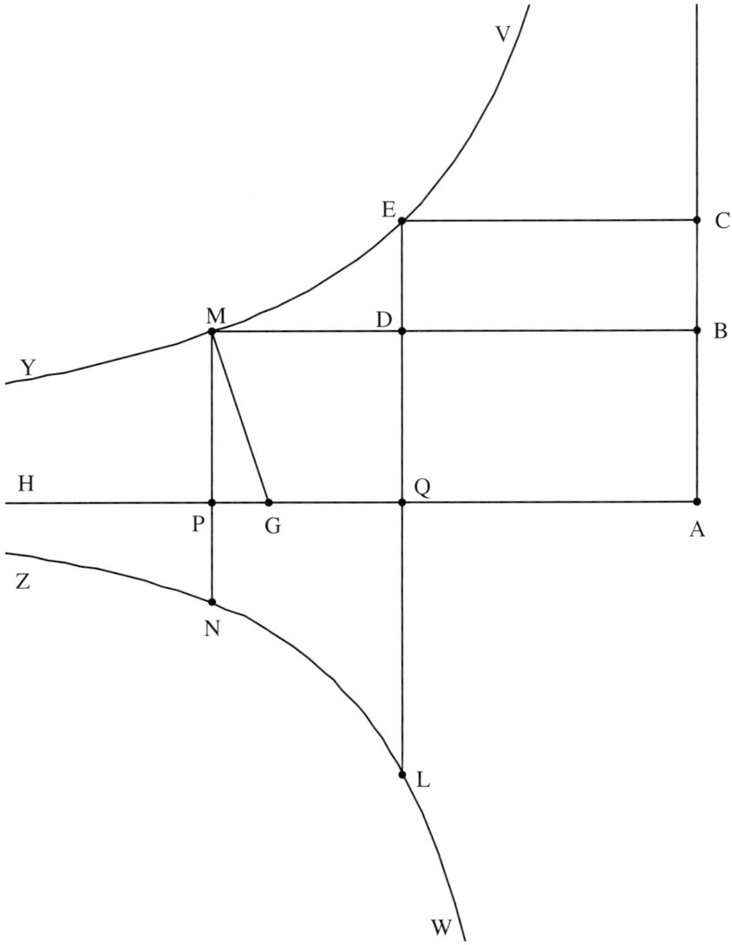

Figure 9.2 Diagram relating to Newton's proof of his version of van Heuraet's theorem

as generated by motion (in the usual sense of this term) and admits that the relations of these figures depends on the relations of the punctual speeds of these motions.

This style of argument will not often be repeated by Newton in his later notes. Still, the central idea will never be abandoned: that of considering related geometric magnitudes as generated by motions whose punctual speeds mutually depend on each other and vary according to an appropriate rate corresponding to the geometric relations of these magnitudes.

As this example shows quite well, Newton uses the term 'motion' and its cognates (overall the verb 'to move') in two distinct senses: to refer both to motions of points or lines in our sense of this term, and to the punctual 'determination' of these motions. The term 'determination' as related to motions had already been used by Descartes, Fermat and Hobbes, in different senses (Descartes 1637, 17–18; Descartes 1644, II, 55–58; Descartes 1647, II, 41, 99; Descartes 1897–1910, letters XCVI, CXI, CCXX, CCXXX, CCXXXIV, DXXI), and Newton will use it later on different occasions (cf. 1967–1981, I, p. 372, for the first occurrence). When motions are rectilinear, their determination, in Newton's sense, reduces to the scalar component of the punctual speed, since their directional component is constant and there is no need to take it into account. But things go in a quite different way when these motions are not rectilinear.

For the time being, let us consider only the simplest case, that of rectilinear motions. I shall come back to the case of curvilinear motions in Section 9.4.

Let x and y be two variable segments generated at the same time by two points moving according to a rectilinear motion. The relation of these segments in any instant of time depends on the (scalar components of the) punctual speeds of these motions. But also the reciprocal is true: for these segments to be related by a certain relation, the (scalar components of the) punctual speeds of these motions have to satisfy some appropriate conditions. Hence, two problems arise quite naturally: (*i*) Given the relation of x and y, to look for the (scalar components of the) punctual speeds of the motions that generate them; (*ii*) Given the (scalar components of the) punctual speeds, to look for the relation of x and y. These are just the two problems that Newton states in the *De Methodis*. But why are they relevant for the solution of geometrical problems concerned with curves?

A first answer comes, implicitly, from a short note probably redacted at the beginning of the Fall of 1665 (1967–1981, I, pp. 343–347), where these problems are stated and the first of them is solved, in the particular case where both the relation between the segments and that between (the scalar components of) their punctual speeds are expressed by polynomial equations in two variables. Here is what Newton writes (1967–1981, I, p. 344):

> 1. If two bodies c, d [Figure 9.3] describe the streight lines ac, bd, in the same time, (calling $ac = x$, $bd = y$, $p =$ motion of c, $q =$ motion of d) & if I have an equation expressing the relation of $ac = x$ & $bd = y$ whose termes are all put equal to nothing. I multiply each terme of the equation by so many times py or $\frac{p}{x}$ as x hath dimensions in it. & also by soe many times qx or $\frac{q}{y}$ as y hath dimensions in it. the sume of these products is an equation expressing the relation of the motions of c & d. . . .
>
> 2. If an equation expressing the relation of their motions bee given, tis more difficult & sometimes Geometrically impossible, thereby to find the relation of the spaces described by these motions.

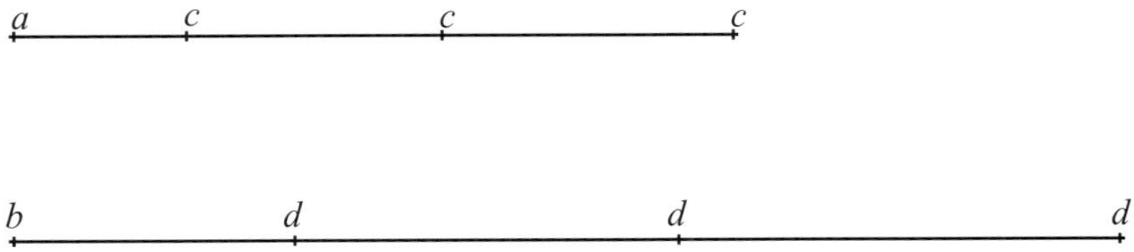

Figure 9.3 Diagram relating to Newton's proof of the algorithm of speeds

The algorithm described in the first proposition is the well-known direct algorithm that leads from any polynomial equation

$$\sum_{i=0}^{n}\sum_{j=0}^{i} A_{i-j,j}x^{i-j}j^{j} = 0 \tag{9.4}$$

to the equality

$$\frac{q}{p} = -\frac{\sum\limits_{i=0}^{n}\sum\limits_{j=0}^{i}(i-j)A_{i-j,j}x^{i-j-1}j^{j}}{\sum\limits_{i=0}^{n}\sum\limits_{j=0}^{i} jA_{i-j,j}x^{i-j}j^{j-1}}. \tag{9.5}$$

This is a particular case of the algorithm described in proposition 7 of the *October 1666 tract.*[23] If we interpret it within the formalism of differential calculus, as we know it today, this algorithm allows us to pass from any polynomial equation $P(x, y) = 0$ to the equality

$$\frac{q}{p} = -\frac{\frac{\partial P}{\partial x}}{\frac{\partial P}{\partial y}}.$$

Still, no compact and general notion equivalent to that of partial derivative of a polynomial is available for Newton at this stage of his researches. Hence, this algorithm is for him nothing but a rule to transform a polynomial $P(x, y)$ into an appropriate ratio of associated polynomials that is taken to express the ratio q/p of (the scalar components of) the punctual speeds of the rectilinear motions generating the segments x and y.

In a note written about one year earlier (1967–1981, I, pp. 236–238), Newton had claimed that the product of y and the ratio of polynomials providing the right-hand side of the equality (9.5) – also described, of course, as the result of an appropriate transformation of an equation like (9.4) – provides the

23 Cf. footnote 19, on p. 228.

sub-normal on the x-axis and at the generic point (x, y) of the curves expressed, with respect to a Cartesian orthogonal system of co-ordinates, by this same equation. From this claim and the equality (9.5), it follows that

$$\frac{q}{p} = \frac{sn._x[P(x, y)]}{y}, \tag{9.6}$$

where $sn._x[P(x, y)]$ is just this sub-normal.

Though Newton did not state explicitly this equality in his note of Fall of 1665, at this date he was certainly aware of it. Once compared with the result about the intrinsic link between the problem of tangents or normals and the problem of quadratures that Newton had obtained some months earlier by modifying the theorem of van Heuraet, this equality provides a way to connect these two geometric problems with the problems of speeds, in the case of curves referred to a system of Cartesian co-ordinates and expressed by polynomial equations.

Suppose that x and y are the orthogonal Cartesian co-ordinates of a curve, and that they remain to each other in a certain relation R. If this relation is expressed by a polynomial equation $P(x, y) = 0$, from the equality (9.6) it follows that the problems of tangents and normals can be solved by passing from this relation to the ratio q/p according to the equality (9.5) and rewriting the right-hand side of this equality in terms of only one of the two variables x and y. Moreover if one sets $\mathrm{AP} = x$, $\mathrm{PM} = y$, $\mathrm{PN} = z$ (Figure 9.1 or Figure 9.2), from the equality (9.6), it follows that the condition (9.3) transforms into

$$z = K\frac{q}{p}.$$

Hence, according to Newton's version of the theorem of van Heuraet, the problem of squaring the curve of orthogonal Cartesian co-ordinates x and z can be solved by passing from the relation R^* that links these co-ordinates to each other to a polynomial equation $P(x, y) = 0$ such that $q/p = z/K$. If this is so, the trapezoid delimited by this curve, taken between the abscissas $x = \xi$ and $x = \kappa$, is indeed equal to $K|y_\kappa - y_\xi|$.[24]

The only difficulty that possibly arises in the solution of the former of these problems, when R is expressed by a polynomial equation $P(x, y) = 0$, is that of rewriting the right-hand side of the equality (9.5) in terms of only one of the two variables x and y. If the relation R^* is given somehow, the difficulty that possibly arises in the solution of the latter problem is that of finding an appropriate polynomial $P(x, y)$, provided that there is one (which is of course not warranted, in general).

Two classical problems concerned with curves – the problem of tangents or normals and the problem of quadratures – are thus reduced, under appropriate

24 Cf. footnote 22, on p. 231.

restrictive conditions, to problems concerned with punctual speeds of rectilinear motions which are, in turn, equivalent to algorithmic problems belonging to the field of Vietian analysis. But if these conditions are not met, are the problem of tangents or normals and the problem of quadratures also connected in some ways with problems concerned with punctual speeds of rectilinear motions?

To answer this question, it is relevant to know how the equality (9.6) is obtained. Did Newton merely get, in two distinct ways, two coincident algorithms (the algorithm of the tangents or normals and the algorithms of speeds)? Or did he understand in general – based on geometrical-mechanical arguments, independent of any equation – that the ratio of the (scalar components of the) punctual speeds of the generative motions of two segments y and x is equal to the ratio of the sub-normal on the x-axis and the ordinate of the curve of orthogonal Cartesian co-ordinates x and y? If the latter possibility obtained, then Newton knew, in the Fall of 1665, that the connection between the problems of tangents, normals and quadratures and problems concerned with punctual speeds of rectilinear motions – which is manifested by the equality (9.6) together with his version of the theorem of van Heuraet – does not depend on the way the relevant curves can be expressed with respect to a system of Cartesian co-ordinates. If the former possibility obtained, then he could not have avoided to wonder if this connection also holds for curves that, though referred to such a system of co-ordinates, cannot be expressed by polynomial equations.

No direct evidence is available for deciding among these possibilities. Still, it is perhaps relevant to remark that, in his *Geometrical lectures* (1670), Barrow proved a theorem equivalent to a generalization of the equality (9.6) to any curve referred to a system of Cartesian orthogonal co-ordinates. In lecture III, he remarked that any curve can be conceived as the result of the composition of two motions: one of a straight line *az* (Figure 9.4) that moves parallelwise from position AZ so as its point a moves along a fixed perpendicular straight line AY, and the other of a point m that moves on the former of these lines so as to describe the curve (1670, pp. 28–29 and Child 1916, pp. 49–51).[25] Then, in lecture IV (art. XI), he proved that the ratio of the (scalar components of the) punctual speeds of these motions at whatever point M of the curve is the same as that of the segments PM and TP, provided that TM is the tangent to the curve at M (1670, pp. 32–33 and Child 1916, pp. 55–57). In Newton's notation, and supposing that the straight line TP is the x-axis, and AP $= x$, PM $= y$, this reduces to the equality

$$\frac{q}{p} = \frac{y}{stg._x[y]}, \tag{9.7}$$

25 For sake of simplicity, I have indicated fixed and moving points with capital and small letters, respectively.

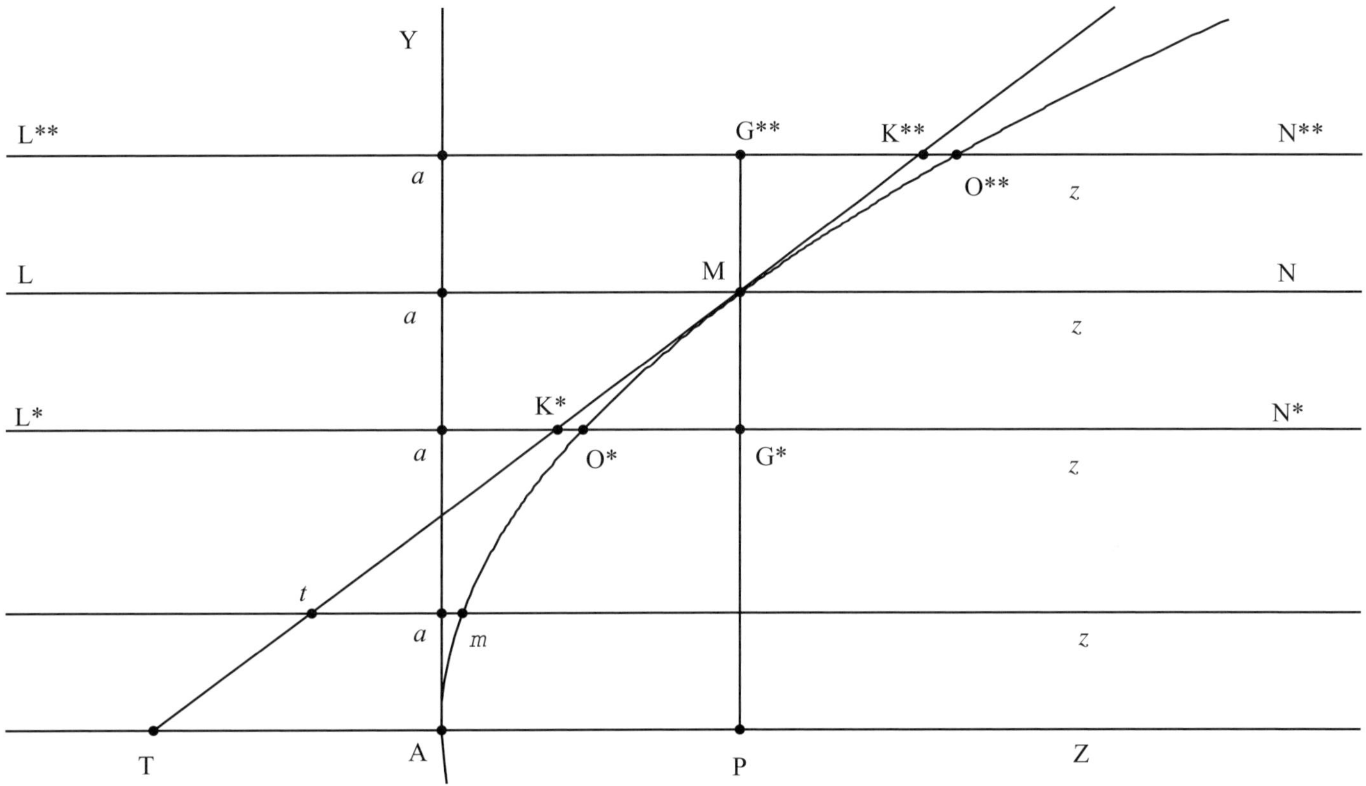

Figure 9.4 Barrow's composition of motions

where $stg._x[y]$ is the sub-tangent of the curve of ordinate y on the x-axis. If the co-ordinates are orthogonal and $sn._x[y]$ is the sub-normal of this same curve on this same x-axis, this equality is equivalent to

$$\frac{q}{p} = \frac{sn._x[y]}{y}, \tag{9.8}$$

which is a generalization of the equality (9.6).

It is possible that Newton had attended a lecture, either at Cambridge University or elsewhere,[26] at which Barrow proved this result. If this is so, he was not only aware that equality (9.6) is only a particular case of the much more general equality (9.8), but he also knew a simple way to prove this last equality. In the printed version of Barrow's lecture this theorem is proved as follows. If it is considered as fixed, the tangent TM also results from the composition of two motions: the same motion of the straight line *az* from which the curve results, and the motion of the point *t* that moves uniformly on the same straight line *az* starting from T and so as to take the position M when *az* takes the position LN. The trajectory of a composed motion such as those that describe both the curve and its tangent depends only on the ratio of the (scalar components of the) punctual speeds of the motions that compose it, and is a straight line if and only if this ratio is constant (1670, p. 28 and Child, 1916 pp. 49–50). Hence, one can suppose, without any loss of generality, that the motions of *az* and *t* are uniform. This being admitted, consider any two positions of *az*: the position L*N* on one side of LN, such that the point *m* takes the position O* which is between the position K* of the point *t* and the point G* at which *az* (in position L*N*) cuts PM; and the position L**N** on the other side of LN, such that the point *m* takes the position O** which is beyond the position K** of the point *t* which is in turn beyond the point G** at which *az* (in position L**N**) cuts PM. Also admit that these positions are such that the curve does not change its concavity and has no extreme between them.[27] When the straight line *az* is in the first of these positions, the (scalar component of the) punctual speed of the motion of *m* along it is smaller than that of the motion of the point *t* also along it, since the former speed is increasing whereas the latter is uniform, and the space O*G* covered by *m* in a certain time[28] is smaller than the space

26 Child (1916, p. 7) suggested that Barrow delivered his *Geometrical Lectures* at Gresham College and that Newton attended them in 1663–1664. This is however far from sure: concerning the relations between Barrow and Newton before 1669, and the possibility that the latter attended some lectures of the former, cf. Newton (1967–1981, I, pp. 10–11, footnote 26).

27 As a matter of fact, Barrow does not make this restrictive condition explicit. Still, such a condition is clearly required by his argument, and, as a consequence, this argument does not apply if M is an extreme or inflection point.

28 Barrow openly considers this time as being 'represented' by the segment G*M.

K*G* covered by *t* in the same time. For an analogous reason (considering the motions as going in the opposite directions),[29] when the straight line *az* is in the second of these positions, the (scalar components of the) punctual speed of the point *m* along it is greater than that of the motion of the point *t* also along it. It follows that, when *az* is in the position LN, these speeds are equal, which is enough to prove the theorem.[30]

9.4 Newton and Roberval's method of tangents

Though no similar proof occurs in Newton's notes, the way he succeeds in showing that the equality (9.6) is nothing but a particular case of a much more general result has a lot of affinities with Barrow's arguments. Still, Newton goes much farther than Barrow, since he also shows that the problem of tangents is intrinsically connected with some appropriate problems concerning motions (either rectilinear or not) and their punctual speeds even when the relevant curves are not referred to any system of Cartesian co-ordinates. This became possible when he became aware of Roberval's method of tangents.[31]

In 1665, this method was known in France by some mathematicians,[32] but had not yet been presented in any published text.[33] This only happened in 1693, when a treatise written by a pupil of Roberval, François de Bonneau, Sieur de Verdus (Roberval 1693), appeared. This treatise certainly communicated notes taken from Roberval's lectures. Though Newton never mentions this treatise, nor the name of Roberval, the content of some of his notes leaves no doubt that he had somehow become acquainted with his method.[34]

29 This condition is implicitly expressed by Barrow through the identification of the relevant time with the segment MG** (as a matter of fact, Barrow, in setting out the second part of his argument, takes MG** to be a time, rather than merely representing it), which is now described in the opposite direction than G*M : cf. the previous footnote.

30 The constant ratio of the (scalar components of the) punctual speeds of the motions of *az* and of *t* on this last straight line is, indeed, equal to the ratio of PM and TP, so that, if the (scalar components of the) punctual speeds of *m* at M is equal to the constant one of *t*, the ratio of the (scalar components of the) punctual speeds of the motions of *az* and of *m* on this last straight line, when this last point is at M, is also equal to the ratio of PM and TP.

31 On Newton and Roberval's method of tangents, cf. Wolfson (2001) which I did not yet know when I wrote my (2005).

32 On Roberval's method and his diffusion, cf. Auger (1962, pp. 58–77), Hara (1965), Pedersen (1968) and Pedersen (1969, pp. 20–23). This method is also studied in detail by A. Warnstel in a chapter of his forthcoming book (in French) on the mathematical works of Descartes.

33 A similar method had been, however, applied by Torricelli to find the tangent of a parabola in Torricelli (1644, pp. 119–121).

34 There is no evidence that speaks to the way this method became known to Newton. It was known by Barrow, who spoke of it in a letter to Collins (Rigand 1841, 34) as a 'method of finding the tangents to curved lines by composition of motions' that had been mentioned

Here is how Verdus presents its 'principe d'invention' (Roberval 1693, p. 70):[35]

> ... in every ... curve, the tangent at whatever point is the direction line of the motion of the movable that describes this same line. Hence, in composing some motions in different ways and in knowing the direction of the composed motion at whatever point of a curve, we shall know, at the same time, its tangent.

The problem with this principle is that it does not make clear how the composition of motions is understood, exactly. In fact in Verdus' treatise at least three different sorts of compositions of motions are considered:

1. A point is submitted to a composed motion if it moves with respect to a system of reference that moves, in turn, with respect to another system of reference.
2. A point is submitted to a composed motion if it is the intersection point of two rigid curves and moves insofar as these curves move separately from each other.
3. A point is submitted to a composed motion if it moves insofar as its distances from two fixed poles, represented by two segments generated by two distinct motions, change at the same time.

Verdus' treatise expounds the method in general in a rather vague way, and then includes different examples each of which is concerned with one or more of these modalities of composition. In each example, we are told how to find the punctual direction of the composed motion, supposing that both the scalar and directional components of the punctual speeds of the two motions that

by Mersenne and Torricelli. This suggests that Barrow became acquainted with it through Mersenne's mention of it in the *Cogitata physico mathematica* (1644, pp. 115–116). But it is also possible that he knew it in some other way, for instance through Hobbes, who was close to Verdus (Skinner 1966) and met Roberval himself in 1642 (Auger 1962, p. 72). It is highly plausible that Barrow mentioned this method in one of his lectures. The third of his *Geometrical Lectures* is, indeed, entirely devoted to the composition of motions which is then used, as we have seen, to investigate tangents. It is possible that Newton was there, learned the fundamental ideas of this method and some of its paradigmatic examples (Newton's notes include many examples occurring in Verdus' treatise), and then elaborated on them by himself.

35 The translation is mine. Here is Verdus' original text: 'en toutes les ... lignes courbes qu'elles puissent estre, leur touchante, en quelque point que ce soit, est la ligne de direction du mouvement qu'a en ce mesme point le mobil qui la décrit. En sorte que composant des mouvemens en diverses façons, et venant à connoistre la direction du mouvement composé en quelque point que ce soit, d'une ligne courbe, nous connoistrons par mesme moyen sa touchante.'

compose it are known. Still, these modalities are not explicitly distinguished and no general procedure or construction is associated with each of them.

Case 1 is that of the motions that generate a cycloid and a spiral, provided that these motions are described respectively as the motion of a point on a wheel that advances by rotating on a straight line (Figure 9.5; this is the motion of a rotating point on a translating plane), and as the motion of a point advancing on a rotating ruler (Figure 9.6; this is the rectilinear motion of a point on a rotating plane).

In the first of these two examples, the second motion is rectilinear. In the second it is not. When it is rectilinear the situation is quite simple: the speed of the point moved according to the composed motion results from the application of the rule of parallelograms to the speeds of the composing motions (Figure 9.5a).

When the second motion is not rectilinear, there is no guarantee that the speed of the point moved according to the composed motion results from the application of the rule of parallelograms, at least if this rule is applied to the speeds of the composing motions. The reason is the following. Suppose that $\mathbf{v}_1$ and $\mathbf{v}_2$ are the punctual speeds of the first and the second motion, respectively. If the second motion is not rectilinear, we have no guarantee that $\mathbf{v}_1$ and $\mathbf{v}_2$ are also the components of the punctual speed $\mathbf{v}$ of the composed motion along their own directions. The same is true also for the two other cases of composition of motions.

An example of case 2 is the quadratrix, described as the trajectory of the intersection point of two rules, one of which rotates around the vertex of a square while the other translates along the direction of a side of this square by remaining perpendicular to it (Figure 9.7).

An example of case 3 is the ellipse, described as the locus of the points such that the sum of their distances from two given points is fixed (Figure 9.8).

Suppose now that a curve $\mathcal{C}$ is the trajectory of a motion $\mathcal{M}$ composed, in one of the previous three ways, by two other motions $\mathcal{M}_1$ and $\mathcal{M}_2$. Suppose also that these two motions are either rectilinear or circular. In both cases the directions of their punctual speeds $\mathbf{v}_1$ and $\mathbf{v}_2$ are known (in the case of a rectilinear motion, this is the same trajectory as the motion; in the case of a circular motion, this is the perpendicular to the radius of this trajectory). Suppose also that the ratio of the scalar components of these speeds is known as well: they can be represented by two segments s_1 and s_2 which are taken in the same directions of these speeds and that are in such a ratio to one other. To find the tangent of $\mathcal{C}$, it is enough to determine the punctual direction of $\mathcal{M}$. The problem is thus to compose $\mathbf{v}_1$ and $\mathbf{v}_2$ in the right way, that is, to find a general construction to be applied to s_1 and s_2 so as to get a straight line that provides such a direction. Once the tangent of $\mathcal{C}$ is known, this curve can be added to straight lines and circles as a trajectory of motions of which other motions are composed so that the tangent of their trajectory can be found

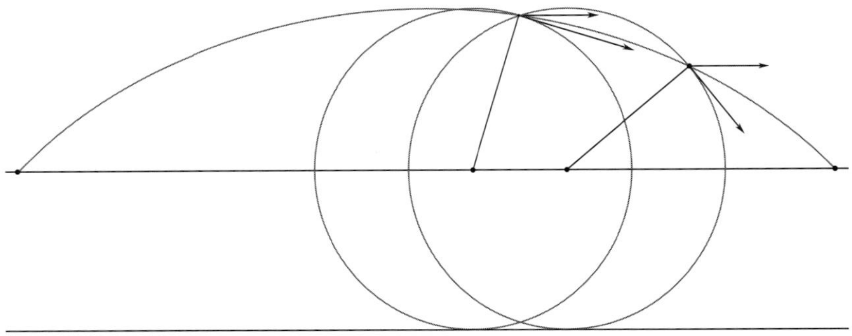

Figure 9.5 The composition of motions in the description of a cycloid

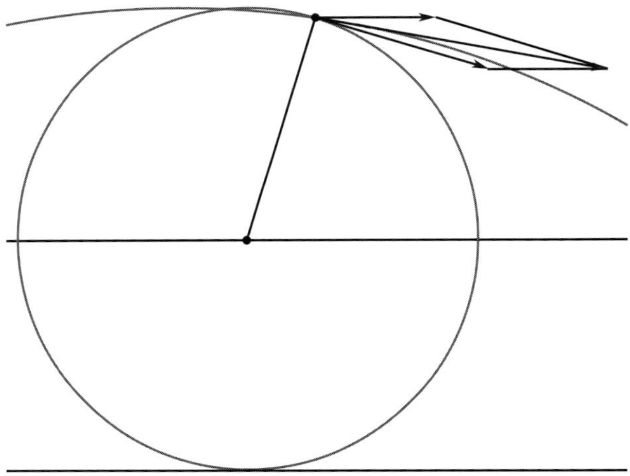

Figure 9.5a The rule of the parallelogram applied to the case of the cycloid

through the same method. And, of course, one can then continue in the same way up to other curves conceived as trajectories of motions composed by other, more and more complex motions.

Roberval treats different cases in different ways. Newton wants, instead, a general principle to be applied in any particular case. Many of the mathematical researches of Newton between the Fall of 1665 and the Spring of 1666 are locating just such a principle. It is finally found, in its definitive and general form, in May of 1665, and it is expounded in two notes (the second of which results from a revision of the first) written on May 14th and 16th (Newton 1967–1981, I, pp. 390–392 and pp. 392–399). This same principle is also expounded in proposition 6 of the *October 1666 tract*. Propositions 1–5 of this treatise are merely used to provide the necessary ingredients of this exposition.

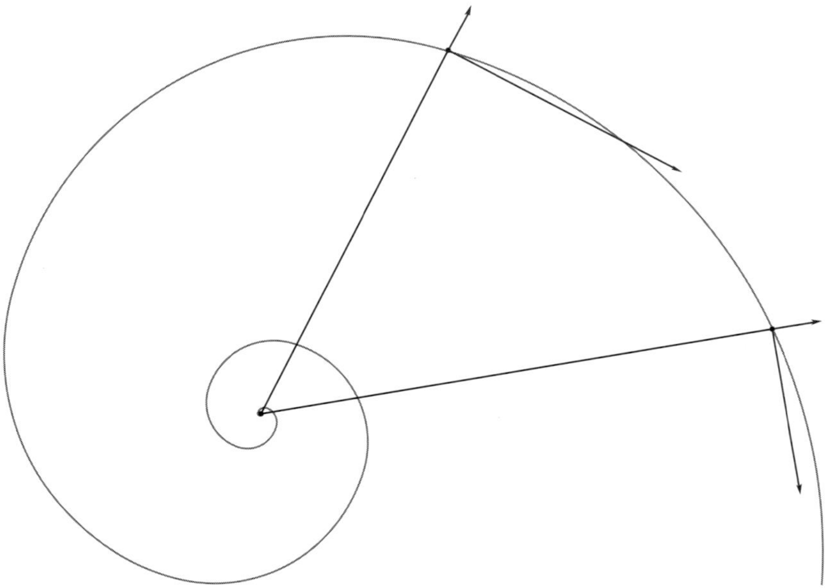

Figure 9.6 The composition of motions in the description of a spiral

It seems that Newton understood that the first and third among the three previous modalities of composition of motions can be reduced to the second one, that is, that there is a way to pass, through appropriate constructions, from the two former cases to the latter (for details, cf. Panza 2005, 412–414). It follows that any composed motion can be viewed as the motion of the intersection point of two rigid curves that moves separately.

If the tangents of these curves and the (ratio of the) punctual speeds of their respective motions are known, it is moreover easy to find the punctual direction of the composed motion, and thus the tangent of its trajectory, as follows.

Suppose that YMX and ZMW (Figure 9.9) are the moving curves and M is their intersection point. Suppose also that UMO and VMP are the tangents to these curves at the point M, and that the punctual speeds of the motions of these curves are represented (scalarly and directionally) by the segments MR and MQ. It follows that the direction of M is provided by the diagonal MT of the quadrilateral MRTQ which is constructed by drawing from R and Q two parallel lines to the tangents UMO and VMP, respectively.

The justification is easy. The point M is affected in fact by four motions: the two motions of the curves YMX and ZMW and the two motions that this point has on these curves in order to continue to be their intersection point. The segments MR and MQ represent, respectively, the punctual speeds of

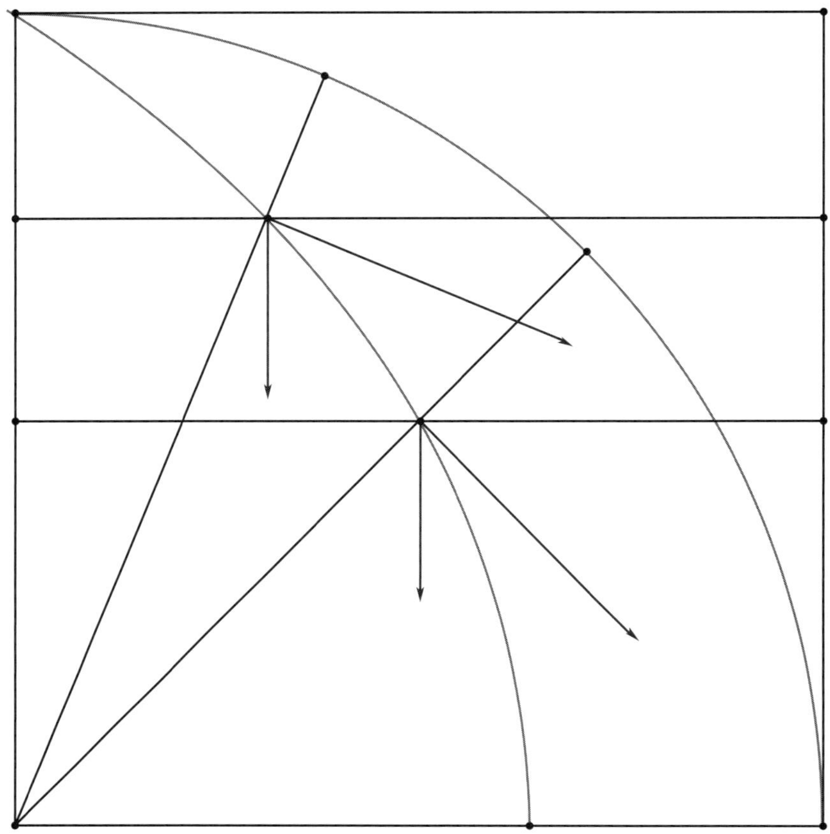

Figure 9.7 The composition of motions in the description of a quadratrix

the two former motions. The segments RT = MT′ and TQ = MT″ represent, respectively, the punctual speeds of the two latter ones. By composing these four motions two by two according to the rule of parallelograms, one gets exactly the direction MT.

Provided that the tangents of straight lines and circles are known, one can easily find, in this way, the tangents of the trajectories of the intersection point of two moving straight lines, two moving circles, or a moving straight line and a moving circle. And again, once this is done, the tangents of the trajectories of the intersection point of two curves corresponding to these trajectories can be found in the same way, and so on.

But, for this to be possible, the ratio of the scalar components of certain speeds has to be determined. And, for that, the algorithm of speeds for segments related by a polynomial equation can be useful.

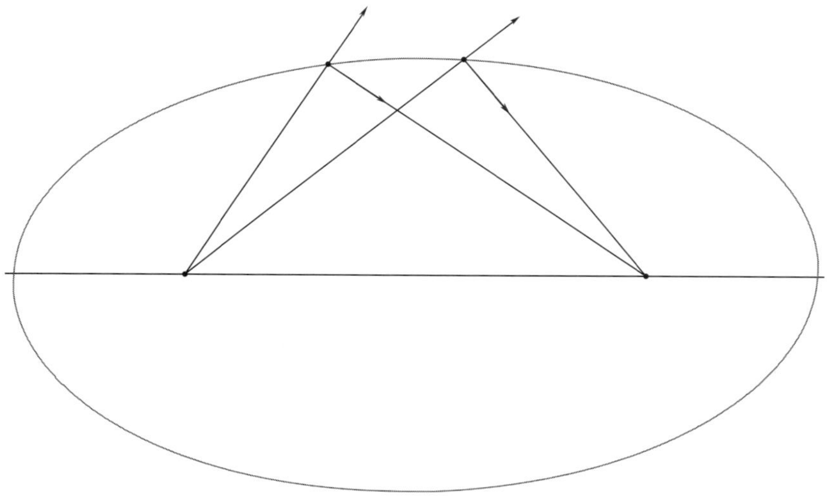

Figure 9.8 The composition of motions in the description of an ellipse

The simplest case obtains when the two curves YMX and ZMW reduce to two straight lines each of which translate along the direction of the other (Figure 9.10). This is just the configuration involved in Barrow's previous proof. But now it is nothing but a particular case of a more general configuration. In such a particular case, each of the two lines provides, then, the punctual direction of the motion of the other and is its own tangent. Newton's general principle reduces, thus, to the rule of parallelograms (which is consistent with the fact that the motion of the point M can also, in this case, be described as the motion of a point with respect to a system of reference that moves rectilinearly with respect to another system of reference). Hence, if the punctual speeds of these straight lines are represented by the segments MR and MQ, to solve the problem it is enough to construct the rectangle MRTQ, for its diagonal MT is the sought after tangent.

Barrow's result – that is, the equality (9.7) – is thus quite easily proved as a particular case of a more general result that concerns tangents of curves independently of any system of co-ordinates to which these curves might be referred.

9.5 Back to the *De Methodis*

With all this in mind, we can now come back to the first reduction of the *De Methodis*, which, recall, consists in reducing geometrical problems concerning curves to two quite general problems concerning motions (cf. p. 228 above).

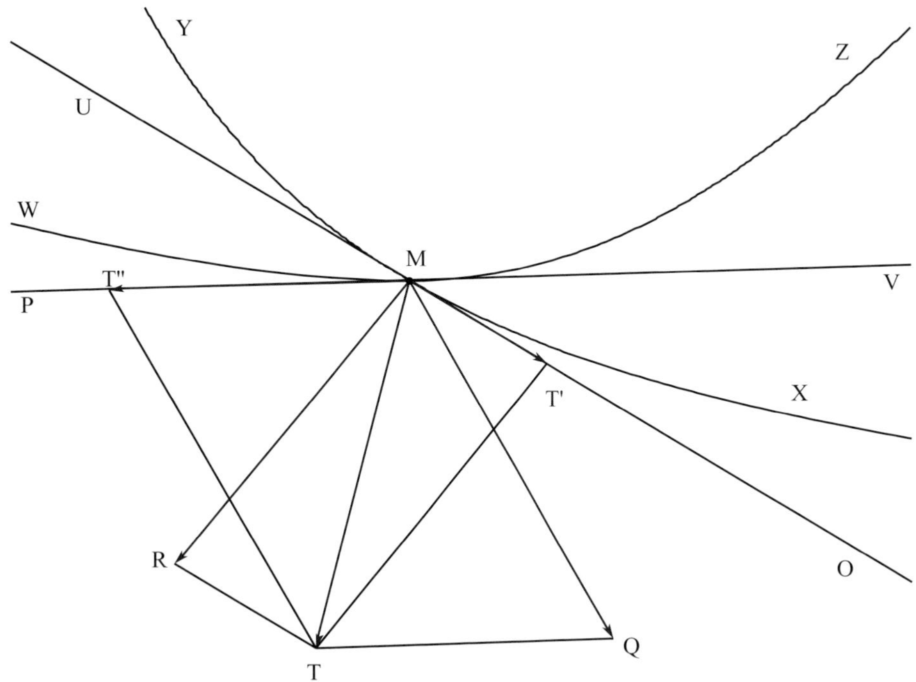

Figure 9.9 Newton's general law for composing motions

If we compare these problems with propositions 1–8 of the *October 1666 tract*, we find that Newton has eliminated both the general context provided by the theory of composition of motions (propositions 1–6) and the assumption that spaces (in the first problem) and speeds (in the second) are linked to each other by a polynomial equation (propositions 7–8: cf. pp. 228–229). So, a question quite naturally arises: What are the spaces and speeds that Newton is speaking of? Are they merely segments generated by rectilinear motions of points and the punctual speeds of these motions (which are nothing but scalar quantities)? Or are they some sort of trajectories of rigid curves and points on these curves and the punctual speeds of them (which cannot be reduced to scalar quantities)?

There is no question that Newton's text is ambiguous. However, Newton offers some clarifications by presenting a quite simple example (1967–1981, III, p. 73):[36]

> So in the equation $x^2 = y$ if y designates the length of the space described in any time which is measured and represented by a second space x as it increases with uniform speed: then $2mx$ will designate the speed with which the space y at the same moment of time proceeds to be described.

The letter 'm' replaces 'p', here. This is a minor change, but it goes together with two other, more relevant ones. Newton openly supposes that: (*i*) the space x is covered by a uniform motion; and (*ii*) this space measures and represents time.

The first supposition is not absolutely new. It had been used by Barrow[37] and Newton himself had at times appealed to it in his earlier notes. And, once it is admitted, the second – also used by Barrow – seems quite natural. Still, the way Newton employs this second supposition reveals that it is not merely a local trick for him. It is rather a symptom of a quite deep change in Newton's conceptions. Time is not here understood as the real time in which motion takes place; it is merely the second term of an analogy (Guicciardini 1999, pp. 19–20). And this is also the case of space. The reason is simple: Newton is no longer referring to the motions of points or lines; he is no longer considering geometrical quantities generated by these motions. He is rather referring to variations of quantities conceived as pure variables. There is no need to insist

36 I quote Whiteside's translation (but maintain the symbol 'm' instead of replacing it with the symbol '$\dot{x}$' as Whiteside does by using a notation that Newton will only introduce in 1691: cf. Whiteside's footnotes 83 and 86). Here is Newton's original (1967–1981, III, p. 72): 'Sic in æquatione $xx = y$ si y designat spatij longitudinem ad quodlibet tempus quod aliud spatium x uniformi celeritate increscendo mensurat et exhibet descriptam: tunc $2mx$ designabit celeritatem qua spatium y ad item temporis momentum describi pergit. . . .'

37 Cf. the previous footnotes 28 and 29.

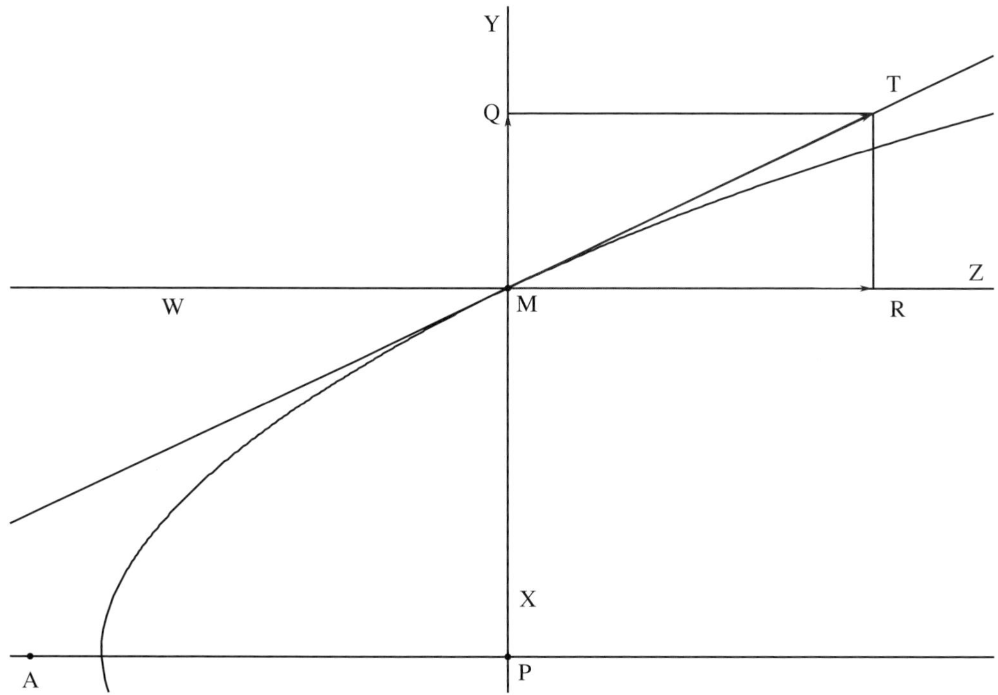

Figure 9.10 Applying Newton's law in the case of the motion of two straight lines each of which translates along the direction of the other

on this point, since Newton himself is quite clear about it (1967–1981, III, p. 73):[38]

> And hence it is that in the sequel I consider quantities as though they were generated by continuous increase in the manner of a space which a moving object describes in its course.

Quantities are, thus, not spaces generated by motions, that is, segments generated by moving points or surfaces generated by moving lines. They are instead that which is 'generated by continuous increase' in the same way as space is 'generated by motion'. But things are even clearer in what follows:[39]

> However, as we have no estimate of time except in so far as it is expounded and measured by an equable motion, and as, furthermore, only quantities of the same kind may be compared one with another [as well as only] their speeds of increase and decrease [can be compared one with another], I shall, in what follows, have no regard to time, formally so considered, but, among the quantities propounded which are of the same kind, I shall suppose some one to increase with an equable flow, and the others to be referred to it as though it were time, so that the name 'time' may, by analogy, be conferred upon it. And so, whenever in the following the term 'time' occurs . . . by that name should be understood not time formally considered but that other quantity through whose equable increase or flow time is expounded and measured.

If Newton does not speak of trajectories and composed motions, it is, thus, because he does not want to refer to real motions, but rather to a more general kind of change. To say it in Aristotle's language, he is no longer interested in displacement of points or local change (*φορά*) as such, but rather in a more general kind of change (*κίνησις*) that includes displacement of points as a particular case. Let us call it 'quantitative variation'. What is this exactly?

From the beginning of 1664 – while studying Wallis' method of quadratures (1967–1981, I, pp. 91–95) – Newton had understood that it is enough, for a

38 I quote Whiteside's translation. Here is Newton's original (1967–1981, III, p. 72): 'Et hinc est quod in sequentibus consideratem quantitates quasi generatæ essent per incrementum continuuum ad modo spatij quod mobile percurrendo describit.'

39 I slightly modify here Whiteside's translation. Here is Newton's original (1967–1981, III, p. 72): 'Cùm autem temporis nullam habeamus æstimatione nisi quatenus id per æquabilem motum localem exponitur et mensuratur, et præterea cùm quantitates ejusdem tantùm generis inter se conferri possint et earum incrementi et decrementi celeritates inter se, eapropter ad tempus formaliter spectatum in sequentibus haud respiciam, sed e propositis quantitatibus quæ sunt ejusdem generis aliquam æquabili fluxione augeri fingam cui cæteræ tanquam tempori referantur, adeoque cui nomen temporis analogicè tribui mereatur. Siquando itaque vocabulum temporis in sequentibus occurrat . . . eo nomine non tempus formaliter spectatum subintelligi debet sed illa alia quantitas cujus æquabili incremento sive fluxione tempus exponitur et mensuratur.'

certain geometric quantity – typically a segment or a portion of space – to be able to be regarded as a variable, that another geometric quantity be available and be such that the value of the former depends on its value. This latter quantity works then as a parameter for the variation of the former. It is the crucial idea of a principal variable.

For quite a long time, Newton seems to have been convinced that the fundamental way to express the relation between a geometric quantity and the parameter of its variation consists in writing an algebraic – typically a polynomial – equation interpreted on these quantities. His work on tangents and quadratures, especially that inspired by Roberval's method, taught him that this same relation can also be expressed in a quite different, and more general and fundamental way, by appealing to generative motions and their compositions.

The previous quotation manifests a new, crucial achievement. It reveals that Newton is not dealing with the variation of geometric quantities – or of any other particular sort of quantities – and with the way of expressing their mutual relations. He is rather dealing with quantitative variation as such, understood as a special kind of change. This is the kind of change characterized by the fact that any particular example of it – let us say $\mathcal{X}$ – is univocally identified and completely determined insofar as the link that connects it to a principal change of this same kind, on which any other one depends, is determined by means of a law that establishes how this principal change is reflected in $\mathcal{X}$. Let $\mathcal{T}$ be the principal quantitative variation. This means that a particular quantitative variation $\mathcal{X}$ is univocally identified and completely determined insofar as an appropriate particular relation $\mathcal{R}(\mathcal{X}, \mathcal{T})$ is determined. The subjects of $\mathcal{X}$ and $\mathcal{T}$ (the entities that are supposed to vary) are not relevant here, and the intrinsic nature of $\mathcal{T}$ is also not relevant, and, as a matter of fact, could not be determined. This is not the principal quantitative variation because it is uniform. Things go the other way around: $\mathcal{T}$ is (taken to be) uniform because it is the principal variation.

One could argue that this idea is not new, since that which is described is (in the language of the Scholastics) nothing but the change of intensive qualities. This is incorrect, however. Newton seems, indeed, to permute the *definiens* and the *definiendum*: quantitative variation is not defined by appealing to the notion of intensive quality; rather a quantity is conceived, in its abstract generality, as that which is submitted to a quantitative variation. Though quantities are designated by atomic symbols – like x or y – they are not the specific objects that these symbols stand for. They are rather that which varies according to the relations that are somehow expressed by appealing to these symbols, for example – but not only – through a polynomial equation. The following passage is quite explicit about this (1967–1981, III, p. 73):[40]

40 I quote Whiteside's translation, with some minor, essentially typographic changes (among which there is that already mentioned in footnote 36). Here is Newton's original (1967–1981, III, p. 72): 'Quantitates autem quas ut sensim crescentes indefinitè considero, quo

> But to distinguish the quantities which I consider as just perceptibly but indefinitely growing from others which in any equations are to be looked on as known and determined and are designated by the initial letters *a*, *b*, *c*, *&c.*, I will hereafter call them 'fluents' and designate them by the final letters *v*, *x*, *y* and *z*. And the speeds with which they each flow and are increased by their generating motion (which I might more readily call 'fluxions' or simply 'speeds') I will designate by the letters *l*, *m*, *n*, and *r*.

Newton is thus ready for the second reduction. The two previous problems about spaces and speeds can now be re-stated as follows (1967–1981, III, pp. 75 and 83):[41]

> Problem 1. Given the relation of the flowing quantities to one another, to determine the relation of the fluxions.
>
> Problem 2. When an equation involving the fluxions of quantities is exhibited, to determine the relation of the quantities one to another.

The explicit reference to equations that occurs both in the previous passage where Newton introduces the terms 'fluent' and 'fluxion' and in the statement of the second problem is confirmed in the solution of these problems. Though in the statement of the first problem, Newton is speaking in general of a relation between fluents and fluxions, he solves the problem (1967–1981, III, pp. 74–82) under the condition that this relation is expressed through an appropriate equation: namely, either an algebraic equation (either polynomial or not) between the relevant fluents, or an algebraic equation including a variable expressing the area or the length of a curve expressed in terms of one of the relevant fluents.[42] Moreover, in solving the second problem (1967–1981, III, pp. 82–112), he supposes that one or more algebraic equations among the relevant fluents and fluxions are given and shows how to determine one of these fluents in terms of another one through an algebraic, possibly infinitary, algebraic expression.

distinguam ab alijs quantitatibus quæ in æquationibus quibuscunque pro determinatis et cognitis habendæ sunt ac initialibus literis a, b, c, *&c* designantur, posthac denominabo fluentes, ac designabo finalibus literis v, x, y, et z, Et celeritates quibus singulaæ a motu generante fluunt et augentur (quas possim fluxiones vel simpliciter celeritates vocitare) designabo literis l, m, n et r.'

41 I quote Whiteside's translation. Here is Newton's original (1967–1981, III, pp. 74 and 82): 'Prob. 1. Relatione quantitatum fluentium inter se datâ; fluxionum relationem determinare.' 'Prob. 2. Exposita æquatione fluxiones quantitatum involvente, invenire relationem quantitatum inter se.'

42 Newton's example (1967–1981, III, p. 78) is the equation $z^2 + axz - y^4 = 0$, where z is supposed to be the area of the circle referred to a system of Cartesian orthogonal co-ordinates of equation $w = \sqrt{ax - x^2}$. This example is tractable, since Newton proves that $r = m\sqrt{ax - x^2}$ (where r and m are the fluxions of z and x, respectively), that is, according to the differential formalism: $dz/dx = \sqrt{ax - x^2}$ or $z = \int_0^x \sqrt{at - t^2}dt$.

It could seem, thus, that the generalization involved in the passage from motions to quantitative variations is immediately thwarted by a new regression to the particularity of algebra merely extended through the appeal to specific geometric quantities as areas and lengths. Things are not so, however. To understand why, it is convenient to briefly reconstruct the development of Newton's ideas described above. Newton began by considering polynomial equations as the privileged way to express curves with respect to Cartesian co-ordinates, and showed that, if curves are so expressed, the problems of tangents or normals and of quadratures can be solved through the consideration of rectilinear motions that are taken as the generative motions of these co-ordinates. He passed then from curves so expressed to curves considered as trajectories of composed motions, independently of any particular system of co-ordinates or any sort of equation expressing them, and showed that the possibility of solving the problems of tangents or normals and of quadratures for curves expressed by polynomial equations, through the consideration of rectilinear motions, is nothing but a particular consequence of a more general relation between these trajectories and the components of the relative motions. Finally, he replaced motions with quantitative variations, rectilinear trajectories with fluents, and punctual speeds with fluxions. Still, in this new quite general context, the specification of any particular variation depends on the specification of relations between fluents and fluxions. And, insofar as fluents are not particular sorts of quantities – being rather quantities insofar as they are related to each other –, there is no way to specify these relations by considering particular geometrical or mechanical configurations. Hence, the formalism of Vietian analysis – that is, algebraic equations – returns to take a central role as a privileged way for specifying these relations and thus identifying particular fluents and fluxions. Even the appeal to areas or the lengths appears, in this context, as an easy way to introduce a purely algebraic relation between fluents and fluxions.[43] Fluents and fluxions are thus, so to say, abstract quantities: quantities conceived as nothing but the subjects of quantitative variations, and Vietian analysis is the tool used to specify these variations.

Of course, the intrinsic limitations of this tool affect the extension of the domain of quantitative variations. Still, through his double reduction, Newton has opened a new field for mathematical investigations. This is what, at the very beginning of the *De Methodis*, he calls 'field of analysis',[44] that is, the general doctrine of abstract quantities, conceived as I have just said. Though

43 Cf. footnote 42, above. The appeal to the area of a circle, in the example considered in this last footnote, is only useful to introduce the following system of equations:

$$\begin{cases} z^2 + axz - y^4 = 0 \\ w = \sqrt{ax - x^2} \\ r = mw \end{cases} .$$

44 Cf. the quotation appended to footnote 16.

the *De Methodis* comes back, after the solution of the two previous general problems, to the usual geometric problems concerned with curves, this field is largely extended after this first definition, and a large part of the history of mathematics, after Newton's *De Methodis*, consisted in efforts to enlarge it, by extending the formalism of Vietian analysis (using, among others things, two crucial ingredients which are already part of the toolbox used in this treatise, namely, infinite series and fluxional – or better, in the language that later became common, differential and integral – equations). Eulerian analysis is just the result of the efforts made to structure this field and to absorb in it other branches of mathematics. Newton's field of analysis can thus be viewed as the original kernel of it.

9.6 After the *De Methodis* (or concluding remarks)

As is well known, Newton will quickly change his mind and devote his mathematical energy to classical geometry and the possibility of extending it without modifying its intrinsic nature by using any extraneous formalism.[45] There are many reasons for this change, some of which are certainly not based on mathematical concerns. Still, the previous story teaches us something that may help us to understand this change, which, as far as I know, has gone unnoticed by commentators.

The theory of composition of motions that Newton elaborated based on his understanding and development of Roberval's method of tangents is a theory according to which punctual speeds are considered as proto-vectorial magnitudes: they have both a scalar and a directional component, which are both relevant for composition. Once real motions are abandoned in favour of quantitative variations, and punctual speeds are replaced by fluxions, only the scalar component is conserved, since, in Newton's theory of composition of motions, the directional one was accounted for only through the positional relations of the motions involved, which were represented by diagrams. Still, the problem of considering directions of motions and speeds in the description of physical phenomena could not be avoided.

Hence, Newton's field of analysis could appear to be the original kernel of an autonomous mathematical theory – like Eulerian analysis will later be – only under the condition that this theory be conceived as a theory of pure scalar relations, capable, at most, of providing a framework for accounting for the relations that physical bodies have to each other because of their intensive qualities. This theory could not provide, as such, a language for describing physical reality, by idealization, but only a tool for calculating intensive relations of magnitudes whose particular nature and other sorts of relations had to

45 Among many other possible references about this matter, cf. Guicciardini (1999, pp. 101–104).

be independently specified. Briefly speaking, the interpretation of the relations between abstract quantities as relations between particular quantities could not develop without a crucial addition of information that this autonomous mathematical theory could not account for. The further development of differential calculus, which allowed for the possibility of changing the principal variable by passing from some differential ratios to others, captured at least part of this information. Together with the introduction of appropriate differential and variational principles, this allowed, during the eighteenth century, the growth of analytical mechanics (Panza 2002). But in Newton's theory of fluxions, these developments were blocked by the presence of a unique independent variable understood by analogy with time. The appeal to classical geometry – on which his theory of composition of motions was ultimately founded – should thus have appeared to Newton as a condition for using mathematics to speak of the physical world up to a sufficient degree of accuracy. This could, perhaps, partially explain the absence of the theory of fluxions in the *Principia*: this could have, at most, provided a local tool to be used there, but it could not have been, as such, the basic device of a new natural philosophy.[46]

Still, Newton's field of analysis, became – mostly thanks to mathematicians who did not share Newton's peculiar geometrical outlook – the nucleus of a new form of pure mathematics, whose applications depended on modalities quite different than those proper to classical geometry. This was just eighteenth-century analytic mathematics. My main aim has been to suggest that Newton has to be considered as one of the main fathers of this form of mathematics – better, as its original, first father.

46 To avoid misunderstanding, I repeat myself: this could be, at most, a partial explanation; other reasons for which I cannot account here, are certainly also relevant.

PART III

The reception of Newton

10

Newton, Locke, and Hume

GRACIELA DE PIERRIS

From the 1690s until at least the second half of the eighteenth century, European scientific and literary circles standardly perceived Newton's and Locke's systems as resting on very similar principles and methods; these systems were commonly blurred together as forming a single vision composed of natural and moral philosophy.[1] Moreover, a long tradition of Locke scholars extending to our time has found close links between Locke's and Newton's views on the methodology of natural science. Indeed there is no denying that Newton and Locke share a similar conception of scientific method, if this is simply described as one based on rational and regular experiments and observations, and the use of generalization and deduction. (These similarities have been emphasized, for example, by G. A. J. Rogers and John Yolton.)[2]

This is essentially an abridged version of my (2006). I am grateful to the editors of *Hume Studies* for permission to publish the present version here. For discussions related to the topic of the present paper, see the articles by Mary Domski and Lynn Joy in this volume.

1 Feingold (1988). This article masterfully weaves together the circumstances surrounding the association of the two thinkers in both the expert and popular minds of the period.

2 Rogers (1978, p. 229), writes, for example: "[W]hat Locke found in the *Principia* was the exemplification of a method to which he himself already subscribed. He already believed that a combination of observation, generalization or induction, and deduction was the only route to knowledge of nature and that the *Principia* exhibited just that method in its most fruitful manner... It confirmed for him all his own methodological conclusions... The *Principia* was for Locke the vindication of a general methodological approach to which he had subscribed for perhaps twenty years." (I thank Mary Domski for bringing to my attention this particular passage as pointedly summarizing Rogers's overall view of a close kinship between Locke and Newton.) In this and other articles, Rogers argues extensively for a close connection between Locke's and Newton's scientific methodology and, in general, for what he takes to be the important influence of Locke on Newton. Other articles where Rogers develops the same view are, for example, (1979a) and (1982). In a similar vein, John Yolton, in a more cautious tone, writes, for example in (1969, p. 193): "In his admiration for Boyle, Newton, and Sydenham, Locke was praising these men for this method of carefully observing and recording the observed coexistence of qualities. In his own scientific interests Locke practiced this method also. Theory and hypothesis must find their place in the context of experience and history. The scientists of the day had been making new discoveries and advances by using the method praised by Locke."

Hume also explicitly associates his work and his method with Newton's, aspiring to be the Newton of a new science of human nature. This is very prominent, in particular, in the Introduction to the *Treatise*[3] and Section One of the *Enquiry*.[4] Thus, if both Locke and Hume are Newtonians, one could plausibly identify Hume's conception of the methodology of science with Locke's. Nevertheless, as I shall argue, there is a clear and sharp distinction to be drawn between Hume's Newtonian inductivism and Locke's conception of the methodology of natural science in the *Essay*.[5] In his conception of both the scientific method and the origin and meaning of our idea of causation, Hume is deeply indebted to what he takes to be the Newtonian inductive methodology for the study of nature. This is not to deny Locke's enormous influence on Hume: as with any historical development of philosophical ideas, Hume's epistemological views would not have been possible without the contributions of his predecessors, especially Locke's.[6] Yet Locke represents a crucial transition between Cartesian rationalism and the full-blown empiricism of Hume, and, in particular, there are very important vestiges of rationalism in Locke's epistemology influencing his conception of scientific methodology: the idea of the containment of the effect in the cause, the postulation of a hidden microstructure of primary qualities or properties of bodies, the attendant notion of a metaphysical

3 All citations of *A Treatise of Human Nature* (abbreviated as *Treatise* or T) are from the David Fate Norton and Mary J. Norton edition (Hume 2000), and thus include the book, part, section, and paragraph numbers. All citations of *An Enquiry concerning Human Understanding* (abbreviated as *Enquiry* or EHU) are from the Tom L. Beauchamp edition (Hume 1999), which includes section and paragraph numbers. All citations of *An Enquiry concerning the Principles of Morals* (abbreviated as EPM) are from the Tom L. Beauchamp edition (Hume 1998), which include sections and paragraph numbers.

4 In the *Enquiry*, at EHU 1.14–15, Hume expresses the hope that his own science of human nature might meet with the same success as Newton's determination of the laws and forces that govern the motions of the planets. Here Newton is not mentioned explicitly by name, but it is obvious that Hume refers to Newton when he writes about "the philosopher" who established the laws and forces of planetary motions: "Astronomers had long contented themselves with proving from the phaenomena, the true motions, order, and magnitude of the heavenly bodies: Till a philosopher, at last arose, who seems, from the happiest reasoning, to have also determined the laws and forces, by which the revolutions of the planets are governed and directed." In the Introduction to the *Treatise*, Hume expresses the same aspiration of modeling his own science of human nature on Newton's method. This is evident in Hume's desideratum, explicitly stated at T Introduction 8, of avoiding conjectures and hypotheses in explaining the most general and certain principles derived from experience. In the present paper I shall dwell on this desideratum, which I take to be central to Newton's and Hume's rejection of the methodological scientific ideal of the mechanical philosophy.

5 All citations from Locke (1975), include the roman numeral numbers of the book and chapter, followed by the Arabic number of the section.

6 In particular, as I argue in my book manuscript, *Ideas, Evidence, and Method: Hume's Skepticism and Naturalism concerning Knowledge and Causation*, Hume adopts and radicalizes the Lockean perceptual and imagist model of apprehension of items before the mind, conceived as the standard of ultimate evidence.

necessary connection between cause and effect, and the (for Locke unattainable) ideal of an a priori demonstrative knowledge or science of nature.

Hume follows Newton in substituting the ideal of inductive proof for the ideal – characteristic of the mechanical philosophy – of a demonstrative science of nature based on a postulated hidden microstructure. Although Hume does not fully do justice to a fundamental aspect of Newtonian methodology – the mathematization of nature[7] – he adopts the basic ideas of Newton's inductive method as presented in the Rules for the Study of Natural Philosophy in *Principia*, Book III, especially as articulated in the crucially important Rule III.[8] Hume's notion of inductive proof, which is at the heart of his conception of causation and scientific methodology, consists in a universalization (whenever possible and subject to future experimental revisions) of our past and present uniform experience, with the attendant assumption that nature is, in Newton's words, "ever consonant with itself." Hume's embrace of Newton's inductive method marks a central point of departure from Locke's conception of scientific methodology, for, as I argue below, the desideratum of achieving Newtonian inductive proofs *replaces* the ideal of demonstrative knowledge of nature and *liberates* scientific method from the a priori hypotheses of the mechanical philosophy. In particular, the inductive derivation of laws from manifest uniform phenomena takes priority over the hypothetical postulation, prior to what experience can teach us, of a hidden microstructure of primary qualities – which, according to the mechanical philosophy (shared by both Descartes and Locke), necessitates the causal relations among bodies and between bodies and our senses. Therefore, despite Hume's extensive debt to Locke, Hume does not have a Lockean conception of causation and scientific methodology.

10.1 Newton and Locke on scientific methodology

The central idea of the Newtonian inductive method, as summarized in Newton's Rules, is that exceptionless or nearly exceptionless universal laws are

7 Neither Hume nor Locke has the resources in their empiricist conception of mathematics and the relationship of mathematics to physics to incorporate the constitutive role of mathematics in Newton's physics. By contrast, precisely this constitutive role is emphasized in Kant's reading of Newton: see, for example, Friedman (1992). I. Bernard Cohen (1980) emphasizes the methodological import of Newton's mathematization of nature; and this approach is further developed by George E. Smith (2002a). According to Cohen and Smith, at the heart of Newton's method is a process of mathematical idealization and successive approximations by revision. There is no doubt that, in his adoption of Newton's inductive method, Hume ignores the role of mathematics and idealizations.

8 Rule III prescribes (Newton 1999, p. 795): "*Those qualities of bodies that cannot be intended and remitted [i.e. qualities that cannot be increased and diminished] and that belong to all bodies on which experiments can be made should be taken as qualities of all bodies universally.*" All citations are from Newton (1999). Cohen, in his guide (1999, p. 200), explains that the terms "intension" and "remission" go back to late-medieval doctrine referring to qualities that "undergo an intension or remission by degrees."

inductively derived from "manifest qualities" or observed "phenomena," and only further observed phenomena can lead us to revise these laws. In his official methodological pronouncements, Newton explicitly and emphatically opposes the purely hypothetical explanations of the mechanical philosophy standing in the way of his inductive argument for the law of universal gravitation.

Newton's Rules III and IV state that his method of inductive universalization must be applied without the interference of hypotheses, and, in his explanations of these Rules, Newton explicitly depicts the hypotheses of the mechanical philosophy as obstructing this method. This is clear in Newton's explanation of Rule III, for example:

> For the qualities of bodies can be known only through experiments; and therefore qualities that square with experiments universally are to be regarded as universal qualities... Certainly idle fancies ought not to be fabricated recklessly against the evidence of experiments, nor should we depart from the analogy of nature, since nature is always simple and ever consonant with itself. The extension of bodies is known to us only through our senses, and yet there are bodies beyond the range of these senses; but because extension is found in all sensible bodies, it is ascribed to all bodies universally. We know by experience that some bodies are hard. Moreover, because the hardness of the whole arises from the hardness of its parts, we justly infer from this not only the hardness of the undivided particles of bodies that are accessible to our senses, but also of all other bodies. That all bodies are impenetrable we gather not by reason but by our senses. We find those bodies that we handle to be impenetrable, and hence conclude that impenetrability is a property of all bodies universally. That all bodies are movable and persevere in motion or in rest by means of certain forces (which we call forces of inertia) we infer from finding these properties in the bodies that we have seen... Finally, if it is universally established by experiments and astronomical observations that all bodies on or near the earth gravitate [*lit.* are heavy] toward the earth, and do so in proportion to the quantity of matter in each body, and that the moon gravitates [is heavy] toward the earth in proportion to the quantity of its matter; and that our sea in turn gravitates [is heavy] toward the moon, and that all planets gravitate [are heavy] toward one another, and that there is a similar gravity [heaviness] of comets toward the sun, it will have to be concluded by this third rule that all bodies gravitate toward one another. Indeed, the argument from phenomena will be ever stronger for universal gravity than for the impenetrability of bodies, for which, of course, we have not a single experiment, and not even an observation, in the case of the heavenly bodies. Yet I am by no means affirming that gravity is essential to bodies.[9]
>
> (Newton 1999, pp. 795–796)

9 Thus, at the end of the exposition of this rule, Newton asks the reader to accept an immense leap: universal gravitation among all parts of matter in the Universe. Alexandre Koyré (1968, p. 268), referring to this last claim, exclaims: "This is an affirmation of an incalculable scope."

Here Newton illustrates the use of his method by first describing the inductive inference to the generalization that all bodies are extended – which proposition, contrary to Descartes, is not taken to be an a priori assumption known by the pure intellect alone. Thus, in a way very congenial to Hume, Newton claims that we inductively infer that all bodies – observed and unobserved – are extended only on the basis of having uniformly observed that the bodies which are in the range of our senses are extended.

Moreover, Newton here explicitly contrasts the strength of the argument for universal gravitation with the case of the impenetrability of the heavenly bodies, for which, as Newton points out, nobody in his time has a single experiment or observation on which to ground an induction. Locke and Boyle – contrary to Descartes – assume that impenetrability is one of the essential (primary) qualities of all bodies. Indeed, for these empiricist mechanical philosophers, the property of impenetrability is the most fundamental grounding of what they – as mechanical philosophers and contrary to Newton – take to be the most intelligible form of causation in physical nature: motion by impact or impulse. Thus, one of the morals of Rule III is that the use of the inductive method is contrary to a procedure which begins from the hypothetical assumption that impenetrability is a primary quality essential to any piece of matter whatsoever. From a Newtonian perspective, laws that might govern the hypothetical impenetrable parts of celestial bodies could gain equal standing with the law of universal gravitation only if one could derive these laws from observations by means of his inductive method. And in Rule IV and its explanation, Newton prescribes that the conclusions of an inductive universalization from observations be regarded as true or nearly true until observed exceptions lead to their revision.[10] The laws inductively derived from phenomena are regarded as truly universal (or very nearly so) – and thus are taken to be exceptionless (or very nearly so) – until more observations lead to restrictions on their accuracy or scope. But no mere mechanical hypothesis (such as the vortex theory) can lead to such restrictions. Only uniform and constant manifest experience can lead to revisions of the inductively established laws of nature; and the goal of this entire process is to lead, eventually, to completely exceptionless universal laws where no further restrictions are necessary.

Newton distinguishes between the status of universal conclusions established by inductive proof and his own procedure of using experiments to show the probability of a conjecture or hypothesis. Newton regards propositions proved or "deduced from the phenomena" and "made general by induction" as having the "highest evidence that a proposition can have in this [experimental] philosophy." By contrast, Newton explicitly denies that his own hypotheses (or

10 See Newton (1999, p. 796): "In experimental philosophy, propositions gathered from phenomena by induction should be considered either exactly or very nearly true notwithstanding any contrary hypotheses, until yet other phenomena make such propositions either more exact or liable to exceptions. This rule should be followed so that arguments based on induction may not be nullified by hypotheses."

conjectures) have the attributes of inductive proofs. Newton makes this point, for example, in a letter to Cotes, March 1713:

> [A]s in geometry the word "hypothesis" is not taken in so large a sense as to include the axioms and postulates, so in experimental philosophy it is not to be taken in so large a sense as to include the first principles or axioms which I call the laws of motion. These principles are deduced from phenomena and made general by induction: which is the highest evidence that a proposition can have in this philosophy. And the word "hypothesis" is here used by me to signify only such a proposition as is not a phenomenon nor deduced from any phenomena but assumed or supposed without any experimental proof.[11]

At the end of a letter to Boyle, February 28, 1678–9, after proposing various conjectures about the ether, including one concerning the cause of gravity, Newton writes: "[B]ut by what has been said, you will easily discern, whether in these conjectures there be any degree of probability, which is all I aim at."[12]

Reading Newton's inductive Rules in combination with the above-quoted passages from the Scholium and some of his letters strongly suggests that Newton is explicitly targeting the mechanical philosophers in his formulation of the Rules. Newton's central target is the rationalist version of this philosophy – as defended by Descartes, Leibniz, and their followers – and he is most concerned to prevent their a priori demonstrative ideal from hindering or "nullifying" his own use of universalizing induction. However, the differences between Newton and Locke are more complicated and subtle than the differences between Newton and these rationalist philosophers, for the obvious reason that both Locke and Newton hold that observations and experiments are, in the end, all the evidence we have in the study of nature. Nevertheless, despite this general common ground, Locke remains wedded to central assumptions of the rationalist mechanical philosophers, and these prevent him from anticipating, in the *Essay*, the key ideas of Newton's inductive method as characterized in his Rules. Indeed, this should not be at all surprising, since the crucial Rule III, where Newton first explicitly emphasizes the tension between his method and the hypotheses of the mechanical philosophy, was first added to the second edition of the *Principia* in 1713, and Rule IV, where Newton completes this polemic by warning of the dangers of "nullifying" the inductive method by hypotheses, was only added to the third edition in 1726 – both long after Locke's death in 1704.[13]

11 See Newton (2004, p. 118). 12 See *ibid.* p. 11.

13 Alexandre Koyré, in 1968, Chapter VI, argues that a comparison of the manuscripts showing the changes throughout the three editions of the *Principia* illustrates the polemical character of the Rules directed against the continental rationalists, especially Descartes and Leibniz. Koyré argues, in particular, that Rule IV was added to the third edition precisely because the defenders of the mechanical philosophy still persisted in rejecting universal gravitation after the second edition.

The first important difference between Locke and Newton is that Locke is an advocate of the mechanical philosophy (Locke is likely to have in mind Boyle's empiricist version) – which he calls the "corpuscularean Hypothesis" – as providing the most intelligible explanation of the operations and qualities we observe in bodies. The most intelligible such explanation, common to *both* rationalist and empiricist mechanical philosophers, conceives all fundamental causal action as communication of motion by impact or impulse. Locke is very explicit about this in the first three editions of the *Essay*, for example (II, VIII, 11): "*Bodies operate* one upon another, and that is manifestly *by impulse*, and nothing else. It being impossible to conceive, that Body should operate on what it does not touch, (which is all one as to imagine it can operate where it is not) or when it does touch, operate any other way than by Motion."[14] Newton, by contrast, is not committed to the privileged intelligibility of the mechanical communication of motion by impact or impulse, and, in particular, he explicitly distinguishes the action of the (so far unknown) cause of gravity from that of all "mechanical causes."[15]

14 In the fourth edition (1700) of the *Essay* II, VIII, 11 and 12, Locke makes some limited changes in response to Newton's theory of gravitation. In Locke's reply to Stillingfleet's second letter – (Locke 1824, vol. III, pp. 467–468) – Locke has announced his intention to change those passages in the *Essay* which assert "that bodies operate by impulse, and nothing else." This is because he has been "convinced by the judicious Mr. Newton's incomparable book, that it is too bold a presumption to limit God's power, in this point, by my narrow conceptions." However, in this letter Locke still claims that he can conceive the operations of bodies one upon another in no other way but by impulse – this conception derives from our idea of body and what we know of matter. Thus, the letter continues: "The gravitation of matter towards matter, by ways inconceivable to me, is not only a demonstration that God can, if he pleases, put into bodies powers and ways of operation, above what can be derived from our idea of body, or can be explained by what we know of matter, but also an unquestionable and every where visible instance, that he has done so." The change affecting *Essay*, II, VIII, 11 in the fourth edition commits Locke solely to the view that the only way *we can conceive* bodies to operate is by impulse, and this is how bodies produce ideas in us – Locke now does not make any claim about the operation of bodies *themselves* upon one another. Similarly, the first sentence of *Essay* II, VIII, 12, which in the previous editions read: "If then Bodies cannot operate at a distance..." (Nidditch edition, Locke 1975, critical apparatus at the bottom of page 136), is simply eliminated. These changes are really very modest; in particular, the new sentences incorporated in *Essay* II, VIII, 11, are still consistent with the belief in the superior intelligibility of motion by impulse or impact. In addition, there is no reference to Newton's theory of gravitation in the *Essay* in any of its editions. Besides the letter to Stillingfleet, other writings in which Locke makes similar favorable pronouncements concerning Newton's theory of gravitation are Locke (1892) and Locke (1693/1989). For a different view of the significance of these favorable Lockean pronnouncements, see Downing (1998).

15 Thus, the famous "hypotheses non fingo" passage from the second edition General Scholium begins as follows (Newton 1999, p. 943): "Thus far I have explained the phenomena of the heavens and of our sea by the force of gravity, but I have not yet assigned a

There is a related aspect of the model of intelligibility shared by the rationalist mechanical philosophers and Locke which has significant methodological implications and thus marks a second important difference between Newton and Locke. This is the assumption that any proper causal explanation of the operations and qualities we observe in bodies reduces to a hidden configuration of the primary qualities of their "insensible Parts."[16] In particular, the microstructure of "insensible Corpuscles" (characterized by their primary qualities) underlying all observable phenomena is supposed to explain and necessitate the effects of bodies on one another and on ourselves (*Essay*, IV, III, 25): "These insensible Corpuscles, being the active parts of Matter, and the great Instruments of Nature, on which depend not only all their secondary Qualities, but also most of their natural Operations, our want of precise distinct Ideas of their primary Qualities, keeps us in an uncurable Ignorance of what we desire to know about them. I doubt not but if we could discover the Figure, Size, Texture, and Motion of the minute Constituent parts of any two Bodies, we should know without Trial several of their Operations one upon another, as we do now the Properties of a Square, or a Triangle."

To be able to penetrate into the exact configuration of the assumed primary qualities of bodies is the guiding methodological desideratum for achieving proper causal explanations, and thus what Locke calls "Knowledge" or "Science" of nature. To be sure, Locke, unlike the rationalists, emphasizes a skeptical gap (suggested here and further explained below) between what our faculties can actually perceive and the particular microstructural configuration of primary qualities underlying the phenomena. Nonetheless, for both the rationalist mechanical philosophers and for Locke, the ultimate causal explanations of what we observe reside in precisely this hypothetical hidden microstructure. By contrast, Newton, as we have seen, is especially concerned that the favored

cause to gravity. Indeed, this force arises from some cause that penetrates as far as the centers of the sun and planets without any diminution of its power to act, and that acts not in proportion to the quantity of the *surfaces* of the particles on which it acts (as mechanical causes are wont to do) but in proportion to the quantity of *solid* matter, and whose action is extended everywhere to immense distances, always decreasing as the squares of these distances." Moreover, when Newton entertains the possibility of explaining the action of gravity by the pressure exerted by an interplanetary ether in Query 21 of the *Opticks*, this pressure is not due to motion by impact (as in the vortex theories of Descartes, Huygens, and Leibniz) but to short-range repulsive forces acting at very small distances.

16 Locke gives several overlapping lists of primary qualities; perhaps the most complete list appears at *Essay*, II, VIII, 9: "These I call *original* or *primary Qualities* of Body, which I think we may observe to produce simple *Ideas* in us, *viz.* Solidity, Extension, Figure, Motion, or Rest, and Number." Locke then often uses "Bulk" to comprise solidity and extension together, whereas "Texture" appears to refer to the way in which the various "insensible Corpuscles," with their individual sizes and figures, are situated with respect to one another (so as, for example, to reflect light of various colors).

hypothetical causal explanations of the mechanical philosophy do not interfere with his use of the inductive method.

A third important difference between Locke and Newton concerns how they conceive of "primary" properties. Whereas Newton does talk of "primary," "original" or "simple" properties, these, for him, are discovered only by observations and experiments – as a product of his inductive method. In a groundbreaking analysis of the extent to which Locke differs from Newton concerning the methodology of science, Howard Stein argues that, unlike Locke, Newton does not take his primary and original properties as constituting an antecedently fixed list, prior to and independently of what experimental inductive investigation may then discover.[17] In particular, Newton uses the terminology of "primary," "original," or "simple" properties of light in the early parts of the *Opticks*, where these include, for example, the intrinsic degrees of refrangibility of differently colored rays revealed in his famous prism experiments. Thus, as Stein points out, Newton's conception of "primary" or "original" properties is always open to what experience can teach us by the application of his inductive method.[18]

In support of Stein's point, I should add, first, that whenever Locke gives lists of primary qualities in the *Essay*, he does not envision the possibility that we might modify this list in light of experience. Second, since Newton's Rules III and IV explicitly oppose the mechanical philosopher's method of starting with hypotheses that cannot be modified by experimental investigation, these Rules also imply the rejection of a hypothetical fixed list of primary properties in advance of what experience can teach us. Third, whereas Locke's primary–secondary quality distinction is associated with a skeptical gap between our perceptions and an underlying hidden reality that explains them, Newton's notion of "primary" or "original" properties is associated with no such gap: these, on the contrary, are continually made accessible to us by the inductive or experimental investigation of manifest phenomena.

The skeptical gap on Locke's account leads to a crucially important fourth methodological difference between himself and Newton: in spite of his skepticism, Locke retains the a priori ideal of knowledge of nature. As we have seen in our discussion of the second point of difference, Locke, at *Essay*, IV, III, 25, claims that *if* we could discover the hidden configuration of primary qualities, then we *would* have a priori demonstrative knowledge ("without Trial") of the operations of bodies. However, since we cannot in fact penetrate to this

17 See Stein (1990). More generally, in my discussion of the methodological differences between Newton and Locke, I am very much indebted to this article. However, I especially emphasize the points that illuminate the closely related methodological differences between Locke and Hume.

18 G. A. J. Rogers, by contrast, consistently takes the primary–secondary quality distinction as a central point of agreement between Newton and Locke. See, for example, (1982, p. 225).

hidden structure, we are left in "an incurable Ignorance." Locke's emphasis on this problem certainly distances him from the rationalist proponents of the mechanical philosophy. For Locke, however, we cannot achieve demonstrative knowledge or science of bodies precisely because we cannot acquire knowledge of what Locke assumes, together with the rationalists and independently of observation or experiment, to be the necessary, intrinsic connection between the primary qualities of bodies and their operations on other bodies and on ourselves. Because of precisely this ignorance, although we can empirically arrive at probable connections and regularities among the secondary qualities we observe, we cannot establish certain and undoubted rules governing them (see, for example, *Essay*, IV, III, 14). We can merely conventionally collect qualities under a general name – the nominal essence of a particular substance – guided by the observation of manifest qualities.[19] Thus, nominal essences allow us to sort individuals into species of things, but we will never know whether the observed qualities we compile under these nominal essences correspond with the real essences of things (see, for example, *Essay*, III, VI, 9).

Therefore, the general prescription to rely on experience and observation in the study of nature, which Locke undoubtedly shares with Newton and Hume, does not lead to a rejection of the explanatory ideal of the mechanical philosophy. On the contrary, it is only by fulfilling this ideal, for Locke, that we could ever attain true "Knowledge" or "Science," of nature; and, in "experimental philosophy," we must instead be content with what Locke calls "Judgment" and "Opinion" (see, for example, *Essay*, IV, III, 26).[20] In sum, Locke's skepticism about the possibility of a genuine "Science" of nature depends on central tenets of the mechanical philosophy, and his view of what experimental inquiry can achieve (mere "Judgment" and "Opinion") is a consequence of his demonstrative ideal of "Knowledge" and "Certainty."[21]

19 For Locke's conventionalism about nominal essences of different kinds of substances, thus about natural kinds, see *Essay*, III, VI.

20 The certain knowledge that for Locke we can in fact attain includes (necessary) intuitive knowledge, (necessary) demonstrative reasoning (as in Descartes, mathematics, not logic, is the paradigm of both intuitive and demonstrative knowledge), knowledge of the existence of God, and sensitive knowledge of the real existence of things without the mind. However, it does not include the specific nature of existing physical things. See, for example, *Essay*, Book IV, Chapters II–IV, VI, and IX–XI.

21 Edwin McCann – for example, in (1983) – has argued against the interpretation I follow in this paper, according to which Locke adopts a geometrical, deductive model of the powers and operations of bodies. On McCann's view, when Locke affirms, for example in passages such as *Essay*, IV, III, 28, that the "mechanical affection" (primary qualities) of bodies have "no affinity at all" with the sensations of secondary qualities produced by them in our minds, and thus "we can have no distinct knowledge of such Operations beyond our Experience; and can reason no otherwise about them, than as effects produced by the appointment of an infinitely Wise Agent, which perfectly surpass our Comprehensions," Locke means that there are contingent, divinely established laws of nature, which do not

We thereby finally arrive at a fifth important difference between Locke and Newton: unlike Newton's conception of his own inductive method, probable opinion in Locke can never result in a truly universal exceptionless law. Locke does not anticipate an experimental method leading to the formulation of inductively established, exceptionless universal laws of the kind envisioned in Newton's Rules III and IV; and he does not arrive, in particular, at the idea that such an inductive method can replace the demonstrative ideal of the mechanical philosophy with an alternative ideal of scientific knowledge. In Locke, experience is merely a source for the modification and revision of what we provisionally regard as nominal essences, but there is no way, as we have seen, that we could ever know whether such nominal essences correspond to – or even approximate – the truly necessary connections determined by the real essences. In devising conventional nominal essences of particular substances, we rank things by using general names in order to class individuals together into species or kinds in accordance with our observations and experiments. However, we could never attain either certainty or necessity or knowable exceptionless universality in this way. For example, the regularities we have observed in making general claims about gold – (we have observed that gold, unlike iron, has always been malleable) – have no knowable connection with the truly exceptionless universality we could obtain only by a knowledge of gold's real essence.

Thus, even if we interpret Locke's nominal essences as generalizations resulting from induction (since the formulation of nominal essences depends on repeated observations), these generalizations could never amount to what Newton calls inductive or experimental "proofs" of truly universal exceptionless laws of nature. Genuine exceptionless universality, for Locke, could only result from certain and demonstrative knowledge, which, in the case of bodies, is for us unattainable. For Locke, therefore, corresponding to the unbridgeable skeptical gap between primary and secondary qualities, real and nominal essences, there is a parallel unbridgeable gap between the regularities we actually observe and the truly universal, absolutely certain laws which must demonstratively flow from the real essences:

follow with necessity from the mechanical nature of bodies. On this view, Locke is not a pure mechanist: the connection of observable qualities and powers with the microstructure or real essence of bodies is not strictly a priori, but contingently established by the arbitrary power of God. In order to have Knowledge of bodies, in addition to knowing their real essence (which we do not know), we need to know the contingent general connections ordained by God, which can only be cognized experimentally. Michael Ayers, in (1991), vol. II, chapter 12, argues against McCann's interpretation (and other more limited voluntarist interpretations, such as Leibniz's and Margaret Wilson's). In this controversy I side with Michael Ayers: not only do I take Locke to endorse the demonstrative ideal of the knowledge of nature, but I also agree with Ayers that Locke's pronouncements about unknown conjunctions possibly established by God are skeptical *epistemological* claims of possibility, not *ontological* claims regarding the contingent character of the laws of nature.

> The more, indeed, of these co-existing Qualities we unite into one complex *Idea*, under one name, the more precise and determinate we make the signification of that Word; But yet never make it thereby more capable of *universal Certainty*, in respect of other Qualities, not contained in our complex *Idea*; since we perceive not their connexion, or dependence one on another; being ignorant both of that real Constitution in which they are all founded; and also how they flow from it . . . Could any one discover a necessary connexion between *Malleableness*, and the *Colour* or *Weight* of *Gold*, or any other part of the complex *Idea* signified by that Name, he might make a *certain* universal Proposition concerning *Gold* in this respect; and the real Truth of this Proposition, That *all Gold is malleable*, would be as *certain* as of this, *The three Angles of all right-lined Triangles, are equal to two right ones.*
>
> (*Essay*, IV, VI, 10)[22]

Thus, Locke identifies truly universal laws of nature with absolutely necessary and demonstratively certain laws, grounded in the (forever unknowable) real constitution or essence of bodies. Because of this conception of the universality and necessity of the laws of nature, Locke's empirical scientific methodology is left with an exclusive emphasis on probable opinion concerning the differences among particular substances, such as the observable differences in the sensible qualities of gold and iron. Locke nowhere envisions the third category of inductive or experimental "proofs" in Newton's sense.

As we have seen, Newton, in his remarks on Rule III, explicitly warns against the danger of allowing mechanical hypotheses to interfere with what he takes to be an inductively established universal law – the law of universal gravitation – and Rule IV goes on to emphasize that restrictions in the accuracy or scope of

22 In a similar passage at *Essay*, IV, III, 25, which I partially quoted before, Locke affirms again the impossibility of deriving truly universalizing conclusions from observations, and suggests that knowledge of universal laws (if *per impossibile* we could have it) would be demonstrative knowledge as in geometry: "If a great, nay far the greatest part of the several ranks of *Bodies* in the Universe, scape our notice by their remoteness, there are others that are no less concealed from us by their *Minuteness*. These insensible Corpuscles, being the active parts of Matter, and the great Instruments of Nature, on which depend not only all their secondary Qualities, but also most of their natural Operations, our want of precise distinct *Ideas* of their primary Qualities, keeps us in an uncurable Ignorance of what we desire to know about them. I doubt not but if we could discover the Figure, Size, Texture, and Motion of the minute Constituent parts of any two Bodies, we should know without Trial several of their Operations one upon another, as we do now the Properties of a Square, or a Triangle . . . But whilst we are destitute of Senses acute enough, to discover the minute Particles of Bodies, and to give us *Ideas* of their mechanical Affections, we must be content to be ignorant of their properties and ways of Operation; nor can we be assured about them any further than some few Tryals we make, are able to reach. But whether they will succeed again another time, we cannot be certain. This hinders our certain Knowledge of universal Truths concerning natural Bodies; and our Reason carries us herein very little beyond particular matter of Fact."

such laws can only be grounded in further inductive evidence. Proceeding in this way, we can successively correct for any exceptions that may inductively be found, so as eventually to approximate closer and closer to a truly universal and exceptionless inductive generalization. Generalizations grounded by this method have the "highest evidence that a proposition can have in this [experimental] philosophy," with which no corpuscularean hypothesis or conjecture can possibly compete. Newton's conception of experimental philosophy, unlike Locke's, is fashioned in explicit opposition to the demonstrative ideal of the mechanical philosophy, and his conception of inductive generalization, in particular, is intended to replace this ideal with a contrary purely inductive ideal. This is precisely the crucial step that Locke never takes and which, as we shall see, is taken by Hume.

10.2 Hume's Newtonian ideal of causation and induction

Newtonian inductivism inspires Hume's own positive account of causation and conception of scientific methodology – both in opposition to the mechanical philosophy of nature which Hume has inherited. As we have seen, in the received view of causation, shared by both Descartes and Locke, nature or reality has an ultimate or intrinsic constitution of primary qualities which underlies the causal relations we can observe – the view of causation is thus intimately related to a view of the necessity in nature. For Locke, in particular, a material necessity independent of both our minds and our available empirical methods explains causal connections and resides in the intrinsic causal powers of bodies – powers with which the primary qualities of substances are endowed. Locke shares the mechanical philosophy's understanding of how a microstructure of primary qualities in one body can necessitate effects in another: the motions of the microscopic parts of one body are communicated by impact to the microscopic parts of another.

This view of causation would explain the causal nexus in a *single case* – if we could penetrate into the intrinsic hidden microstructure of the bodies involved. In my view, Hume's argument against the very notion of singular causation is the first instance where we can appreciate Hume's moves against the mechanical philosopher's conception of efficient causation. Indeed, it is precisely in the context of arguing against singular causation that Hume famously argues against the demonstrative derivation of effects from causes, and this argument is also explicitly directed against the received view that we have grounds for claiming that there really is, independently of our observation of causes and effects, a necessary connection between them. In advancing objections to taking a singular observation of a relation between objects as causal, Hume is not merely preparing the ground to claim, following the model of Newtonian induction, that the central ingredient in our idea of causation is constant conjunction, that is, uniform experience of like causes followed by like effects.[23] He is raising, in addition, several interconnected skeptical doubts regarding the mechanical philosophy's model of causal explanation: the containment of the effect within the cause, the ideal of demonstrative knowledge of causation, and the reality of a necessary connection between cause and effect – all dependent on the postulation of a hidden configuration of primary properties.

Contrary to the mechanical philosopher's conception of the containment of the effect in the cause, Hume argues that the ideas of cause and effect are *distinct* ideas, entirely separable and thus independent from one another (T 1.3.3.3): "[A]ll distinct ideas are separable from each other, and as the ideas of cause and effect are evidently distinct, 'twill be easy for us to conceive any object to be non-existent this moment, and existent the next, without conjoining to it the distinct idea of a cause or productive principle."[24]

Hume also makes it explicit that his rejection of the conception of the containment of the effect in the cause goes hand in hand with his rejection of an a priori, demonstrative model of causal explanation:

> 'Tis easy to observe, that in tracing this relation, the inference we draw from cause and effect, is not deriv'd merely from a survey of these particular objects, and from such a penetration into their essences as may discover the dependence of the one upon the other. There is no object, which implies the existence of any other if we consider these objects in themselves, and never look back beyond the ideas which we form of them. Such an inference wou'd amount to [demonstrative] knowledge, and wou'd imply the absolute contradiction and impossibility of conceiving any thing different. But as all distinct ideas are separable, 'tis evident there can be no impossibility of that kind. When we pass from a present impression to the idea of any object, we might possibly have separated the idea from the impression, and have substituted any other idea in its room.
>
> (T 1.3.6.1)

To "consider these objects in themselves" is to consider only the meager evidence of impressions of sensation or objects before the mind. Any further ideas that we take to have been inferred from these direct presentations might have been erroneously inferred, for there is no intrinsic connection in terms of the content of distinct presentations before the mind such as those we regard as a cause and an effect. It follows from this that it is intelligible to regard anything as the cause of anything else; and the reference to the postulation of an essence, in particular, makes it clear that Hume is here targeting the mechanical philosophers, including Locke.

23 Hume adds constant conjunction at T 1.3.6 and EHU 4, Part II, after he has argued in both works against singular causation.

24 See also T 1.3.6.1; EHU 4.6; and so on.

By emphasizing that all we observe are distinct, separable events, Hume implies that the only clue to any legitimate postulation at the microscopic level is provided by the macroscopic observation of separate events. Any connection that we would find at the microscopic level, if we were to advance by means of observation to it, would thus be an inductively derived connection, not an intelligible necessary connection of the intrinsic content of the (ideas of) cause and effect. Hume is not precluding the attempt to advance by the inductive method into the microscopic level. Yet, because our only guide is what we observe, we can only inductively generalize from the observed to the unobserved, and thus claim that if we could penetrate into the microscopic level, we would still only observe separate, distinct events, just as we do at the macroscopic level. The postulation of a hidden microstructure prior to what we can observe not only interferes with the inductive method, but it is also entirely idle, since claims about the unobserved microscopic level can only be inductively inferred from regular and constant experience at the macroscopic level. In particular, the most we can claim concerning laws of impact is that the same inductively inferred laws holding at the macroscopic level between distinct and separable events hold for the not-yet-observed microscopic level as well.

In *Enquiry*, Section 6, Hume places the laws of impact and gravitation (and common-sense generalizations based on constant conjunctions) on an entirely equal footing with respect to their legitimacy and intelligibility – which, in all cases, is based on the inductive method and thus the observation of constant conjunction. Unlike Locke, in the *Essay*, who explicitly claims that the only intelligible explanation of motion is by impulse and suggests that Newton's *Principia* is a work in pure mathematics,[25] Hume unreservedly accepts universal gravitation as a law of nature, and takes Newton's theory to articulate a fundamental law of nature on a par with all other inductively established laws (EHU 6.4): "There are some causes, which are entirely uniform and constant in producing a particular effect; and no instance has ever yet been found of any failure or irregularity in their operation. Fire has always burned, and water suffocated every human creature: The production of motion by impulse and gravity is an universal law, which has hitherto admitted of no exception." Notice the qualification, in accordance with Newton's Rules III and IV, that these laws have *hitherto* been observed to be exceptionless, thus the suggestion that they are open to revision by experience.

Hume takes the laws or principles of elasticity, gravity, cohesion of parts, and communication of motion by impulse to be completely equivalent with respect to legitimacy and intelligibility, for, in all these cases, we have inductively discovered laws of nature arising from the observation of constant conjunction.

25 At *Essay*, IV, VII, 3, Locke writes: "Mr. *Newton*, in his never enough to be admired Book, has demonstrated several Propositions, which are so many new Truths, before unknown to the World, and are farther Advances in Mathematical Knowledge."

And again, the power or necessary connection, which hypothetically might be taken to be involved in action by contact, is as unintelligible as gravitational action at a distance (EHU 7.25): "We surely comprehend as little the operations of one [the Supreme Being] as of the other [the grossest matter]. Is it more difficult to conceive, that motion may arise from impulse, than that it may arise from volition? All we know is our profound ignorance in both cases." In the immediately preceding paragraphs, Hume has argued against occasionalism, and, before addressing occasionalism, Hume has also argued against Locke's view that we acquire the idea of power from the actions of the mind. A footnote to the last quoted words then stresses the equal unintelligibility attending the idea of a power that allegedly operates in inertia, motion by impact and gravitational action at a distance: we have to limit our claims to "facts," that is, to observed constant conjunctions and the inductively inferred conclusions from such observations. As Newton's Rule III prescribes, all we can rely on are observations, and if the observations are sufficient in number, uniform and constant, we can generalize by induction to unobserved cases of the same kind.

At T 1.3.6, in the course of answering a question he has posed at T 1.3.2.15 ("Why we conclude, that such particular causes must *necessarily* have such particular effects; and what is the nature of that *inference* we draw from the one to the other, and of the *belief* we repose in it?"), Hume adds as an essential ingredient of the notion of causation the observation of constant conjunction – that like objects have been always placed in like relations. This experience amounts to the observation of uniformities of the kind Newton illustrates, as we have seen, in the explanation of Rule III. But how and why can we generalize from what we have hitherto uniformly observed to the unobserved? Hume asks (T 1.3.6.4): "[W]hether we are determin'd by reason to make the transition [from the experience of the constant conjunction of events of a first kind with events of a second kind to unobserved events of either kind, or to universal laws of nature] or by a certain association and relation of perceptions." In view of the previous contrasts Hume has drawn between natural and philosophical relations (at T 1.1.4–5), and between knowledge and probability (at T 1.3.1–2), Hume's question here can be taken to be whether we are determined to draw an inference by natural principles of association of the mind (by natural relations), or, instead, we base the causal inference on philosophical relations, where the latter, in my view, involve reflective comparisons that yield legitimizing reasons.[26] Hume then writes (T 1.3.6.4): "If reason determin'd us, it wou'd

26 In De Pierris (2002), I discuss Hume's distinction between two kinds of philosophical relations at *Treatise* 1.3.1–2 – those established either by intuitive and demonstrative reasons, or by experience itself – and I render it in terms of two methods for justifying claims about relations: either solely on the basis of comparisons of intrinsic features of the relata, or on the basis of their extrinsic relations. I shall leave for another occasion my defense of the view that in asking whether we are determined by reasoning to draw causal

proceed upon that principle, *that instances, of which we have had no experience, must resemble those, of which we have had experience, and that the course of nature continues always uniformly the same.*"

As we have seen, in Rules III and IV Newton formulates the basic idea of his inductive scientific methodology, and the explanation of Rule III includes the guiding principle of the inductive method that nature is always consonant with itself. In my view, the principle "*that instances, of which we have had no experience, must resemble those, of which we have had experience, and that the course of nature continues always uniformly the same*" in T 1.3.6 is Hume's own version of the Newtonian principle, which licenses the universalization of completely uniform experience into exceptionless laws. As Alexandre Koyré emphasizes, for Newton this methodological principle has a *justificatory* role, since it is precisely on its basis that we are licensed to go beyond the data of the senses and attribute to all bodies whatsoever – observed and unobserved – the qualities we have observed so far (1968, p. 267): "And it is just because nature is consonant to herself that we can generalize the data of experience and attribute to *all* bodies the properties that experience shows us in those which are within our reach."

Although Hume does not explicitly refer to Newton's Rule III in his epistemological writings, he does explicitly cite it in the *Enquiry concerning the Principles of Morals*, in support of his empirical generalization regarding how we are determined to approve of the social virtues. Hume suggests that because the principle of usefulness or utility has been found to have a great force or energy as the (sole) source of the moral approbation paid to the (more artificial) social virtue of justice, it must have a considerable force or energy in the case of such (more natural) social virtues as humanity, benevolence, and so on:

> The necessity of justice to the support of society is the SOLE foundation of that virtue; and since no moral excellence is more highly esteemed, we may conclude that this circumstance of usefulness has, in general, the strongest energy, and most entire command over our sentiments. It must, therefore, be the source of a considerable part of the merit ascribed to humanity, benevolence, friendship, public spirit, and other social virtues of that stamp . . . It is entirely agreeable to the rules of philosophy, and even of common reason; where any principle has been found to have a great force and energy in one instance, to ascribe to it a like energy in all similar instances. This indeed is Newton's chief rule of philosophizing[1].
>
> (EPM 3.48)

The footnote reads: "Principia, Lib. III."

inferences, Hume raises the question of whether we have legitimizing or justificatory reasons (either a priori or a posteriori) for causal inductive inferences to the unobserved. For the view that at T 1.3.6 Hume does not raise skeptical doubts about induction see Garrett (1997, chapter 3).

In order to understand the full import of this example, it is important to see that there are, both in Newton and in Hume, two ways of applying Newton's methodological principle that nature is consonant with itself. In the first place, this principle licenses the inductive inferences from particular constant conjunctions of the same type to unobserved cases of, or universal generalizations over, these same types. The formulation of Newton's Rule III articulates this kind of straightforward induction, and, in the explanation of this Rule, Newton mentions the principle that nature is consonant with itself as guiding such inferences. He illustrates how such a principle guides the inferences from the experience that particular bodies or parts of bodies are extended to the generalization that all bodies are extended, from the experience that particular bodies are impenetrable to the generalization that all are impenetrable, and the same inductive procedure is followed to conclude that all bodies are movable and endowed with the power of inertia. However, in his most important application of Rule III, Newton shows that induction and its justificatory principle of the uniformity of nature have an even more far-reaching application. At a second stage, a higher level of generalization can proceed by the same methodological rule: different generalizations obtained by induction and the principle of the uniformity of nature in different realms can now be unified with one another under an even more comprehensive generalization. Thus, in the explanation of Rule III, Newton illustrates this second type of application of induction and the principle of the uniformity of nature with his own inference to the law of universal gravitation. The law of gravitation holds in the realm of objects close to the Earth; it holds in the different realm of the relation between the planets and the Sun; it again holds in the different realm of the interaction between the Moon and the sea, and so on. The law of universal gravitation is generalized to all bodies whatsoever starting from these more restricted lower-level generalizations.

In this way, the principle of gravity, which initially has explanatory power in one realm (bodies close to the Earth) is shown also to have explanatory power in another realm (the motions of the planets). Inductive generalization and the assumption of the uniformity of nature unify all these lower-level generalizations under a higher-level generalization. Similarly, in Hume, the principle of public interest and utility is inductively shown to have "sole" explanatory power concerning the (more artificial) social virtue of justice. Then, in a second stage, public interest and utility can be further generalized so as to explain a "considerable part" of the (more natural) social virtues of humanity, benevolence, and so on. The result is a higher-level generalization that unifies all of the virtues under a single law of public interest and utility. This unification can be achieved precisely because of a second-level assumption that human nature is uniform. Hume thus takes Newton's Rule III as a model for his own inductive investigation of human nature, and he thereby models the moral sciences on the Newtonian method in the natural sciences.

This general point is further substantiated by Hume's own "rules by which to judge of causes and effects" presented at T 1.3.15. The first three rules comprise the definition of cause in terms of constant conjunction (uniform experience). Rule four states that we can turn into a rule an operation that we naturally follow, namely, that when we have (inductively) discovered causes and effects on the basis of uniform experience, we usually extend our observation to every phenomenon of the same kind. The fifth rule also depends, as the previous ones, on the assumption that we establish causes and effects on the basis of uniform experience: when several different objects produce the same effect it must be by the same (more properly, by a resembling) quality present in all of the causes. And the sixth rule is parasitic on the fifth: the difference in the effects of two resembling causes must proceed from the particular qualities on which they differ (T 1.3.15.8): "For as like causes always produce like effects, when in any instance we find our expectation to be disappointed, we must conclude that this irregularity proceeds from some difference in the causes." Rule six thereby registers *irregularities*, but prescribes what reasoning might lead us to refine them in the direction of perfect uniformity. It is precisely by assuming the uniformity of nature – that like causes produce like effects – that we can then undertake this process of refinement.

In rule seven, however, the crucial Newtonian background stands out even more clearly. At issue are compound causes of "compounded" effects, and Hume warns against misapplications of this rule with an example from the moral sciences.[27] Hume illustrates exactly what he has in mind in his earlier discussion of "the probability of causes":

> We may establish it as a certain maxim, that in all moral as well as natural phaenomena, wherever any cause consists of a number of parts, and the effect encreases or diminishes, according to the variation of that number, the effect, properly speaking, is a compounded one, and arises from the union of the several effects, that proceed from each part of the cause. Thus because the gravity of a body encreases or diminishes by the encrease or diminution of its parts, we conclude that each part contains this quality

27 See T 1.3.15.9: "When any object encreases or diminishes with the encrease or diminution of its cause, 'tis to be regarded as a compounded effect, deriv'd from the union of the several different effects, which arise from several different parts of the cause. The absence or presence of one part of the cause is here suppos'd to be always attended with the absence or presence of a proportionable part of the effect. This constant conjunction sufficiently proves, that the one part is the cause of the other. We must, however, beware not to draw such a conclusion from a few experiments. A certain degree of heat gives pleasure; if you diminish that heat, the pleasure diminishes; but it does not follow, that if you augment it beyond a certain degree, the pleasure will likewise augment; for we find that it degenerates into pain."

> and contributes to the gravity of the whole. The absence or presence of a part of the cause is attended with that of a proportionable part of the effect. This connexion or constant conjunction sufficiently proves the one part to be the cause of the other. As the belief, which we have of any event, encreases or diminishes according to the number of chances and past experiments, 'tis to be consider'd as a compounded effect, of which each part arises from a proportionable number of chances and experiments.
>
> (T 1.3.12.16)

Hume here has in mind the conclusion – central to the law of universal gravitation – that each part of a gravitating body (like the Earth) also gravitates (so that the gravity of the whole is the sum of the gravities of the individual parts), and Hume proceeds to apply this Newtonian model, once again, to an example from the moral sciences (T.1.3.12.16): "As the belief, which we have of any event, encreases or diminishes according to the number of chances and past experiments, 'tis to be consider'd as a compounded effect, of which each part arises from a proportionable number of chances and experiments."

Newton establishes the property of gravity in question in Propositions 6 and 7 of Book III of the *Principia*, which crucially depend on Rule III.[28] We first show, by experiment, that all bodies falling toward the Earth are attracted by gravity in proportion to their quantity of matter. We then observe, by the equality of action and reaction, that all these bodies must attract the Earth as well, and we conclude, by Rule III, that the latter attraction (for which we do not yet have experiments) must also take place in proportion to the Earth's quantity of matter. Therefore, the Earth's gravity arises from, and is compounded out of, the individual gravitational attractions of its parts.[29] It

28 I am especially indebted to Michael Friedman here.

29 Cotes's summary in his Preface to the second edition explains this point especially clearly (Newton 1999, p. 387): "The attractive forces of bodies, at equal distances, are as the quantities of matter in these bodies. For, since bodies gravitate toward the earth, and the earth in turn gravitates toward each body, with equal moments, the weight of the earth toward each body, or the force by which the body attracts the earth, will be equal to the weight of the body toward the earth. But, as mentioned above, this weight is as the quantity of matter in the body, and so the force by which each body attracts the earth, or the absolute force of the body, will be as its quantity of matter. Therefore the attractive force of entire bodies arises and is compounded from the attractive force of the parts, since (as has been shown), when the amount of matter is increased or diminished, its force is proportionally increased or diminished. Therefore the action of the earth must result from the combined actions of its parts; hence all terrestrial bodies must attract one another by absolute forces that are proportional to the attracting matter."

therefore appears overwhelmingly likely that Hume not only has Newton's argument for universal gravitation clearly in mind in the *Treatise*, but he also takes Newton's Rule III as the fundamental guide for his own inductive investigations of the moral sciences there – just as he explicitly does so later in the second *Enquiry*.[30]

Yet Hume's own positive notion of causation modelled on Newton's inductive method is not limited to the justificatory role of the principle of the uniformity of nature in inductive inference. That we must not assume anything before (or beyond) experience, that the premises of our causal inductive inferences consist of uniform experience, and the universalizing character of the laws resulting from such inferences to the unobserved are also essential ingredients of Newton's inductive method. As I have shown, Locke did not anticipate the import of Newton's Rules III and IV, and he certainly did not envision replacing the ideal of demonstrative knowledge of nature by the Newtonian inductive method. In Locke, there is no serious consideration of the merits or limits of a principle of induction to ground genuinely universal laws – a principle which legitimizes the formulation of universal exceptionless causal laws concerning all corporeal substances, or even all corporeal substances of a particular kind, beyond the narrow scope of our experiments. What might be taken to be inductive generalizations in the formulation of nominal essences of substances need not correspond with the ultimate metaphysical explanation of causal relations involving these substances, which lies at the level of their hidden microstructure.

According to Hume, as we have seen, the mechanical philosophy's a priori model of the causal relation in terms of the intrinsic necessary structure of substances is not a correct but unattainable ideal model; rather, it provides a completely misguided model of the causal relation. This ideal is entirely misplaced when applied to matters of fact, not because there is an unknowable inner microstructure of necessary connections that explains the regularities we observe, but rather because the very idea of necessary connection as an intrinsic quasi-geometrical containment is itself completely incorrect. As Hume's characterization of causation (as a philosophical relation) at T 1.1.4–5 revealingly puts it, causation does not concern relations that we can ascertain by merely

30 Newton himself makes the essential role of Rule III perfectly explicit in the course of Propositions 6 and 7. According to Corollary 2 to Proposition 6 (Newton 1999, p. 809): "All bodies universally that are on or near the earth are heavy [or gravitate] toward the earth, and the weights of all bodies that are equally distant from the center of the earth are as the quantities of matter in them. This is a quality of bodies on which experiments can be performed and therefore by rule 3 is to be affirmed of all bodies universally." Corollary 1 of Proposition 7 then concludes (p. 811): "Therefore the gravity toward the whole planet arises from and is compounded of the gravity toward the individual parts."

inspecting the intrinsic features of presentations before the mind. Instead, causation concerns *extrinsic* relations among separate, distinct presentations before the mind (most notably, the relation of constant conjunction), so that there can be no relation of containment between the two distinct events we call cause and effect.[31]

However, from his radical skeptical standpoint, at the last stage of his skeptical argument concerning causation, necessity, and induction (in both the *Treatise* and the *Enquiry*), Hume argues against the possibility of grounding his own positive notion of causation in terms of constant conjunction and inductive inference. In particular, he focuses on the very principle of the uniformity of nature presupposed by such inferences. Hume thus argues, in particular, that there is neither an a priori nor an inductive justification of the general principle guiding Newton's inductive method. Therefore, from Hume's radical skeptical standpoint, the Newtonian inductive leap from what we have hitherto observed to an exceptionless generalization including the unobserved is also ultimately ungrounded. Hume's inquiry into the justification of the guiding methodological principle of Newtonian induction shows, in my view, that at T 1.3.6 and EHU 4.2 Hume targets what he himself takes to be the best possible form of causal inference. Nevertheless, the confident use of this method within the natural standpoint of science and common life is not affected by his own unsustainable radical skeptical argument. That Hume's skepticism about the best possible inductive method is directed at his own Newtonian model of scientific inference sharply brings out the mutual autonomy of Hume's two standpoints.[32]

31 See footnote 26 above. As I explain in the paper cited there, Hume's distinction between two types of philosophical relations (roughly, intrinsic and extrinsic) is essentially the same distinction he later makes in the *Enquiry* between relations of ideas and matters of fact. From Hume's point of view, therefore, to conceive of causation in terms of quasi-geometrical containment goes against the very distinction between relations of ideas and matters of fact.

32 I have developed this theme of the mutual autonomy of Hume's two standpoints in detail in my (2001), and also in my (2002). In the latter article I first argued, in particular, that Hume fully endorses, outside his radically skeptical standpoint, the normativity of inductive proof with its attendant principle of the uniformity of nature, which he models on Newton's Rule III. I also argued, in this way, against Annette Baier, who takes Hume's negative conclusion concerning the ultimate justification of the principle of the uniformity of nature as an argument against rationalist, deductivist attempts at grounding the inductive inference. In my view, on the contrary, precisely because Hume argues at T 1.3.6 that there is no inductive justification of the principle of induction, Hume raises radical skeptical doubts about his own Newtonian inductivist model of scientific inference. Nevertheless, according to Hume, within the natural standpoint (as opposed to the radically skeptical standpoint), our best inductive method (following Newton's Rule III) enables us to formulate "well-established" and exceptionless (albeit revisable) causal laws of nature.

There is, therefore, a fundamental asymmetry between Hume's skepticism concerning the a priori model of causal explanation of the mechanical philosophy and his skepticism concerning Newton's (and his own) inductive method. Within the naturalistic standpoint of science and common life, Hume emphatically endorses the inductive method but not the a priori model of causal explanation. Inductive inference, for Hume, even before it is refined into the ideal Newtonian inductive method by a reflection compatible with common life and science, follows natural principles of association of the mind and relies only on experience. But the a priori reasoning of the mechanical philosophers, involving the postulation of a hidden configuration of primary qualities and powers demonstratively necessitating effects, does not follow such natural operations. This asymmetry also reveals itself in the character of Hume's positive notion of necessity, within the standpoint of common life and science. Hume explains (but does not ultimately justify) such a positive notion of necessity as a projection from our inductive inferences,[33] whereas the necessity of hidden powers postulated by the mechanical philosophy has no analog, for Hume, in any natural operations of the mind. This inherited notion of necessity is simply rejected once and for all – from *both* standpoints.

33 Hume writes (T 1.3.14.20): "Tho' the several resembling instances, which give rise to the idea of power, have no influence on each other, and can never produce any new quality *in the object*, which can be the model of that idea, yet the *observation* of this resemblance produces a new impression *in the mind*, which is its real model. For after we have observed the resemblance in a sufficient number of instances, we immediately feel a determination of the mind to pass from one object to its usual attendant, and to conceive it in a stronger light upon account of that relation . . . Necessity, then, is the effect of this observation, and is nothing but an internal impression of the mind, or a determination to carry our thoughts from one object to another."

11

Maupertuis on attraction as an inherent property of matter

LISA DOWNING

11.1 Introduction

I begin with a caveat. This paper examines Maupertuis from the very particular perspective of two issues that come together in his *Discours*: (1) The history of regularity-based defenses of Newtonian gravity, that is, the tradition, originating in Newton himself and prominent in the early eighteenth century, of defending the law of gravity as a mere regularity, not requiring any account of underlying causes.[1] Maupertuis, in the *Discours*, is one of the most influential exemplars of this tradition, but he also steps firmly beyond it, as will be detailed in what follows. (2) Lockean Newtonianism, that is, the uses of Locke's thought, especially his skepticism about knowledge of real essences, as a resource for the defense of Newton. Such Lockean themes are prominent too in the *Discours*.[2] In this paper I examine the *Discours* from both of these perspectives, hoping to illuminate both the text itself and a delicate episode in the history of philosophy of science.

First, however, I examine an important point of comparison: Willem Jacob Van s'Gravesande's 1720 text, *Physices elementa mathematica, experimentis confirmata*. I should, therefore, say a bit about the special importance of these *two* texts in the context of the defense of Newton. Newton appeared to require defense because Newtonian gravity was widely perceived to face a problem:[3] it

1 Of course, the question of Newton's considered view of the status and source of gravity is a tortured one, and even the seemingly simpler question of his position in the *Principia* is subtle, as has been emphasized recently in the work of Stein, Smith, and Janiak (among others). There is no doubt, however, that Newton's own words inspired a regularity defense. Notably, Newton states that he treats forces, including attraction, mathematically, not physically (Newton 1999, pp. 408, 589), suggests that this leaves questions about the causes of attraction open (589), and disavows hypotheses about the cause of gravity (943, this in the General Scholium, where he is responding to criticism).

2 In "Locke's Newtonianism and Lockean Newtonianism" (Downing 1997) I discuss some of this same material towards the end of better articulating Locke's complex attitude towards the natural philosophy of his day.

3 Indeed, it was perceived to face multiple related problems.

could not be straightforwardly linked to an acceptable metaphysics, for treating attraction as a genuine physical quality violated standard doctrines about the passivity of body. One prominent strategy for defending Newton took its initial cue from the Newton of the *Principia*, who eschewed speculation as to the cause of gravity, and thus apparently treated gravity as a manifest effect and avoided linking his dynamics to metaphysics.[4]

This defense of Newton, however, left many readers unsatisfied. Firstly, since many passages in the *Principia* seem to describe gravity as an attractive force, intrinsic to bodies, allowing them to affect distant bodies, it is natural to take such talk literally, lacking any developed alternative interpretation.[5] Furthermore, as Cotes pointed out to Newton himself in correspondence, Newton's application of his third law to gravitational attraction between distant bodies (Book III, prop. V, cor. 1) appears to require that bodies be able to act directly on one another at a distance.[6] Secondly, Newton's evasive tactics raise pointedly the question of whether scientific explanation can be had in the absence of causal explanation. Both these issues were, of course, highlighted by Newton's critics, who argued, in effect, that Newton was attributing gravity to bodies as an (unintelligible) intrinsic attractive power, or he was proposing a perpetual miracle, or he was offering up a radically incomplete theory.[7]

A number of early Newtonians saw that a way out of this impasse might be provided by the following basic strategy: Treat physics and metaphysics as separate domains, the former concerned with regularities among the phenomena, the latter with underlying causes. Reconceive scientific explanation accordingly: if an occurrence can be deduced from a more general principle or system of principles, it has been given a scientific explanation. This strategy begins

4 Again, there are many issues of Newton interpretation lurking here. One question is whether Newton is consistent in this position, even within the *Principia*. (The General Scholium, for example, seems to verge on speculation about the cause of gravity.) A further question is what it is to treat gravity as a manifest effect. Stein and Janiak have in effect argued that this doesn't rule out some modest ontology: that for Newton gravity is a natural power of bodies (Stein 2002), that gravity is itself a cause, though one not yet physically characterized (Janiak 2008).

5 See Heilbron (1982, p. 40) and McMullin (1978, p. 116).

6 See Newton (1850/1969, p. 153). Newton's reply apparently neglects the problem raised by Cotes and simply asserts that universal gravitation is proven from the phenomena. For more discussion, see Koyré (1968, pp. 273–283), Janiak (2008, pp. 168–174).

7 Leibniz, Newton's most philosophically distinguished critic, in effect assumes that Newton must be giving some sort of causal account and thus makes the first two criticisms (Leibniz's letter to Hartsoeker, published 1711, in Newton 2004, pp. 111–112). He implicitly holds that without a causal account, it is not a genuine piece of physics. This last charge can be found explicitly in the early anonymous review of the *Principia* in the *Journal des Sçavans*, wherein the author complains that Newton writes as a geometer, not a physicist, and thus supplies a mere mechanics, but not, as yet, a physics (Cohen 1971, pp. 156–157). Versions of all three charges can be found in Fontenelle's much later *Théorie des Tourbillons Cartésiens* (1752).

from Newton's remarks in the *Principia*, but isolates this tendency, develops it, strengthens it, and makes it explicit. Two enormously influential examples of this strategy were Willem Jacob Van s'Gravesande's 1720 text, the *Elementa*, and Pierre Louis Moreau de Maupertuis's 1732 tract, *Discours sur les différentes figures des astres.*

s'Gravesande's book appeared in two different English translations in 1720, immediately following its first publication in Latin.[8] It was widely disseminated and treated as an authoritative exposition of Newtonian physics and Newtonian methodology.[9] It was, in particular, a strong influence on Maupertuis, whose *Discours* famously marked an official introduction of Newtonian gravity into the Cartesian bastion of the Académie Royale des Sciences in Paris.[10] Moreover, aspects of their approach to the defense of Newton can be found, to one extent or another, in a wide range of Newtonians, including Pemberton, Keill, MacLaurin, Voltaire, and d'Alembert. Further, I will show that the views put forward by s'Gravesande and Maupertuis in this context are of significant philosophical interest in their own right. Although the fact that early Newtonianism contains a "positivistic" streak has often been remarked upon,[11] not enough has been done by way of philosophical analysis of the relevant texts: s'Gravesande's *Elements* and Maupertuis's *Discours* are the most philosophically rich examples of this tendency and repay serious study. Indeed, Maupertuis's *Discours* contains the best developed defense of attractionism per se of any published work in the period. Moreover, as I will argue below, the philosophy of science presented in it is complex and points beyond the more straightforwardly positivistic line carried over from s'Gravesande.

11.2 s'Gravesande

The Dutch natural philosopher s'Gravesande's Newtonian credentials were impeccable, having been cemented by a 1714 trip to England, which resulted in his election to the Royal Society and personal acquaintance with Newton. After returning to Holland and taking up a professorship at Leiden, where he himself had been educated, he turned to the composition of the *Elementa*, subtitled (in translation) "An Introduction to Sir Isaac Newton's Philosophy."

8 I quote from Desagulier's 1721 (second edition) translation below. The other translation is credited to Keill.

9 See Heilbron (1982, p. 45), Thackray (1970, p. 102), Ruestow (1973, pp. 113–139). A. Rupert Hall has called it the most influential book of its kind, at least before 1750 (Hall 1970, p. 510). On s'Gravesande's importance more generally, see Cassirer (1951, pp. 60–64).

10 As briefly discussed below, aspects of Newton's work were already influential in France.

11 E.g. Heilbron (1982), Hankins (1970, p. 3).

11.2.1 Defense of the necessary priority of laws

The preface to s'Gravesande's *Elementa* is in no way an explicitly polemical work. Nothing about its measured tones suggests a response to controversy, or even an acknowledgment of controversy. Nevertheless, this introduction to the introduction to Sir Isaac Newton's philosophy clearly functions as, and was intended to function as, a defense of Newton. Although 'attraction' occurs not at all in the preface, and gravity is only mentioned once, the preface contains, as we will see, the ingredients of a powerful and influential defense of attractionism and, more generally, significant developments in philosophy of science.

The preface begins by articulating two recurring themes: (1) The subject-matter of physics is the laws that God has prescribed to the Universe. (2) An epistemic modesty becomes us in questions of natural philosophy, for "What has led most People in Errors, is an immoderate Desire of Knowledge, and the Shame of confessing our Ignorance," but "there is a learned Ignorance that is the Fruit of Knowledge, and which is much preferable to an ignorant learning" (*Elements*, I, p. viii). These two themes are quickly joined, however, as s'Gravesande attempts to justify his claim that the natural philosophers must seek empirical laws, rather than pursuing a more Cartesian method. This justification is based on an account of our limited epistemic situation:

> *What Substances are, is one of the things hidden from us, We know, for instance, some of the Properties of Matter; but we are absolutely ignorant, what Subject they are inherent in.*
>
> *Who dares affirm that there are not in* Body *many other Properties, which we have no Notions of? And who ever could certainly know, that, besides the Properties of* Body *which flow from the Essence of Matter, there are not others depending upon the free* Power *of* GOD, *and that extended and solid Substance (for thus we define Body) is endowed with some Properties without which it could exist. We are not, I own, to affirm or deny any Thing concerning what we do not know. But this Rule is not followed by those who reason in Physical Matters, as if they had a compleat Knowledge of whatever belongs to Body, and who do not scruple to affirm, that the few Properties of Body which they are acquainted with, constitute the very Essence of Body.*
>
> (*Elements* I, p. xi)

Our knowledge of some properties of bodies does not amount to knowledge of the essence of body. Moreover, even if we did know the essences of bodies, we could not rule out that bodies had other, inessential, qualities, bestowed upon them by God.[12]

12 A direct appeal to superaddition is also found in Keill (1809, p. 419) (originally published in 1708), where the topic is a short range attractive force distinct from gravitational attraction.

These reflections sound, of course, thoroughly Lockean. They begin from the undeniably Lockean point that we are ignorant of the real essences of things, including the essence of body in general. Interestingly, they proceed in a way that suggests two common misreadings of Locke. First, s'Gravesande suggests that God might superadd to bodies qualities which do not flow from their essence. Although many, including Leibniz, have understood Locke to have suggested as much in his correspondence with Stillingfleet, in my view he was a consistent essentialist; his talk of superaddition is not meant to suggest that God might attach to bodies qualities that do not derive from their actual real constitution.[13] The problem, as we will see later, lies in s'Gravesande's failure to cleanly distinguish between nominal and real essence. Further, the paragraphs that follow suggest that s'Gravesande has conflated essences with logical subjects, in a way that could easily be inspired by a reading (or misreading) of 2.23 of Locke's *Essay*:[14]

> *What do they mean by saying that the Properties of Substance constitute the very Substance?*
>
> *Can those Things subsist when joyn'd together that cannot subsist separately? Can Extension, Impenetrability, Motion, &c. be conceived without a Subject to which they belong? And have we any Notion of that Subject?*
>
> (*Elements* I, p. xi)

Nevertheless, we can see here a reasonably effective, if rather basic, argument against a Cartesian-style strict mechanism. s'Gravesande assumes that we have no insight, no intellectual intuition, into the natures of things. Given this, there cannot be any grounds for a demand that all the properties we observe bodies to have be reducible to a mechanistic short list of preferred, supposedly essential, qualities.

s'Gravesande's further goal, however, is to defend the general irreducibility of laws in natural philosophy:

> *It is past doubt, for instance,* That a Body once mov'd continues in Motion: that Reaction is always equal and contrary to Action. *And several other such Laws concerning Body have been discovered: which can no way be deduced*

13 See Downing (2007).

14 Michael Ayers has argued convincingly that Locke did not think of "substance" as an entity distinct from all properties, i.e. as an in principle unknowable logical subject. However, s'Gravesande was certainly not the last to read Locke in this way. (See Ayers 1977, especially p. 78.)

Of course, I don't mean to suggest that s'Gravesande was particularly interested in Locke interpretation. Nevertheless, it is worth examining how Lockean doctrines appear, sometimes altered or distorted, in many Newtonians. For an interesting discussion of a variety of ways in which s'Gravesande is influenced by and responds to Locke, see Schuurman (2004, pp. 129–155).

> *from those Properties that are said to constitute Body; and since those Laws always hold good and upon all Occasions, they are to be look'd upon as general Laws of Nature. But then we are at a loss to know, whether they flow from the Essence of Matter, or whether they are deducible from Properties, given by* GOD *to the Bodies, the World consists of; but no way essential to Body; or whether finally those Effects, which pass for Laws of Nature depend upon external causes, which even our Ideas cannot attain to.*
>
> (*Elements* I, p. xii)

Not even the most basic laws of motion, including the inertial law that a body in motion continues in motion unless opposed by some force, can be deduced from the qualities some deem to be essential to bodies. The aim and end of science, then, can only be the articulation of such laws:

> *It appears then sufficiently, what is the End of Physics, from what Laws of Nature the Phænomena are to be deduc'd, and wherefore when we are once come to the general Laws, we cannot penetrate any further into the Knowledge of Causes.*
>
> (*Elements* I, p. xiii)[15]

We should pause, at this point, to note some possible objections, not from the perspective of a Cartesian who would defend the real use of the intellect in identifying essences, but from the perspective of an influential set of early Newtonians: Whiston, Bentley, and Clarke. s'Gravesande's position that we can make no claims about the essence of body brings him into fundamental disagreement with this trio. They unanimously maintained that we *can* draw conclusions about the nature of body from experience. Furthermore, they would have rejected s'Gravesande's claim that the law of motion according to which every body in motion remains in motion cannot be shown to follow from the essence of body. On the contrary, Whiston and Clarke both held that this law flows directly from the passive nature of matter – as a passive

15 It is interesting to contrast MacLaurin (1748/1968) here, who is similarly empiricist and anti-metaphysical (for example, attributing to Newton the view that "metaphysical considerations . . . had often misled philosophers, and had seldom been of real use in their enquiries" [p. 8]) but who nevertheless would dissent from this last statement. Like s'Gravesande, MacLaurin defends Newton as not having attempted to give the cause of gravity, but suggests that "the tracing the chain of causes is the most noble pursuit of philosophy; but we meet with no cause but what is, itself, to be considered as an effect, and are able to number but few links of the chain" (p. 17). Thus, science aims at tracing the chain of causes, although at any point in time, we will have to stop somewhere. MacLaurin treats the ether as one speculative hypothesis about the cause of gravity, and implies that further progress on this issue is not ruled out.

D'Alembert, by contrast, positions himself much more closely to s'Gravesande (as well as to Berkeley) by treating mechanics as "the science of effects, rather than the science of causes" (D'Alembert 1743/1967, p. xxiii; see also Hankins 1970, p. 153).

entity it has this one negative power.[16] More importantly, in their view, from our observations of the passivity of matter, we can conclude that attraction *could not* flow from its essence. Given that attraction cannot simply be due to bodies as they are in themselves, it must be due to God. However, if it is asked how God could bestow such a quality upon bodies, all three of these authors conclude that he must do so by a continual activity.[17] Thus, Clarke, Bentley, and Whiston conclude not with metaphysical agnosticism but with a particular metaphysical account of attraction as God's action. s'Gravesande's failure to convincingly address the issue – Why can we not draw conclusions about essences from experience?– leaves him vulnerable to attack from this quarter. Maupertuis's case, as we will see, is somewhat different, both because his discussion of essence is more nuanced and because he does not attempt to close off connections between physics and metaphysics.

11.3 Maupertuis

Pierre Louis Moreau de Maupertuis was a member of the Académie Royale des Sciences in Paris from the age of 25. His most celebrated early work, the *Discours sur les différentes figures des astres*, was, like s'Gravesande's *Elementa*, written not long after an influential trip to England. Although historians such as Thackray (1970, pp. 83–101) and Guerlac (1981, pp. 41–73)[18] have made clear that many aspects of Newton's thought were widely discussed, even accepted and transformed, among French natural philosophers quite soon after their initial publication, Maupertuis's discourse is still a remarkable historical document, for it represents the first public defense of attractionism in the Paris Academy, where Cartesian ideology still dominated. It thus represents a crucial stage in the early career of Newtonianism in the Cartesian stronghold of France.[19]

16 Clarke (1738, II, p. 697), Whiston (1696/1978, p. 6). Interestingly, s'Gravesande sounds much more like Clarke or Whiston in his later commentary on the first law of motion:

> We see that Bodies by their Nature are inactive and incapable of moving themselves; wherefore unless they be moved by some extrinsical Agent, they must necessarily remain for ever at rest.
>
> (*Elements* 1: 49)

17 See Clarke (1738, II, p. 601), Whiston (1696/1978, p. 218), Bentley (1838/1996, III, p. 168). I thus disagree with John Henry's (1994a) contention that Bentley proposes (and Newton accepts) an account of gravitational attraction as a superadded quality. Unfortunately, I cannot provide a full discussion of Henry's interesting case here.

18 And more recently J. B. Shank (2004).

19 See Beeson (1992), Brunet (1931), Thackray (1970, p. 96). It should be noted that the views Maupertuis expresses so ably in this early work are not necessarily representative of those he held later in his career. In particular, although Maupertuis retains a Lockean skepticism about knowledge of the essence of body, his views about the relation of physics and metaphysics clearly evolve from those suggested in the *Discours*. For example, when it came to his principle of least action, Maupertuis seemed willing to allow that metaphysical

Maupertuis's discourse is also notable for its philosophical content: it contains one of the best developed defenses of attractionism of the period. The second chapter of Maupertuis's discourse is devoted to a "discussion métaphysique sur l'attraction," in which Maupertuis seeks to identify and defuse sources of resistance to Newton's theory of gravity. Thus, unlike s'Gravesande, Maupertuis specifically acknowledges the existence of a dispute "which divides the greatest philosophers" (*Discours*, p. 10).[20] He strategically underplays, however, his own role as a polemicist in that dispute, claiming that he will not "pronounce" on the question but only "compare the ideas" of the two (*Discours*, p. 10).

11.3.1 First facts vs. causal explanations

As Maupertuis depicts it, the central dispute between Cartesians and Newtonians concerns the question of whether gravity ought to be regarded as the effect of circulating vortices of matter, or whether it may be treated "as if it were an inherent property of bodies" "without looking for its cause" (*Discours*, p. 10). Maupertuis's initial defense of the Newtonian position emphasizes this last proviso, stating that Newton himself officially treats universal attraction or gravitation as a fact, not a cause, leaving open the possibility of a deeper causal explanation in terms of subtle matter, perhaps even a fully mechanistic one (*Discours*, p. 12).[21] This in itself, of course, as we have already seen, is not an original point; indeed, the observation that Newton did not claim to have settled the causes of gravity was a sort of Newtonian piety, found, e.g., in the writings of Keill (1758, p. 4), Desaguliers (1734, pp. 6, 21), MacLaurin (1748/1968, p. 10), and Voltaire (1741, p. 186).[22]

This strategy, however, motivates the following question: Does a theory which fails to provide an acceptable causal explanation of the phenomena it discusses count as an acceptable piece of natural philosophy? In returning a positive answer to the question, Maupertuis follows s'Gravesande, but Maupertuis's handling of the question is more direct and is specifically focused on the question of gravity. Whatever gravity may be, he argues, it is always a "first fact," from which one can depart in order to explain the other facts which depend on it (*Discours*, p. 12). "Every regular effect, though its cause be unknown, may be the object of the Mathematicians" (*Discours*, p. 12), and the

argument might have direct implications for natural philosophy. This presumably reflects the increasing Leibnizian influence on his later thought.

20 Translations of Maupertuis's text are my own. All references to the *Discours* are to the original 1732 edition, unless otherwise noted.

21 As Janiak's work reminds us, Maupertuis's interpretation of Newton, while hardly idiosyncratic, may not be entirely correct. The claim that the *Principia* does not aim to settle the causes of gravitational attraction does not entail that gravitational attraction may not itself be regarded as a cause.

22 Also, it is mouthed by Whiston, Bentley, and Clarke.

resulting theory is indeed explanatory: it explains the phenomena which can be deduced from it:

> Galileo, without knowing the cause of the gravitation of bodies towards the earth, did not fail to give us a very beautiful and very sure theory on this gravity and to explain the phenomena which depend on it.
>
> (*Discours*, p. 12)

Maupertuis is thus making the methodological point that universal attraction may be taken as a first principle for physics, whether or not it is metaphysically primary, that is, whether or not gravitational attraction can be causally reduced to some more fundamental properties of bodies. Maupertuis buttresses this position by arguing that ultimate causal explanations elude us in any case, so it would be a mistake to insist on them when it comes to gravity: "I do not believe that it is permitted to us to ascend to first causes, nor to comprehend how bodies act upon one another" (*Discours*, p. 13). He concludes this part of his case for Newtonian gravity by suggesting that the search for the cause of this force might be left "to more sublime Philosophers" (*Discours*, p. 13), implying that it is not a task for natural philosophers.

Again, the suggestion that the pursuit of physics can be divorced from metaphysical questions about underlying causes, coupled with and supported by an agnosticism about ultimate causes, is highly reminiscent of s'Gravesande. What is most interestingly different about Maupertuis is that, unlike his Dutch colleague, he is drawn back into the question of the causes of gravity, albeit framed in terms of possibility rather than actuality.

11.3.2 Attraction as intrinsic quality, real vs. nominal essences, and primary qualities

As we have seen, Maupertuis's first defense of Newtonianism invokes agnosticism about causes. Nevertheless, his next step is to address the question of whether a causal account which makes gravity the effect of an inherent attractive power in matter can be ruled out a priori as a "Monstre métaphysique" (*Discours*, p. 13). A first question to ask here is why Maupertuis felt compelled to address this question. If regularities may be taken as first principles, why is any further defense of attraction required? Maupertuis's willingness to answer this challenge at length suggests that he believes that if we are in a position to rule out a priori the possibility of an inherent property of attraction in matter, the Newtonian is in trouble. The difficulty is two-fold: (1) If the question of the existence of intrinsic attractive powers can be definitely settled, then progress can evidently be made on this relatively metaphysical front, which suggests that neglecting it may not be a legitimate strategy. (2) If the question is settled in the negative, then the insistence that some other explanation (whether Cartesian impact or God's action) must be available and ought to be sought looks correspondingly compelling. This Newtonian predicament is neatly flagged by Maupertuis in his subtle first characterization of Newtonianism: the Newtonians treat gravity *as if it were an inherent property*. For example, the rather frequent Newtonian protest that, for all they knew, gravity might be produced by impulse,[23] was pretty clearly disingenuous: for a mechanical model which worked by simple impact would make gravity proportional to surface area,[24] and, while ether hypotheses were floated to explain gravity, the ether invoked was typically elastic, i.e. characterized by interparticulate attractive and/or repulsive forces.[25] Maupertuis seeks to *legitimate* the "as if" of attractionism by arguing that the possibility that attraction is an inherent property[26] of bodies cannot be eliminated.[27]

We would be in a position to definitively rule out or affirm attraction, Maupertuis asserts, were our epistemic situation quite different from our actual one:

23 An example is provided by Maupertuis, although he seems to attribute the claim to Newton, rather than endorsing it himself:

> Newton . . . often stated . . . that it might even be that this tendency was caused by some subtle matter emitted by bodies, and was the effect of a veritable impulsion.
>
> (*Discours*, p. 12)

24 Newton made this point against the Cartesians in the General Scholium (Newton 1999, p. 943). It is echoed, for example, in Voltaire (1738/1967, p. 201) and MacLaurin (1748/1968, p. 387).

25 This was certainly the case with the ether of Newton's 1717 Queries to the *Opticks*. See McGuire (1977, p. 117) and Heimann and McGuire (1971, p. 242).

26 Maupertuis uses the terminology of inherent property, "propriété inhérente." I understand this as meaning the same as "intrinsic"; thus, the question at issue is whether attraction can be regarded as if it were a property seated in each body. The contrast would be an extrinsic quality, externally imposed, e.g. by an aether or by God's continuous action. If attraction were intrinsic, this still leaves open the question of whether or not it is primordial, that is, an ultimate quality irreducible to more fundamental qualities (which is one thing that might be meant by calling attraction an essential quality). It also leaves open the question of whether or not we would call something matter/body only if it possessed attraction (which, following Boyle and Locke, is another thing that might be meant by calling attraction an essential quality). Newton famously disavows the claim that gravity is essential to matter (1999, p. 796). Maupertuis does not pronounce on this question here, though as we will see below he allows that for all we know gravity could be a primordial quality.

Eric Schliesser (2010b) has argued that Newton himself, in his (posthumously published) *Treatise of the System of the World*, treats gravity as an interaction, thus something relational, but an interaction partially grounded in an intrinsic property of all matter. Maupertuis does not specifically consider this position, but I suspect he would include it under his broad "as if" characterization, since it holds that gravity derives from an intrinsic property of all matter.

27 s'Gravesande in effect does this as well by arguing that there is no reason to demand reducibility to mechanist qualities, but the argument is considerably less systematic and satisfying.

> If we had complete ideas of bodies, such that we well understood what they are in themselves, and what their properties are to them, how and in what number they reside in them; we would not be at a loss to decide whether attraction is a property of matter. But we are very far from having such ideas; we only know bodies by a few properties, without knowing at all the subject in which these properties are united.
>
> (*Discours*, pp. 13–14)

The counterfactual situation described by Maupertuis here is one in which we would know the real essences of bodies – that which they are in themselves and that which gives them their properties. This is clearly another version of the same Lockean point about our ignorance of real essences that we saw above in s'Gravesande's "Preface."[28] Maupertuis's version is more thoroughly Lockean, however.[29] For one thing, Maupertuis explicitly includes a version of Locke's doctrine of nominal essences, noting that our actual situation is one in which we know not the real but only the nominal essences of bodies; that is, we know what co-existent observable properties we take to be characteristic of such and such a body (e.g. Rover) or such and such a type of body (e.g. gold):

> We perceive some different collections of these properties, and that suffices for us to designate the ideas of such or such particular body.
>
> (*Discours*, p. 14)

Furthermore, while s'Gravesande's grounds for asserting the unknowability of real essences seem to center on the uncharacterizability of the logical subject, Maupertuis's argument, as we will see, is quite different. Indeed, his initial way of putting the point, in terms of our inability to understand how a thing's observable properties hang together, bears a striking resemblance to some of Locke's formulations in the earlier drafts of the *Essay*:

> Hence it comes to passe that we have noe Ideas nor notion of the essence of matter, but it lies wholy in the darke. Because when we talke of or thinke on those things which we call material substances as man horse stone the Idea we have of either of them is but the complication or collection of those particular simple Ideas of sensible qualitys which we use to find united in the thing cald horse or stone . . . which because we cannot apprehend how they should subsist alone or one in an other we suppose they subsist & are united in some fit & common subject . . .
>
> (Locke 1990, pp. 129–130)

28 A similar point is made by Voltaire (1738/1967, p. 182): "we know nothing at all of what Matter is; we know only some few of its Properties."

29 Maupertuis's admiration for Locke's *Essay* was later made explicit in his 1743 address to the Académie française. He there describes Locke as having shown that "grammar" (what Locke calls the "doctrine of signs" at 4.21.4, which includes both words and ideas) lies at the foundation of the other sciences (Maupertuis 1756, III, p. 264), a belief that fuels his own *Réflexions philosophiques* (1740).

What we are missing, according to Locke, and what requires us to employ the obscure idea of substance in general, is access to real essences that would show us why particular sets of observable properties accompany each other as they do. In order to properly characterize *Maupertuis*'s argument, however, we will need to examine its development in some detail.

Maupertuis begins by introducing a notion of primary or primordial property. Having accumulated sufficient experience of bodies to collect properties into nominal essences usable for distinguishing particular bodies or types of bodies, our next step, as human knowers, is as follows:

> We advance one step further, we distinguish these properties into different orders. We see that while some vary in different bodies, some others are always the same; and from that we regard the latter as primordial properties and as the grounds of the others.
>
> (*Discours*, p. 14)

The *universality* of extension and impenetrability, Maupertuis continues, leads us to put them in the order or category of primordial properties, and thus to regard them as intrinsic and irreducible qualities. He then distinguishes other properties which are less universal, belonging to bodies only when they are in a certain state, e.g. the property of moving other bodies at impact, which is found in all bodies in motion. Maupertuis argues, however, that these experience-based distinctions that we make among properties do not allow us to *exclude* any properties from bodies, other than those which are actually contradictory to universal properties:

> We are right to exclude from a subject only the properties contradictory to those which we know are found in it: mobility being found in matter, we can say that immobility is not; matter being impenetrable, is not penetrable.
>
> (*Discours*, p. 16)

At this point, however, we can pose the challenge to Maupertuis that we posed, on behalf of Clarke, Bentley, and Whiston, to s'Gravesande: Why can we not draw conclusions about essences from this uniform experience? Here we reach the core of Maupertuis's argument. He argues that, without an understanding of how the primordial properties stick together, so to speak, we cannot require that all other properties obviously reduce to them:

> But again, was the collection of these properties necessary? And do all the general properties of bodies reduce to them? It seems to me that it would be to reason badly to wish to reduce them all to them.
>
> (*Discours*, p. 15)

If we saw necessary connections among the known properties of body, e.g. if we apprehended that a body cannot be extended without being impenetrable, we might have some grounds to suppose that we had grasped the real essence of body. This too, I think, is a genuinely Lockean thought: part of what appeals to Locke about mechanist natural philosophy is that the primary qualities it posits seem to be internally connected to one another. Maupertuis contends, however, that while this sort of understanding is not ruled out as a matter of logic, we clearly do not have it:

> But is there some necessary connection between these properties? Could extension not exist without impenetrability? Should I foresee through the property of extension which other properties accompany it? That is what I do not in any way see.
>
> (*Discours*, pp. 14–15)[30]

Lacking this, we must be more modest in our claims:

> It would be foolish to wish to assign to bodies properties other than those which experience has taught us are found in them; but it would perhaps be more foolish to wish, with a small number of properties scarcely known, to pronounce dogmatically the exclusion of all others; as if we had the measure of the capacity of the subjects, when we are acquainted with them only by this small number of properties.
>
> (*Discours*, pp. 15–16)

Thus, we cannot suppose that we have a knowledge of the real essence of body which would allow us to proclaim that attraction is excluded from the nature of bodies.

Maupertuis's next step in his defense of attraction is to consider whether the notion of attraction as an intrinsic property of bodies is somehow incoherent or "less conceivable" than the properties commonly acknowledged to belong to bodies. He addresses this question by comparing the strict mechanist notion of impulse with its Newtonian competitor, attractive force:

> Common people are not at all astonished when they see one body in motion communicate this motion to others; because they are accustomed to seeing this phenomenon, they are prevented from perceiving the marvelousness of it; but Philosophers . . . take care not to suppose that impulsive force is

30 It is interesting that Maupertuis does not attempt to argue in the other direction, that impenetrability can exist without extension; thus he has not provided an effective argument that there are no necessary connections to be found here at all. It seems that he has in his sights especially the Cartesian claim that the essence of body is extension and all further properties follow from extension. Descartes of course maintained in his correspondence with More that extension does entail impenetrability (Descartes 1985–1991, III, pp. 362, 372).

> more conceivable than attractive. What is this impulsive force? How does it reside in bodies? Who would have been able to divine that it resided in them before having seen bodies collide?
>
> (*Discours*, pp. 16–17)

Maupertuis acknowledges the naturalness of mechanism and mechanist explanations, but maintains that this consideration ultimately ought to carry little weight. He argues that impulse is no more intelligible than attraction; experience has made the phenomenon of impulse familiar, but philosophers find that impulsive force is no more conceivable than attractive. Here Maupertuis again expands on a point made by Locke, namely, that impulse itself is not ultimately intelligible, for we cannot comprehend the communication of motion at impact.[31] This led Locke to include the communication of motion, along with cohesion and the production of sensation, on the list of phenomena which we cannot explain except by appealing to God's omnipotence.[32] Maupertuis concludes that impulse and attraction are on the same footing.[33] In doing so, he was followed by Voltaire, in his influential popularization of the Newtonian system, *Elémens de la philosophie de Neuton.*[34]

One possible response to the perceived problems with impulse, of course, is the occasionalist one put forward by Malebranche. Maupertuis, however, considers and neatly rejects this tactic, if it is employed for anti-attractionist ends:

31 Locke, of course, was not the first to discuss this problem. Malebranche, for example, uses it as one basis from which to argue for occasionalism.

32 See *Essay* 4.3.29:

> the coherence and continuity of the parts of Matter; the production of Sensation in us of Colours and Sounds, *etc.* by impulse and motion; nay, the original Rules and Communication of Motion being such, wherein we can discover no natural connexion with any *Ideas* we have, we cannot but ascribe them to the arbitrary Will and good Pleasure of the Wise Architect.

33 Indeed, one might wonder why Locke never explicitly draws this same conclusion. Nevertheless, I believe that this was in the end Locke's view. The only privilege ultimately assigned to impulse over attraction is its peculiar naturalness, i.e., the fact that it coheres (via the all-important notion of solidity, which itself has dynamic implications in Locke's view) with the commonsense conception of body that we derive from ordinary experience.

34 See Voltaire (1738/1967, p. 85):

> We ought to suppose, that we know no more of the Cause of Impulsion, than we do of that of Attraction. We even have not a greater idea of the one than the other of these Powers; for no-body can conceive why a Body has Power by them to move another from its Place.

Right below this remark, Voltaire recommends "Mr. *Maupertuis*'s Metaphysical Discussion upon Attraction."

> But perhaps someone will say that bodies do not have impulsive force at all. A body does not impress movement on the body that it strikes; it is God himself who moves the struck body, or who has established some laws for the communication of motions . . . If bodies in motion do not have the property of moving others; if when a body strikes another, the latter is only moved because God moves it, and has established some laws for this distribution of motion; by what right could one affirm that God could not wish to establish parallel laws for attraction [la Tendance]? As soon as it is necessary to appeal to an all-powerful agent whom only a contradiction stops, one must say that the establishment of parallel laws includes some contradiction: but that is what one will not be able to say; and so is it more difficult for God to make two distant bodies tend or move towards each other, than to wait to move them until one body has been struck by another?
>
> (*Discours*, pp. 17–18)

While Maupertuis shows little sympathy for this sort of appeal to God's action,[35] he rightly observes that the attractionist has no difficulty telling the same story.[36]

The last anti-attractionist argument considered by Maupertuis is billed by him as the most substantive ("le plus solide") that can be made against attraction (*Discours*, pp. 18–19). This argument seeks to show that gravity is less intelligible than contact action by establishing that we see the necessity of some sort of contact action, since it logically follows from motion and impenetrability, two established properties of bodies, whereas we do not see the necessity of gravity. As Maupertuis puts it, if bodies are impenetrable, and one body moves against another, it cannot continue to move without penetrating it, therefore God *must* establish some law of impact (*Discours*, p. 18). However, it is not clear that God must establish a law of attraction. To this Maupertuis responds:

> But if gravity were a property of the first order; if it were attached to matter, independently of the other properties; we would not see that its establishment was necessary, because it would not owe its establishment to the combination of other properties.
>
> (*Discours*, p. 20)

Maupertuis's basic point is that the fact that attraction is not evidently necessary in the way that contact action arguably is, i.e. logically derivable from

35 Despite Maupertuis's clear distancing of himself from the occasionalist element of Malebranche's system, J. B. Shank (2008, p. 287) has argued that the *Discours* invokes a Malebranchian skepticism about human understanding.

36 This point has its parallel in Berkeley, for whom impulsive forces are no less problematic than attractive, while ideas can obey laws of attraction as easily as laws of impact.

uncontroversial properties of bodies (impenetrability and motion), does not count against its being a primordial property/property of the first order.[37]

This passage is crucial to understanding Maupertuis's conception of a primordial property. What it demonstrates is that the primordial properties are not simply the universally experienced ones; i.e. the concept of a primordial property is not the concept of a universally experienced property. If it were, there would be no open question as to whether gravity is a primordial property or not: if it is universally experienced, it is, if not, not. Rather, the primordial properties are properties that are genuinely basic to body, i.e. irreducible to other properties. Gravity, Maupertuis suggests, may for all we know be one such property. In the above cited passages (*Discours*, p. 14) where universality is invoked, Maupertuis's point is to explain how it is that we come to take certain properties as primordial: we suppose that the properties we universally experience in body are its basic and irreducible properties. While it seems that Maupertuis regards this as an acceptable working assumption, it is clear from the example of gravity that he does not suppose that it settles the question.

Here again, Maupertuis's thought tracks Locke's with remarkable subtlety. Like Maupertuis's, Locke's prose suggests, at some points, that he is conflating epistemic and metaphysical versions of the primary/secondary quality distinction. However, both make an implicit distinction between the two. Both hold, moreover, that our epistemic version of the distinction, that is, our commonsense view about what the metaphysically primary qualities of bodies really are, is determined by uniformities in experience. Both agree, however, that these uniformities do not suffice to definitively identify the metaphysically primary qualities. What our universal experience gives us is the nominal essence of matter itself. The real essence of matter might in fact be quite different.

This gives us more than one possible metaphysical status for attraction. It might be a primary quality, part of the (otherwise unknown) real essence of matter. It might flow as a *consequence* from the unknown real essence of matter. Maupertuis's prose here suggests further that, unlike Locke, he would

37 In making this point he follows Cotes, who in his preface to the second edition of the *Principia* (1713) addressed the opponents of attraction as follows:

> For either they will say that gravity is not a property of all bodies – which cannot be maintained – or they will assert that gravity is preternatural on the grounds that it does not arise from other affections of bodies and thus not from mechanical causes. Certainly there are primary affections of bodies, and since they are primary, they do not depend on others. Therefore let them consider whether or not all these are equally preternatural, and so equally to be rejected, and let them consider what philosophy will then be like.
>
> (Newton 1999, p. 392)

not foreclose the anti-essentialist hypothesis that the irreducible properties of bodies in fact do not come united into internally connected categorical properties, i.e. real essences, and, thus, that attraction might be basic without being connected to such a real essence. Here, one might suggest that Maupertuis expresses a more Lockean view than Locke himself does. I've argued elsewhere that Locke never questions the essentialist metaphysics that he takes to be our natural metaphysics, required for the world to be in principle intelligible to us.[38] However, given that he held that our natural physics, Boylean mechanism, had in fact been defeated by experience, he should have regarded this metaphysical assumption as itself defeasible. Interestingly, when Maupertuis returns briefly to this issue again in a later edition of the *Discours*, in the concluding chapter, he hews slightly closer to Lockean essentialism, suggesting that "apparently" "if attraction has a place in Nature" "to the eyes of someone who understood the whole essence of bodies, attraction would be a necessary consequence of that essence." In our current epistemic situation, however, we can do nothing but refer to the will of God, who has somehow spread out ("répandre") attraction in matter (Maupertuis 1756, 1: 161).[39]

11.3.3 Implications of Maupertuis's defense of attractionism

We have seen that Maupertuis's sophisticated defense of Newtonian gravity/attraction trades on the following thoroughly Lockean points: (1) A general knowledge of the natural world based on a grasp of ultimate causes eludes us; natural philosophy must therefore settle for experience-based regularities. (2) We know the nominal essences of bodies, but not their real essences; i.e. there are regularly recurring observable properties through which we identify bodies, but we don't comprehend the causal nexus of those properties. (3) Impulse itself is not fully intelligible, for the communication of motion at impact is inexplicable by us, given our corporeal concepts. However, in Maupertuis's hand they are mobilized towards a new end, the defense of attractionism.

But what, in the end, are the implications of that account? In particular, how well does Maupertuis's defense of the possibility of attraction as an intrinsic quality fit with the apparent metaphysical agnosticism of his opening remarks? It is worth remembering that Maupertuis had in fact billed this chapter as a sort of metaphysical interlude by titling it as he did. Nevertheless, he concludes by emphasizing that he does not claim to have provided a metaphysics of attraction and that he wishes to consider *questions de fait*:

38 Downing (2008, pp. 115–116), Downing (2007, pp. 378–380). Some amendment to this thesis might be required to take full account of *Essay* 4.10.18.

39 On the history of the *Discours* and the differences among its editions, see Terrall (2002, especially p. 76). This new final chapter dates from 1752.

> All which we have just said does not prove that there is attraction in Nature; I do not have any further ambition to prove it. I only set out to examine whether attraction, even when one considers it as an inherent property of matter, was metaphysically impossible. If it were so, the most urgent phenomena of nature could not make it be received; but if it does not contain any impossibility or contradiction, one can examine freely whether the phenomena prove it or not. Attraction is no more, so to speak, than a question of fact; it is to the System of the Universe that one must look in order to find whether it is a principle which really has a place in Nature, to what extent it is necessary in order to explain the phenomena, or finally whether it is uselessly introduced to explain facts which are well explained without it.
>
> (*Discours*, p. 21)

Maupertuis's position is cautiously stated, but nevertheless a view can be discerned here which is in some tension with his apparently forthright initial endorsement of s'Gravesande's law-based model of the aims of science.[40]

Maupertuis's concluding remarks suggest that what he takes himself to have established with his "discussion métaphysique" is the following: (1) Abstract philosophical arguments *are* relevant to the question of the *possibility* of attraction as an inherent quality. (2) The balance of argumentation favors the position that such qualities *are* possible. (3) Given this, the question of the actual existence of attraction as an inherent quality must be settled by experience. This implies, then, that experience is in principle *capable* of settling this question. That is, we might determine (with probability, if not certainty) that the true cause of gravity is the intrinsic attractive powers of bodies.[41] Of course, Maupertuis might still wave off questions of ultimate causes and manners of action as pertaining to "more sublime philosophers," and he can still retain his previous claims that reduction to regularity *suffices* for satisfactory explanation. Nevertheless, it seems that Maupertuis would not in the end endorse s'Gravesande's claim that "when we are once come to the general Laws, we *cannot* penetrate any further into the Knowledge of Causes" (*Elements* 1: xiii, my emphasis). Although Maupertuis is careful to end by framing the question of the existence of attractive powers as a mere question of fact, the fact in question seems no

40 A somewhat similar tension might be seen in Voltaire, who on the one hand calls attraction merely a "constant phenomenon," and on the other a property with which every atom of matter in the Universe is invested, and perhaps a primary cause or first principle (Voltaire 1738/1967, pp. 85, 236, 239). In the end, however, Voltaire is in this work closer to s'Gravesande than Maupertuis, for his notion of first principle is crucially vague, he uses 'property' loosely, and he states forthrightly that "this Attraction, is not, nor can be, the simple Power of one Body to draw another to itself" (Voltaire 1738/1967, p. 237).

41 His view thus recalls Cotes's in the preface to the *Principia*, but it is less radical. Cotes not only suggests that gravity is a primary property of bodies but implies that it is an essential one: "we should not conceive of any bodies that are not heavy" (Newton 1999, p. 391).

longer to be one of the "first facts" invoked in his initial defense, i.e. a mere regularity, but to be a fact about the nature of bodies. The *Discours*, in the end, veers closer to a genuine dynamicism or realism about attraction than at first appears.[42] And, while it maintains that physics can *function* separately from metaphysics, it suggests that each may still have implications for the other.[43,44]

42 This aspect of the *Discours* is commonly neglected; e.g. it seems to be missed by Hankins (1970, pp. 159–160).

43 A line that is in some respects similar had been taken by Pemberton (1728), who defends the finding of intermediate causes as a legitimate natural philosophical activity (p. 12) and notes that "it is not easy to determine, what properties of Bodies are essentially inherent in the matter, out of which they are made, and what depend upon their frame and composition" (p. 19). Nevertheless, he clearly takes this latter issue as a genuine question for natural philosophy, if a difficult one which we may not be able to resolve.

44 Thanks to Eric Schliesser and Andrew Janiak for very helpful comments. Thanks also to audiences at the Center for Philosophy of Science at the University of Minnesota and at HOPOS 2008 (Vancouver). Some of the research that went into this paper was supported, at different points, by the Huntington Library, the Institute for the Humanities at the University of Illinois at Chicago, and the National Endowment for the Humanities. This support is gratefully acknowledged.

12

The Newtonian refutation of Spinoza

Newton's Challenge and the Socratic Problem

ERIC SCHLIESSER

> What I have in my eye is another passage, where, having mentioned David's fool, who said in his heart there is no God, this great philosopher observes, that the Atheists nowadays have a double share of folly; for they are not contented to say in their hearts there is no God, but they also utter that impiety with their lips, and are thereby guilty of multiplied indiscretion and imprudence. Such people, though they were ever so much in earnest, cannot, methinks, be very formidable.
>
> (Philo, quoting Bacon, in Hume's *Dialogues Concerning Natural Religion*)

> This pernicious bigotry, of which you complain, as so fatal to philosophy, is really her offspring, who, after allying with superstition, separates himself entirely from the interest of his parent, and becomes her most inveterate enemy and persecutor. Speculative dogmas of religion, the present occasions of such furious dispute, could not possibly be conceived or admitted in the early ages of the world.
>
> (Hume, *Enquiry Concerning the Principles of Human Understanding*)

12.1 Introduction and summary

In this chapter and other papers I seek to make precise a new view of the ferment within philosophy between the publication of Newton's *Principia* and

I thank audiences at the Southwest Seminar in Early Modern Philosophy held at SFSU, February 2008, especially Michael Friedman, Marleen Rozemond, Gideon Manning, and Donald Ainslie, at Santa Cruz, department of philosophy, especially Paul Roth, and HOPOS 2008 in Vancouver, especially Lisa Shapiro, and BSHS at Oxford, especially Dan Garber, for helpful comments. I also received detailed comments on a penultimate draft from Ryan Hanley, Mary Domski, Yoram Hazony, and Andrew Janiak. Finally, this project was inspired by Abe Stone, who – despite his misgivings about these concepts – helped me distinguish among different variants of the Socratic Problem and Newton's Challenge, and carefully read and commented on the whole paper. The views expressed here are solely the author's.

Kant's *Critique of Pure Reason.* I understand Scottish and French Enlightenment thought prior to Kant's Copernican Revolution in light of philosophers' attempts to come to grips with what I call "Newton's Challenge" by which I refer to the fact that in the wake of the *Principia*'s success the authority of science is used to settle debates within philosophy. The story is complicated because many thinkers are also struggling with versions of what I call the "Socratic Problem," by which I mean that social forces (including religious, political, and moral) can threaten the independence and authority of philosophy. My name ("Socratic Problem") for this is meant to suggest that the submission of philosophy's freedom to other political/moral/religious authority is a perennial issue for philosophy; the Galileo affair is a famous example from the seventeenth century.[1]

Much of the philosophical and exegetical excitement stems from the creative interplay between Newton's Challenge and the Socratic Problem over the independent authority of philosophic reflection. For example, I read Voltaire (and many of the French *philosophes* that followed him) as *embracing* Newton's Challenge by using Newtonian science as a club to beat those religious forces that threaten the independence of the philosopher into submission. I read Rousseau and Hume as offering diverging attempts to reassert (a secular) moral control over natural philosophy.[2]

In this chapter I discuss the philosophic and historical significance of Colin MacLaurin's attacks on Spinoza's metaphysics in his posthumously published, *An Account of Sir Isaac Newton's Philosophical Discoveries* (1748/1968; hereafter *Account*). The main point of the chapter is to illustrate how the Socratic Problem and Newton's Challenge are debated at the start of the eighteenth century. Recognizing the importance and nature of these debates can help us both to understand the partial origin of some canonical versions of our philosophical history and, if we wish, to correct them in favor of more revealing ones. Finally, the mere existence of MacLaurin's treatment undermines a widely accepted historiographic myth that members of the Scottish Enlightenment (Hume, Adam Smith, Reid, etc.) only knew and thought of Spinoza through Bayle's treatment.

The chapter has two main sections. First, I use some texts by Euler, Berkeley, and Newton to introduce the notion of "Newton's Challenge." Second, I use MacLaurin's criticism of Spinoza to flesh out this concept. I distinguish among four different but closely related versions of this challenge: (NC1) a

1 The status of Socrates is ambiguous in the eighteenth century, so the label may be anachronistic. But Condorcet had no doubt, "the death of Socrates is an important event in human history. It was the first crime that marked the beginning of the war between philosophy and superstition, a war which is still being waged amongst us between this same philosophy and the oppressors of humanity and in which the burning of the Pythogorean school was such a significant event" (p. 45).

2 See, for example, Schliesser (2009) and Schliesser (2007).

philosopher claims that mechanics/physics must be consulted in the process of doing metaphysics; (NC2) a philosopher claims that mechanics/physics is epistemically prior to metaphysics; (NC3) a philosopher appeals to the authority of a natural science to settle argument over doctrine, method (etc) *within* philosophy;[3] (NC4) a philosopher claims that natural philosophy/science is immune to metaphysical challenge.

Moreover, in discussing MacLaurin's attack on Spinoza, which is treated in Section 12.3 below, I distinguish among five variants of the Socratic Problem: (SP1) a philosopher claims that practical philosophy takes precedence (in some way) over theoretical philosophy (for example, because the right/duty to do theoretical philosophy must be deduced in practical philosophy); (SP2) a philosopher explains how statements of traditional religious texts (etc.) can be understood as expressions of his philosophical doctrines; (SP3) a philosopher appeals to non-philosophical (political, religious, social) sources as authoritative; (SP4) a philosopher is forced (or threatened) by outside authorities to adjust his views; (SP5) a philosopher is held accountable for the impact of his teachings on his students. These five variants may occur simultaneously or be blended in various ways.[4]

12.2 Newton's Challenge

In this section I discuss some exemplary passages by Euler, Berkeley, and Newton in order to introduce the concept of "Newton's Challenge." I argue that in the aftermath of Newton's phenomenal successes, physics came to be seen as authoritative by some in settling metaphysical questions. Others (Berkeley, Leibnitzians, etc.) attempted to contest this authority. While I do not make the case here (and would not know how to do so), it is important for my larger project[5] that "Newton's Challenge" came to be felt ca 1700. Moreover, I argue that this issue goes beyond Newton's intentions and writings.[6]

3 In this paper I allow that natural science is itself within philosophy. I have a more fine-grained analysis of Newton's Challenge in Schliesser (2011), which overlaps with some of the argument of this chapter.

4 Three caveats: first, here I do not explore the extent to which Newton's Challenge could be treated as a species of the Socratic Problem rather than as an alternative kind. Second, I presuppose for the sake of argument that even if the nature of philosophy or metaphysics are contested and undergo radical changes over time, there are a set of practices, attitudes, and commitments that are distinctly philosophical or metaphysical (etc.). Here I cannot confront the suspicion that this assumption is question-begging. Third, while my taxonomy of the Socratic Problem and Newton's Challenge is by no means exhaustive, it is a bit more fine-grained than necessary for the arguments that follow.

5 Schliesser (manuscript).

6 This chapter, thus, revisits an old theme articulated by A. E. Burtt, although I differ from Burtt by treating Newton's philosophic views non-dismissively. I thank Yoram Hazony for calling my attention to Burtt.

12.2.1 From natural philosophy to mechanics vs. metaphysics, evidence from Euler

When Newton published the first edition of the *Principia* (1687), the fields of science and philosophy were part of a unified enterprise, which included a wide swath of learning and topics. The terms "science," "philosophy," "physics," and "natural philosophy" were often used interchangeably. A part of this enterprise was metaphysics, traditionally the science of "being as such," "the first causes of things," or the "things that do not change."[7] Of course, within the enterprise different sub-disciplines (mixed mathematics, geometry, mechanics, etc.) competed for status and attention; there were also serious, even intense, conceptual and institutional debates over the relative merits of the various methods suitable to and the relative prestige of the various branches of the tree of learning (compare, e.g., Descartes, Bacon, and d'Alembert) even within fairly homogenous intellectual communities like the Royal Society or the French Academy.[8] By the early nineteenth century, science and philosophy were becoming clearly distinguished and sometimes even mutually opposed or indifferent enterprises. There are exceptions to this narrative, because in Victorian Britain one can find occasional individuals (Maxwell, Mill) who have large visions reminiscent of their seventeenth- and eighteenth-century predecessors, and in the German-speaking world Kantian philosophy sometimes provided an overarching unity between philosophy and physics and psychology. But nevertheless the clear trend overall was towards separation. This chapter contributes to our understanding of the intellectual and conceptual elements of the process by which physics came to be seen as a replacement of and competitor to traditional philosophy in the aftermath of Newton's achievements. Here my own focus is on the period between Newton and Kant, the time of the flourishing of the French and Scottish Enlightenment.

A clear statement of the attitude that I have in mind can be found in an influential piece by one of the leading and influential mathematical thinkers of the eighteenth century, L. Euler: "[T]he knowledge of these truths [of mechanics] is [A] capable of serving as a guide in these intricate researches [of metaphysics]. For one would be right [B] in rejecting in this science [metaphysics] all the reasons and all the ideas, [B*] however well founded they may otherwise be, which lead to conclusions contrary to these truths [of mechanics]; and [B**] one would be warranted in not admitting any such principles which cannot agree with these same truths" ("Reflections on Space and Time," 1748b, p. 376).[9] In context Euler is criticizing Idealist metaphysicians' attempts to

7 van Inwagen (2007). 8 Feingold (2000).

9 "Reflexions sur l'espace et le temps." This passage has been quoted by John Stachel in (1977); Stachel credits Arnold Koslow with calling attention to the passage. Rob DiSalle also mentions the passage (and translates it more accurately, although misidentifies the

disagree with Newton's treatment of body, space, and time. Euler's first claim, [A] that mechanics is a guide to metaphysics (which represents a version of NC1 and NC2), even if controversial in some quarters, is not unusual, with variants of it having a long pedigree going back to Plato. Euler's second claim, [B] that with knowledge of mechanics, understood now as having its own privileged (un-philosophical) method, one can authoritatively settle debates within metaphysics (NC3), is an expression of a new, post Newtonian attitude.[10] Thinkers in earlier ages would have probably ruled out Euler's second claim as absurd or barely intelligible – how could the corruptible, visible, changing world allow one to settle facts of the eternal, invisible, unchanging world?

Now Euler stacks the rhetorical decks by speaking of the "truths of mechanics"; after all, (almost) nobody would want to oppose truth. He starts the essay by claiming that "the principles of Mechanics are now solidly established"; he goes on to assert that one cannot say the same about the "general principles of Metaphysics." But he does not give an evidentiary defense of this claim. In context, it's clear he has Newton's first two laws in mind. Nevertheless, Euler is clear that one can wield the authority of mechanics within first philosophy in order to reject (i) a certain privileged access to content ("ideas"), (ii) justification ("reasons"), and (iii) "principles." (I understand Euler's "principles" as explanatory foundations, e.g., axioms, conceptual commitments, general laws, etc.) Moreover, Euler is clear that mechanics allows one also to rule out (iv) competing ways of knowing ("however well founded they otherwise may be,") when they contradict mechanics – this fits nicely with the rejection of alternative forms of justification. Finally, Euler insists that (v) one cannot merely reject content that is contradicted by mechanics, but even possible claims that might follow from extrapolations from principles that one is committed to on un-mechanical grounds. In this context, Euler is silent on the source of mechanics' authority – if it resided in a particular privileged method or, say, in superior empirical accuracy. Euler applied these strictures to himself in public (he found Newtonian attraction unintelligible and privately was attracted to Cartesian vortices).[11] It would be interesting to explore further Euler's grounds for his claim, but here I introduce Euler as illustration of what I call "Newton's Challenge," that is, (to speak directly and a bit anachronistically) that the authority of physics is used to settle debates within philosophy (i.e., NC3). I do not mean to suggest, however, that this was an uncontested authority.

page) in a footnote to his (2002, p. 55 n 31). I have adjusted their translations. For evidence of the influence of this piece, see Friedman (1992). For context on Euler's remarks see Ahnert (2004).

10 For the sake of argument, I am going to ignore how Euler would distinguish between the "truths" of mechanics and, say, the (auxiliary) hypotheses of mechanics and how such a distinction might complicate how one adjudicates claims within metaphysics.

11 See Smith (forthcoming b).

12.2.2 *The authority of Newton contested, Berkeley*

In fact, I became aware of the significance of Newton's Challenge by reading Berkeley's determined and relentless opposition to the authority of mechanics within philosophy. As early as 1710 Berkeley writes:

> The best grammar of the kind we are speaking of, will easily be acknowledged to be a treatise of mechanics, demonstrated and applied to nature, by a philosopher [Newton – ES] of a neighboring nation whom all the world admire. I shall not take upon me to make remarks, on the performance of that extraordinary person: *only some things he has advanced, so directly opposite to the doctrine we have hitherto laid down, that we should be wanting, in the regard due to the authority of so great a man*, did we not take some notice of them.
>
> (Berkeley, 1710, Part I, section 110)

In the British context, Berkeley cannot afford to ignore the authority of Newton and his mechanics, which is mathematical in nature ("demonstrated") and has empirical adequacy ("applied to nature"). I view this as an instance of concern over NC4, that is, Berkeley wants to forestall appeals to the authority of Newtonian science which he takes to be improper metaphysics. Berkeley removed this passage in later editions but as I have documented elsewhere, he frequently returns to discussing the appropriate relationship between mechanics and metaphysics.[12] For example, in *De Motu*,[13] Berkeley deplores that "today [natural philosophy] is almost entirely confined to experiments and mechanics." By contrast, "to treat of the good and great God, creator and preserver of all things, and to show how all things depend on supreme and true being, although it is the most excellent part of human knowledge, is, however, rather the province of first philosophy or metaphysics and theology" (§ 34; see also the complaint about "some modern readers" at *Siris*: § 297).

Elsewhere, I have written about and rationally reconstructed Berkeley's diagnosis and criticism of an (implicit) indispensability argument that supporters of the new natural philosophy would give (Berkeley's third *Dialogue*, pp. 241–242, see also objection 6 at *Principles*, I.50 and objection 10 at I.58):[14]

12 I have discussed this in Schliesser (2005a). For more on Berkeley and Newton's Challenge, see Schliesser (2011).

13 References to *De Motu* (Of Motion) or *The Principles and Nature of Motion and the Cause of Communication of Motions* are to the section numbers of the translation by A. A. Luce in Berkeley (1948–57). While *De Motu* does not explicitly present Berkeley's immaterialism, Berkeley cites it approvingly in his criticism of mechanics in *The Analyst*, query 9, (Berkeley 1992, p. 78).

14 See Schliesser (2005a) "On the origin of modern naturalism"; see Fogelin, (2001, p. 94), who also calls attention to the importance of these objections. One might argue that

(i) Natural philosophy is successful in predicting the empirical phenomena.
(ii) The supposition of "matter" is indispensable to the predictive, empirical success of natural philosophy.
(iii) Hence we should suppose the existence of matter.

We can think of the argument as an abductive inference to the best explanation. We can also think of premise (i) as missing an added thought (ib): "and therefore, Natural philosophy is authoritative today." The reason why I focus on this argument is that it is an early example of somebody discussing the authority of mechanics in attempting to settle metaphysical matters.

Of course, Berkeley would like to *reform* the existing state of affairs: "And it is the searching after, and endeavouring to understand those signs instituted by the Author of Nature, that *ought* to be the employment of the natural philosopher, and not the pretending to explain things by corporeal causes" (*Principles*: I.66; emphasis added; see also I.107, where philosophers are exhorted to look for "final causes of things"). A discerning reader will notice that Berkeley's immediate target here is not Newton, who also embraced the search for global final cause(s), but what I have elsewhere called the "pre-Newton mechanical philosophy," viz., one that tries to explain in terms of (rational reconstructions) of collision of small particles; in the wake of Descartes's and Spinoza's rejection of final causes in physics, it's a science of reductionist and physicalist efficient causation.

12.2.3 Newton's role in setting up Newton's Challenge

The papers by Wallis, Wren, and Huygens of 1668 and 1669 that settled on a widely shared and recognized mathematical treatment of the laws of collision made post-Galilean analysis of motion an autonomous practice relatively insulated from metaphysical and theological concerns. This is why Newton singles them out for praise ("the greatest geometers of our times") in the Scholium to the Corollaries of the Laws of Motion in the *Principia*. Huygens's 1672 *Horologium Oscillatorium* is the paradigmatic work of this sort before the publication of Newton's *Principia*. So, the process I am describing with the label, "Newton's Challenge," is broader than the intentions and aims of the individual, Isaac Newton.

Berkeley is not worried about any such issues of encroachment, or epistemic priority; on his view, all that mechanics does (when properly understood, as instrumentalist) is put forward mathematical principles helping us to predict future phenomena. On this reading, Berkeley's worry is not about NC, but is instead concerned to correct misunderstandings about the nature of mechanics. I thank Lex Newman for putting the objection to me. While this objection has merit, I would claim that Berkeley is motivated to correct misunderstandings about the nature of mechanics *because of NC1*.

Nevertheless, I focus on Newton because, not only does he play a crucial rhetorical role in eighteenth-century debates especially in Britain (and thanks to Voltaire and Mme Du Châtelet in France, too), but by the 1750s, after Maupertuis measured the shape of the Earth and Halley's comet returned, Newton's achievements were perceived to be unprecedented (see, for example, Adam Smith's treatment in "The History of Astronomy"). Moreover, while here I care more about how Newton's achievements were used in subsequent generations, Newton did play a significant role in generating the framework of "Newton's Challenge." To this I turn next.

When Newton publishes the first edition of the *Principia* (1687), he is careful to call it the *Mathematical Principles of Natural Philosophy.* But while the mathematical method matters a great deal to Newton, we should not read Newton's use of "natural philosophy" as in itself a limiting claim (somehow excluding other philosophical projects); in the "Preface" he refers to his work as "mathematical principles of philosophy," and in context he is clear that he has a traditional, broad vision for his enterprise. (Newton's affirmation of the "Mathematical Principles" is meant to be distinctive, in that he proposes a superior, new mathematical-empirical method in addressing traditional concerns.) In the "Preface," he discusses Book III as offering an "example" of his general approach to "Nature"; explicating "philosophy" in terms of "natural powers" and general "forces."[15] So, it appears that Newton is still using "philosophy" and "natural philosophy" interchangeably. However, after the polemics with Leibniz *politicized* matters, Newton emphasized the "experimental" nature of his philosophy, as did Cotes in his influential preface to the second edition of the *Principia.*[16]

A query was added to the second Latin edition of the *Opticks* (1706) that is significant for our purposes: "And if, natural Philosophy in all its Parts, by pursuing this Method, shall at length be perfected, the Bounds of Moral Philosophy will be also enlarged. For so far as we can know by natural Philosophy what is the first Cause, what Power he has over us, and what Benefits we receive from him, so far our Duty towards him, as well as that towards one another, will appear to us by the Light of Nature" (Newton 1979, p. 405). It was maintained in the English translation (and subsequent editions), and I view it as a genuinely Newtonian text.[17] It accords well with the *inductive* argument for God's existence in the "General Scholium," added to the second edition (1713) of the *Principia*: "to treat of God *from* phenomena is certainly a part of natural philosophy" (emphasis added).[18] For Newton the study of motion, duty, and

15 For a terrific introduction to these matters, see Stein (2002). 16 See Shapiro (2004).

17 This is not to say that in any writings after (say) 1700, Newton would have expressed his full views or would not have tried to obscure them; see Snobelen (1999) on Newton's esotericism.

18 See the discussion in Stein (2002, 261). See also Hurlbutt (1965). I quote from Newton (1999). In Hume's *Dialogues* (Part II), Cleanthes concedes that the a posteriori argument only offers probable evidence.

unchanging, first causes are part of a shared enterprise (see also his claim in the *Principia*'s "General Scholium" that although we will know nothing of God's substance, we can "have ideas of God's attributes"). So, Newton maintains the unified picture outlined in the "Preface." Nevertheless, in the passage from the *Opticks*, Newton anticipates the first part [A] of Euler's claim that natural philosophy can guide the search for first causes, or metaphysics (NC2). Moreover, natural philosophy is clearly the more secure, foundational enterprise to other forms of knowledge, so Newton approaches the position of Euler's second claim [B] (that is, commitment to NC3 because Euler appeals to the authority of a natural science in order to settle argument over philosophic doctrine). There is a further way in which Newton anticipates elements of Euler's second claim; his infamous rejection of hypotheses ("General Scholium") anticipates Euler's rejection of reasons and ideas "however well founded they may otherwise be."

Moreover, Newton facilitated "Newton's Challenge" by allowing Cotes (the editor of the second edition of the *Principia*) to publish a highly influential, lengthy preface (1713), in which two competing approaches to philosophy, the Scholastic and Mechanical, are severely criticized from the point of view of "observations and experiments." Within a generation opposing schools would be ridiculed. In part 1 of Hume's *Dialogues Concerning Natural Religion*, Cleanthes, the spokesperson for Newtonian natural religion, leaves no doubt about his willingness to argue from the intellectual authority of Newton's natural philosophy in order to trump philosophical objections against it:

> In reality, would not a man be ridiculous, who pretended to reject Newton's explication of the wonderful phenomenon of the rainbow, because that explication gives a minute anatomy of the rays of light; a subject, forsooth, too refined for human comprehension? And what would you say to one, who, having nothing particular to object to the arguments of Copernicus and Galileo for the motion of the earth, should withhold his assent, on that general principle, that these subjects were too magnificent and remote to be explained by the narrow and fallacious reason of mankind?
>
> . . .
>
> In vain would the sceptic make a distinction between science and common life, or between one science and another . . . Many principles of mechanics are founded on very abstruse reasoning; yet no man who has any pretensions to science, even no speculative sceptic, pretends to entertain the least doubt with regard to them. The Copernican system contains the most surprising paradox, and the most contrary to our natural conceptions, to appearances, and to our very senses: yet even monks and inquisitors are now constrained to withdraw their opposition to it.[19]

19 Somewhat surprisingly, Cleanthes does not appeal to Newton's dynamics in securing the Copernican system.

So, while Newton may not have originated nor intended the full version of "Newton's Challenge," Berkeley rightly discerned where things were heading as Newton's fame and authority was spreading. I now turn to one of Berkeley's antagonists, Colin MacLaurin, whose views formed the inspiration for Hume's portrayal of Cleanthes.[20] I shall focus on MacLaurin's treatment of Spinoza, because it is the most significant discussion of Spinoza's views among Scottish Newtonians and we know that Hume and Adam Smith carefully read the piece.

12.3 Colin MacLaurin's articulation of Newton's Challenge

Colin MacLaurin (1698–1748) was one of the leading British mathematicians and most influential expositors of Newton in the second quarter of the eighteenth century. In his work on fluxions, for example, he tried to respond to Berkeley's criticisms of the calculus.[21] MacLaurin's posthumously published *Account* was not only used by Hume to model Cleanthes' arguments; it was also critically discussed by Adam Smith in his highly regarded "The History of Astronomy."[22] In this section, I discuss how MacLaurin's treatment of Newton illustrates three themes central to my understanding of eighteenth-century philosophy: first, I show how MacLaurin offers an instrumental defense of the freedom of philosophizing, thus, instantiating what I call the "Socratic Problem" (that is, SP1); second, I extend this discussion of the "Socratic Problem" when analyzing MacLaurin's criticisms of Spinoza, where we'll see the "Socratic Problem" co-mingle with "Newton's Challenge"; along the way, I clarify how for MacLaurin a "free" philosopher is to be understood as a *responsible* thinker; finally, and most briefly, I argue that MacLaurin's style of philosophizing offers us new reasons to reject the standard dichotomy between Rationalists and Empiricists when analyzing early modern philosophy. In fact, I show that MacLaurin invented part of the structure of our canon for his own reasons. In what follows I ignore – for brevity and for the sake of argument – how Spinoza might respond to these criticisms.[23]

12.3.1 The Socratic Problem and MacLaurin

Before I document how MacLaurin's philosophy illustrates "Newton's Challenge" in the British context, I should make clear that MacLaurin understands his own project in terms of another set of guiding principles; these reveal instances of the "Socratic Problem." He starts the *Account* as follows: "To describe the phenomena of nature, to explain their causes, to trace the relations

20 See Hurlbutt (1965, pp. 141–145). 21 See Niccolò Guicciardini (1989, p. 47ff).

22 Schliesser (2005b); in the middle of the nineteenth century Maxwell was still citing MacLaurin as an authority, for details see Schliesser (2010b).

23 See Schliesser (forthcoming a) and also Schliesser (forthcoming b).

and dependencies of those causes, and to enquire into the whole constitution of the universe, is the business of natural philosophy... natural philosophy is subservient to purposes of a higher kind, and is chiefly valued as it lays a sure foundation for natural religion and moral philosophy; by leading us, in a satisfactory manner, to the knowledge of the Author and Governor of the universe" (p. 2). MacLaurin has a *unified* project between physics and metaphysics. He connects natural philosophy to religious and moral purposes. More pertinently, MacLaurin justifies natural philosophy, at least in part, in terms of its utility, i.e., its religious and moral pay-off. This is a version of what I have dubbed the "Socratic Problem" (that is, SP1). In the *Theological Political Treatise* (1670), Spinoza had offered a forceful defense of the freedom of philosophizing in speculative matters and insisted that philosophy should not be the handmaiden of theology (Spinoza 1737, preface, p. 19). Reason is and ought to be self-justifying. (To be sure Spinoza promised technological, political, and commercial payoffs as a consequence of this freedom.) As a result, Spinoza inspired much freethinking literature.[24]

One reason to separate philosophy and theology is to reduce the stakes for philosophy. As the second epigraph to this chapter makes clear, Hume thought that Christianity's embrace of philosophy had made philosophy a dangerous enterprise. Hume's encouragement of fideism in his readers, for example, can be understood as reducing Christianity's entanglement with speculative philosophy. This would permit genuine freedom of thought and the possibility of amicable disagreement: "An Instance of true liberty, of which ancient times can alone afford us an example, is the liberty of thought, which engaged men of letters, however different in their abstract opinions, to maintain a mutual friendship, while they agreed in inclinations and manners. Science was the subject of disputations, never of animosity. *Cicero*, an academic, addressed his philosophical treatises, sometimes to *Brutus*, a Stoic; sometimes to *Atticus*, an epicurean" (Hume 1757, emphasis added).

MacLaurin's project is a response to attempts to remake philosophy without appeal to a higher authority. MacLaurin's enterprise appears to fit squarely in the tradition of physico-theology,[25] popularized, for instance, among the Boyle lecturers (who are approvingly mentioned by MacLaurin on p. 62). This tradition makes natural philosophy a handmaiden to theology. For while MacLaurin is quick to claim "an entire liberty must be allowed in our enquiries," it is an instrumental freedom: "that natural philosophy may become subservient to the most valuable purposes [i.e., religion] ... but we ought not to abuse this liberty by *supposing* instead of *enquiring*, and by imagining systems, instead

24 See the widely read Israel (2001); Spinoza's TTP was translated into English in 1737 in London.

25 See Israel (2001, pp. 456–471) – (MacLaurin is not mentioned). A better treatment is Hurlbutt (1965), see especially p. 65ff.

of learning from observation and experience the true constitution of things," (p. 6; emphasis in original). I return to this notion of "liberty" below. But one risk attached to MacLaurin's strategy, which will be exploited by David Hume, who also appeals to "observation" and "experience,"[26] is that natural philosophy may not have the moral or social utility claimed for it. This is by no means an idle concern because throughout the seventeenth and early eighteenth century, there are debates over the moral, religious, and practical utility of the study of nature, especially in Boyle's and Bentley's concerns over the utility of astronomy.[27] Another risk to MacLaurin's strategy is that the findings of empirical sciences stop supporting the claims and practice of Deistic or Theistic metaphysics. But we shall leave that aside for now.

12.3.2 MacLaurin's criticism of Spinoza

Here I explore the details of several of MacLaurin's criticisms of opposing systems of thought. While Descartes and Leibniz come in for more extensive criticism, I focus on MacLaurin's treatment of Spinoza for two reasons. The first is to undermine a historiographic myth: it is often said that in the Scottish Enlightenment, Spinoza was only known through Bayle's discussion (this claim appears to originate in Kemp Smith, 1941). MacLaurin is clearly familiar with Bayle's article (see *Account*, p. 78, the long footnote), but although he approves of Bayle's criticism of Spinoza, he was not much impressed by Bayle – the "weakest" of Spinoza's system's "adversaries" (p. 78, footnote) –, whom he suspected of "a disposition toward scepticism, in relation to the foundations of natural religion" (p. 78). MacLaurin also cites Spinozistic passages that are not in Bayle, and he is clearly familiar with the contents of Spinoza's *Opera Posthuma* – the *Letters*, the *Ethics*, and *Treatise on the Emendation of the Intellect* are mentioned and quoted by MacLaurin. (Somewhat surprisingly, there is no mention of or allusion to the *Theological Political Treatise*, which had been reprinted in London in 1689 and in 1737.)

The second reason for narrowing my focus on MacLaurin's treatment of Spinoza is that Spinoza's "system" provokes an unusually direct and vehement reaction in MacLaurin; he returns to criticizing Spinoza throughout the first hundred pages of his *Account*. MacLaurin sums up his motives for treating Spinoza thus: "Our view in giving some account of it, was [I] not only to shew the absurd consequences to which Des Cartes' system leads, but likewise [II] to

26 When MacLaurin comes to describe Newton's "experimental philosophy" he emphasizes Newton's use of geometry; for MacLaurin, Newton's theory is founded on "experiment and demonstration" (p. 8).

27 See Hurlbutt (1965, p. 86), and the discussion of Richard Bentley in Sermon VIII "A Confutation of Atheism," reprinted in Bentley (1838/1996, pp. 175–177) – see Hurlbutt (1965, p. 58 n 44). For Locke, see Domski's treatment of Locke's ambivalent response to Newton in this volume.

trace Spinoza's doctrine to its source (for the sake of some who may have been misled into a favorable opinion of it), which is no other than the Cartesian fable," (78, footnote). Spinoza's system is offered as a *reductio* of the Cartesian system, and Spinoza is understood strictly as an extension of Descartes's philosophy. MacLaurin approvingly quotes Leibniz who "calls spinozism *un Cartesianisme outré*; and it is apparent that his method, and many of his doctrines, were derived from this source" (*Account*, p. 75). In my final section, I return to MacLaurin's characterization of Descartes, Spinoza, and Leibniz as having a shared enterprise. In what follows I first explore the details of how MacLaurin treats Spinoza as an exemplar of a whole dangerous style of doing philosophy.

12.3.3 MacLaurin on Letter 15

The first mention of Spinoza in the *Account* is in a footnote that MacLaurin adds to his claim that "In every kind of magnitude, there is a degree or sort to which our sense is proportion'd, the perception and knowledge of which is [i] of the greatest use to mankind. The same is the [ii] ground-work of philosophy," (*Account*, p. 17). The footnote reads:

> If we were to examine more particularly the situation of man in nature, we should find reason to conclude, perhaps, that it is well adapted to one of his faculties and inclinations, for extending his knowledge, in such a manner as insight be consistent with other duties incumbent upon him; and that they have not judged rightly who have compared him in this respect (*Spinoz.* Epist. 15) with the animalcules in the blood discovered by Microscopes. He must be allowed to be the first being that pertains to this globe, which, for anything we know, may be as considerable (not in magnitude, but in more valuable respects) as any in the solar system, which is itself, perhaps, not inferior to any other system in these parts of the vast expanse. By occupying a lower place in nature, man might have more easily seen what passes amongst the minute particles of matter, but he would have lost more than he could have gained by this advantage. He would have been in no condition to institute an analysis of nature, in that case. On the other hand, we doubt not but there are excellent reasons, why he should not have access to the distant parts of the system, and must be counted at present with a very imperfect knowledge of them. The duties incumbent upon him, as a member of society, might have suffered by too great an attention to them, or communication with them. Had he been indulged in a correspondence with the planets, he next would have desired to pry into the state of the fixed stars, and at length to comprehend infinite space.
>
> (*Account*, p. 17, footnote)

In the footnote, MacLaurin admits that we have only a partial knowledge of the Universe. This is echoed in the conclusion of Book I of the *Account*: "a compleat system indeed was not to be expected from one man, or one age, or perhaps from the greatest number of ages; could we have expected it from the abilities of any one man, we surely should have had it from Sir *Isaac Newton*: but he saw too far into nature to attempt it," (p. 96). Newton's self-limitation is treated as a virtue. It is to be contrasted with the "folly of philosophical presumption and pride" that MacLaurin associates with the "Cartesian" approach (p. 79); in context it is clear that MacLaurin has Spinoza in mind because he had just spent the previous three pages explicitly criticizing the details of Spinoza's philosophy.

We might want to ask why MacLaurin introduces Spinoza in the footnote. The long sentence which does so is, in fact, puzzling without further clarification. ("[A] If we were to examine more particularly the situation of man in nature, we should find reason to conclude, perhaps, that [B] it is well adapted to one of his faculties and inclinations, for extending his knowledge, in such a manner as insight be consistent with other duties incumbent upon him; and that [C] they have not judged rightly who have compared him in this respect (*Spinoz.* Epist. 15) with the animalcules in the blood discovered by Microscopes.") It is by no means obvious why the part [C] after the semicolon would tell in favor of the main claim of the first half of the sentence [B], unless either more is known about the way animalcules in the blood are said to relate to their environment or the mere mention of Spinoza signals the point of the passage.

In fact, MacLaurin's treatment is a thinly disguised account in terms of the necessity and desirability of final causes. This is the point of the sentence to which the footnote is added – a dual *telos* is even mentioned, viz. (a) "the greatest use to mankind" and (b) the use of "philosophy." (I return to this dual *telos* below.) In particular, man is designed in such a way as to allow for just enough knowledge to perform one's duty and to prevent man from extending enquiry too far; MacLaurin seems to think that (I) important parts of potential knowledge of microscopic and inter-stellar domains are closed off to humans, and (II) this is a good thing. MacLaurin's instrumental defense of the freedom of philosophizing emphasizes, thus, that pursuit of knowledge should not even hope for completion and, more importantly, ought not come at the expense of one's moral or religious obligations.

This latter point is presumably what MacLaurin has in mind when in criticizing Spinoza's (and Descartes's) treatment of the Deity near the end of *Account* (p. 380), he substitutes a characterization in terms of "licentiousness" for "great freedom" in describing their mode of inquiry. In MacLaurin's terminology "free" enquiry is pursued when natural philosophy is subservient to the purposes of natural religion and moral philosophy (SP1). The "licentious" enquiry is not constrained in such a manner – that is, the *free* philosopher does not lose himself in, say, the comprehension of "infinite space" without regard to his religious and moral duties. For MacLaurin the free philosopher is *responsible.*[28] This fits nicely with his already quoted remarks about liberty that should not be abused.

In order to get clear on how to distinguish between the free and the licentious mode of inquiry, it is instructive to turn to the details of Spinoza's fifteenth letter (to Oldenburg), which MacLaurin cites in the long footnote we're discussing. Not unlike MacLaurin's Newton, Spinoza also affirms that he is in "ignorance" of "knowledge of the whole" (see also Spinoza's *Theological Political Treatise*, chapter 4, p. 1). Moreover, it is precisely an attempt to illustrate this *epistemic modesty* that induces Spinoza to introduce an analogy between humans and a worm in the blood. Recall: "This little worm would live in the blood, in the same way as we live in a part of the universe, and would consider each particle of blood, not as a part, but as a whole. He would be unable to determine, how all the parts are modified by the general nature of blood, and are compelled by it to adapt themselves, so as to stand in a fixed relation to one another" (Letter to Oldenburgh, November 1665). Nevertheless, Spinoza introduces the worm analogy, by insisting that "I do not attribute to nature either beauty or deformity, order or confusion. Only in relation to our imagination can things be called beautiful or deformed, ordered or confused." This is entirely consistent with Spinoza's well-known criticism of final causes (in Appendix to *Ethics* I), which, after all, are inferred from perceived order or beauty in nature (see, for example, Newton's argument in the General Scholium) because of people's fears and desires. If nature just *is* (Newton's "blind fate") and cannot be said to be really ordered or intrinsically beautiful, then arguments from or to design or beauty do not get off the ground. Spinoza and MacLaurin both emphasize human epistemic limitations, but because of their other commitments they end up at very different conclusions about final causes.

Later in the *Account* we find evidence for thinking that Spinoza's attack on final causes is motivating MacLaurin's remark in the footnote because, when he turns to some more detailed discussion of Spinoza's system, MacLaurin is explicit about his objection to Spinoza's claim that final causes "are nothing but human fiction," and distaste for how Spinoza "laughs at those who imagine that the eyes were designed for seeing, or the sun for giving light" (*Account*, p. 76; in his notes MacLaurin cites Appendix to Prop. 36 of *Ethics* I).[29] In his footnote (17), MacLaurin, thus, objects to Spinoza's strategy of making our knowledge of nature unavailable as evidence for certain kinds of arguments about our (moral/religious) duty and God's plan for us.

Thus, MacLaurin's free philosopher does not pursue truth independent of other considerations. Philosophic inquiry should be mindful of moral and religious considerations and, when necessary, self-limiting. This is a version

28 This is not the place to investigate the roots of this notion of freedom.

29 Thus, MacLaurin appears to accept particular and general/providential final causes; Newton (following Boyle) only accepts general providence.

of the Socratic Problem (SP1), because the autonomy of philosophy is put at risk. One might object that there need not be a conflict between religious or moral duty and pursuit of philosophical truth. Certainly, MacLaurin would deny there is a problem at all. One might deny this on various grounds, e.g., that truth is one (if one believes that our duties track truths) or that the category 'duty' is simply never in conflict with truthful claims. Here I am not going to explore these options; as mentioned above, MacLaurin clearly believes that in nature there is a dual *telos* in which "the greatest use to mankind" and the use of "philosophy" (understood as the needs of inquiry) coincide. For all I know, this may be a metaphysical necessity or a contingent empirical fact (or evidence that we live in the best of possible worlds). It wishes away a version of the Socratic Problem which is not articulated in terms of truth, but more in terms of how the philosopher relates to the polity/society.

12.3.4 MacLaurin's attacks on Spinoza

Let's now turn to the heart of MacLaurin's treatment of Spinoza. In the span of three pages MacLaurin ridicules (he calls them "absurd," p. 78) a number of characteristic Spinozistic positions: (a) that the existence and essence of substance are eternal truths; (b) Spinoza's rejection of final causes; (c) Spinoza's claims that the Universe is infinite, necessary and "endowed always with the same quantity of motion, or (to use his inaccurate expression[30]) always having the same proportion of motion and rest in it, and proceeding by an absolute natural necessity; without any self-mover or principle of liberty," (*Account*, p. 77); (d) Spinoza's denial of the vacuum; (e) Spinoza's claim that there is "one substance in the universe, endowed with infinite attributes, (particularly, infinite extension and cogitation) that produces all other things, in itself, necessarily, as its own modifications; which alone is, in all things, cause and effect, agent and patient, in all respects physical and moral" (*Account*, p. 78). In all cases, MacLaurin goes out of his way to mention the Cartesian origins of Spinoza's views. Rather than discussing these in detail, I focus on MacLaurin's own summary conclusion:

> In all of these, *Spinoza* has added largely from his imagination, to what he had learned from *Des cartes*. But from a comparison of their method and principles, we may beware of the danger of setting out in philosophy so high and presumptuous a manner; while both pretend to deduce compleat systems from the clear and true ideas, which they imagined they had, of eternal essences and necessary causes. If we attend to the consequences of such principles, we shall the more willingly submit to experimental philosophy, as the only sort that is suited to our faculties. It were unreasonable

30 In a footnote here, MacLaurin quotes Spinoza's Latin text from Letter 15 and *Ethics* II, proposition 13, Lemma 3.

> to charge upon *Des Cartes* the impious consequences which *Spinoza* may have been led into from his principles: but we cannot but observe, to the honour of Sir *Isaac Newton*'s philosophy, that it altogether overthrows the foundation of *Spinoza*'s doctrine, by showing that not only there may be, but that there actually is a *vacuum*; and, instead of an infinite, necessary, and indivisible, plenitude, matter appears to occupy but a very small portion of space, and to have its parts actually divided and separated from each other." (*Account*, p. 77)

There are a number of themes here important to the various arguments I am presenting in this paper. Initially I ignore MacLaurin's motives of tying Spinoza so closely to Descartes. First note that Descartes and Spinoza are presented as aiming for a complete system in "high and presumptuous manner." Recall that this "folly of philosophical presumption and pride" (p. 79) is contrasted later with Newton's virtue and wisdom because Newton did not "attempt" completeness (p. 96). The rejection of completeness and overall systematicity is reminiscent of Euler's rejection of alternative forms for justification within philosophy.[31]

Incidentally, MacLaurin has hit on an unappreciated feature of Descartes's inheritance. Without arguing the case here, we can interpret Spinoza and Newton as presenting us with a fundamental juncture in the history of philosophy; both recognize more or less the same conceptual problems in Descartes's treatment of substance, motion, and individuation: to simplify greatly, Spinoza chooses to correct Descartes in the direction of coherence, simplicity, and completeness, while Newton reforms and narrows Descartes's approach in order to provide (among other things) an account of motion that will permit of exact measurement and further open-ended discovery. Newton's success in doing so generates "Newton's Challenge." Spinoza's achievement can be understood as defusing the grounds on which the "Socratic Problem" has been lodged by Christian theology. From my vantage point by rejecting Spinoza's separation between philosophy and theology MacLaurin instantiates "Newton's Challenge" (NC4) and recreates the "Socratic Problem" (SP1 and SP3).

Second, MacLaurin attacks Spinoza's method, principles, and conclusions. The principles that are attacked are "eternal essences and necessary causes," but MacLaurin says little about these. In MacLaurin's reconstruction Spinoza derives an "infinite, necessary, and indivisible, plenitude," (p. 77) from these principles. He offers extensive quotes (on p. 75) of the Scholium to *Ethics* I, Proposition 15 in English and Latin as well as quotes from Spinoza's fourth letter to Oldenburg about Spinoza's rejection of the vacuum. This is no coincidence. Of the five doctrines (a)–(e) attributed to Spinoza, the denial of the vacuum

31 This theme of completeness and systematicity has a very important afterlife in German Idealism, see Franks (2005).

is the only one that would allow for (plausible) "experimental" refutation. MacLaurin needs it to make his main methodological claim stick.[32]

Throughout the treatment of Spinoza, MacLaurin attacks and ridicules the method of deriving "all our knowledge . . . from true ideas." (Just before MacLaurin had explained that for Spinoza "true idea" is "a clear and distinct idea, as he himself explains it." See also the rest of 78–79.) According to MacLaurin this method of clear and distinct ideas is really the product of the "imagination." (Given that Spinoza sharply distinguishes between the confused ideas of the imagination from the true ideas of the intellect, this is an *ad personam* dig!) MacLaurin concludes his treatment of Spinoza with "He pretends, indeed, to proceed in the geometrical method and style; but while he assumes a definition of substance and of its attributes at his pleasure, and passes from his definitions as true ideas (as he calls them) to the necessary existence of the thing defined, by a pretended immediate consequence, which he will not allow to be disputed, his whole superstructure appears a mere *petitio principii* or fiction. By his way of proceeding, any system whatsoever might be established," (p. 78).

Between all the name-calling we can discern a serious issue; I reconstruct the main gist of MacLaurin's argument as follows:

(I) One can derive empirically false and impious consequences from the method of "clear and distinct," or true, ideas.

(II) In fact, the method of clear and distinct ideas is too unconstrained – one can derive anything from it – it is, thus, a useless method.

(III) Empirical adequacy (*not* intelligibility, completeness, 'clarity and distinctness' etc.) as revealed by experimental science is a more fruitful criterion for evaluating philosophical theories (e.g., in discussing Leibniz, McClaurin claims: "The *criterion* of truth is usually placed in clear and evident perception; but some philosophers seem to value doctrines in proportion as they are obscure. Who would imagine that, in natural philosophy, such arguments should be preferred to the plainest facts and experiments for determining the question concerning a *vacuum*?" (*Account*, p. 82))

(IV) By "empirical" MacLaurin means experimental observation with the aid of "sound" (p. 90) geometry[33] (p. 95).

(V) Moreover, "experimental philosophy" is the "only" method "that is suited to our faculties" (p. 77, see also p. 17, footnote).

32 For an interesting treatment, see Schmaltz (1999). Unfortunately, Schmaltz does not discuss Spinoza's potential response to empirical objections.

33 The context of the quote (from p. 90) is criticism of Leibniz. Thus far, I have given no evidence that for MacLaurin geometry plays a crucial rule within natural philosophy; for more evidence, see (e.g.) his comments on Galileo (63) and his comments on collecting data within astronomy (237).

(VI) Thus, the method of clear and distinct or true ideas (and other philosophical favorites) should be rejected within philosophy and replaced by Newtonian geometric "experiment" and "observation."

MacLaurin, thus, articulates a version of "Newton's Challenge" (NC3) in the service of the Socratic Problem (SP1 – see the first premise where Euler attempts to put mechanics in the service of metaphysics precisely so that he can maintain what he sees as the proper marriage of philosophy and morality). His position is very similar to Euler's. But he goes beyond Euler's way of articulating Newton's Challenge by stressing methodological concerns. The method of clear and distinct ideas is too unconstrained; while given cognitive limitations of human beings, the geometric, experimental philosophy provides the right kind of constraints on our search for truth. One could do worse than look at Hume's writings for seeing what the full argument for the fifth premise would look like. It should be emphasized that the method of true ideas is not the only target of MacLaurin. For example, MacLaurin greatly admires Spinoza's fellow Dutchman, Huygens, describing him as "one of the greatest geometricians and astronomers that any age has produced." Nevertheless, MacLaurin uses Huygens's youthful, Platonic insistence that because there are only six secondary planets (which form a perfect number) in the solar system there is thus no need for further inquiry as a cautionary tale about "borrowed" principles or hypotheses "when applying them with too much liberty to natural enquiries" (51). Again, for MacLaurin, by contrast, the Newtonian method permits responsible freedom.

So, on my reading MacLaurin offers a nice illustration of Newton's Challenge in practice. MacLaurin argues from the empirical success of Newtonian natural philosophy to the *rejection of* alternative positions, methodologies, and foundations within philosophy. (Recall Berkeley's diagnosis of how an indispensability argument might be used to argue inclusively *for* a metaphysical thesis.) At the same time, MacLaurin argues for a certain form of self-limitation: aiming for completeness is likely to get us into trouble. In MacLaurin's hands Newtonian science also means a lowering of expectations. Finally, MacLaurin subordinates his application of Newton's Challenge to his religious and moral outlook; his "free" philosopher is an instance of the Socratic Problem in action.

12.4 MacLaurin, the philosophical historian

Here I conclude by discussing the importance for our historiography and self-understanding of a final feature of MacLaurin's strategy – his simultaneous rejection of the way of ideas and his construction of a Rationalist tradition at the time of the first appearance of "Newton's Challenge."[34] If we leave aside MacLaurin's advocacy of final causes in the name of scientific progress, there is much in his narrative which seems very familiar to the modern reader. In particular, he appears to be the first author to systematically tie together Descartes, Spinoza, and Leibniz as belonging to a shared project. While MacLaurin goes into far greater detail in discussing Descartes and Leibniz, mention of Spinoza's "extravagancies" (94), "impieties" (95; 380), and "most monstrous height" (380) often completes discussion of the wayward Continental threesome. He is probably the first philosophical historian to link the style of these three thinkers' thought together in such a tight manner and at the exclusion of other candidates. MacLaurin's focus on these three reflects his judgment that Cartesian "doctrines still prevail so much" (63); this should alert us to the fact that even in the British Isles the adoption of Newtonian principles and attitudes was a slow, contested process.

Moreover, while we are familiar with reading empirical criticism of the "big three" Continental Rationalists originating in the British Isles, it would be misleading to view MacLaurin as a (minor) representative of British Empiricism. For one, MacLaurin fiercely criticized Berkeley.[35] Also, he seems to have had no interest in Locke (who is unmentioned in the *Account*). Moreover, his attacks on the method of clear and distinct ideas do not just threaten Descartes, Spinoza, and Leibniz; one could marshal similar arguments against Locke, Berkeley, and even Hume (who relies on a notion of adequacy in the inspecting of ideas in the *Treatise*, and there argues from the inconceivability of, say, the vacuum to its impossibility). That is, while MacLaurin defends an empirical approach, his arguments cut across the Rationalist–Empiricist distinction because he gives up on the analysis/inspecting of our ideas as a (privileged) way to make progress in philosophy. MacLaurin himself is better understood as belonging to a tradition that starts with Newton and Cotes (perhaps anticipated by Galileo), and which is deeply critical of much of what we would now perceive as the then-mainstream of Modern philosophy. In his influential preface to the second edition of *Principia*, Cotes had divided the philosophical world into three camps: the Scholastics, the pre-Newtonian Mechanical philosophers, and the Newtonian mathematical-experimental philosophers.[36]

34 My thesis allows that one can find occasional precursors of Newton's Challenge. In my view Berkeley also targets the autonomous nature of the mechanical philosophy after the 1668–9 papers on the laws of collision.

35 MacLaurin (1742, volume II, chapter 12 "Of the Centre of Oscillation," p. 41) acknowledges Berkeley's criticism (the reference is to Query 9 of the *Analyst*, which itself cites Berkeley's "a Latin Treatise, *De Motu*, published at London in the year 1721"). As Niccolò Guicciardini pointed out to me, MacLaurin (1742) treats mathematics as knowledge of ideas of relations not of ideas of things, see volume I, p. 52. So, a fuller treatment of these matters requires more careful articulation.

36 See page 43 of Newton (2004). My terminology is anachronistic.

I claim that it is the signature achievement of the Newtonian philosophers to put the familiar figures of eighteenth-century philosophy (Leibniz, Berkeley, Hume, Wolff) on the defensive about method and the authority of philosophical principles not derived from the success of mathematical-experimental sciences.[37] Once we see this, we can also recover how the philosophers responded and help explain how they eventually gained canonical status and (in the process) not only came to overshadow the Newtonians in the philosophical imagination, but even displaced such tremendous and ambitious thinkers as Diderot, d'Alembert, Condorcet, Adam Smith, and even Hobbes.[38] Moreover, I hope I have prepared the way for a recovery of understanding Berkeley, Hume, Buffon, and, say, Diderot as distinctly reserved about Newton's challenge.

37 For Locke's place in this tradition, see Domski's paper in this volume.
38 For more reflections on this theme, see Schliesser (2010) and Schliesser (2011).

13

Dispositional explanations

Boyle's problem, Newton's solution, Hume's response

LYNN S. JOY

13.1 Introduction

In his *Origin of Forms and Qualities* (1666), Boyle famously proposed a dispositional explanation of how a body's sensible qualities such as color come to be manifested when that body interacts with a human observer. Using his lock-and-key metaphor, he described the sensible qualities of a body in terms of its dispositions to manifest such qualities whenever an observer's body unlocks the relevant dispositions in it. Qualities like color, he argued, have a merely dispositional existence which depends on the size, shape, motion, and combined texture of the colored body's atoms and their causal interaction with atoms of the observer's body.

However, the *very idea of a disposition* itself underwent a major conceptual change when Newton analyzed dispositional properties such as impenetrability in his *De gravitatione* (c. 1664–85) and mass in his *Principia* (1687). Newton turned Boyle's philosophical theory of dispositions on its head by showing that mass could be conceived as an exclusively dispositional property of bodies without requiring that mass be causally grounded in the categorical properties of Boyle's matter. Indeed, as I will argue in this chapter, Newton's science and philosophy were open to the revolutionary possibility that the disposition of mass, when conceived as a natural force acting according to certain mathematical laws, constitutes an existence more fundamental than that of Boyle's matter.

13.2 Boyle's problem concerning the dispositional explanation of sensible qualities

When Boyle undertook his wide-ranging attack on Aristotelian metaphysics in his *Origin of Forms and Qualities*, he challenged in particular the Aristotelian view that there exist, in natural bodies, sensible qualities that not only are real entities distinct from matter but also are entities that can exist separately from matter.[1] Against this view, he offered his own dispositional explanation of how a body's sensible qualities such as color come to be manifested when the body interacts with another body, either one composed of unthinking matter or one that is a man or sensitive animal. Invoking the relationship between a lock and its key, he described a quality like color as a body's disposition to manifest a color whenever another body serves as the key that unlocks the relevant color disposition in the first body.[2] Boyle was notably reluctant to define the sensible qualities themselves beyond giving such dispositional explanations of their production. "I have chosen to Declare what I mean by Qualities, rather by Examples, then Definitions," he said, "*partly* because being immediately or reductively the Objects of sense, Men generally understand pretty well what one another mean when they are spoken of."[3] His examples ranged widely from saline or sour tastes, to melodious or shrill sounds, to more recondite qualities such as the capacity of aqua regis to dissolve the metal gold and the incapacity of aqua fortis to dissolve gold.[4]

Boyle applied his lock-and-key metaphor to this variety of cases, but a serious problem arose from his basic account. In the case of aqua regis dissolving the metal gold, one can indeed regard this mixture of nitric and hydrochloric acids as the key that unlocks the disposition of gold to dissolve. Likewise one can regard the gold as the key that unlocks the action of the solvent disposition of the acid mixture. But, in the case of a human being perceiving a colored body, the relationship between the two lock-and-key explanations is problematic. It is unclear whether the colored body serves as a key by unlocking the human being's disposition to have the sense idea of a color. Or does the human perceiver serve as a key that unlocks the body's disposition to appear colored? Or do both dispositions – that of the colored body and that of its human perceiver – activate each other? Because Boyle was pre-occupied with his attack on the Aristotelians' real qualities, he did not directly address this problem of how the disposition of the colored body and the disposition of its perceiver are related to each other. Instead he concentrated on explaining how the colored body's disposition, and not the disposition of its perceiver, can be reduced to a causal basis composed of atoms. It was this one-sided reductive explanation which provided his main reason for denying the existence of the Aristotelians' real quality of color:

> There is in the Body, to which these Sensible Qualities are attributed, nothing of Real and Physical, but the Size, Shape, and Motion, or Rest of its component Particles, together with that Texture of the whole, which results from their being so contriv'd as they are.[5]

Of course Boyle did recognize that his dispositional explanation of qualities like color would be open to challenge by any critic who attended carefully to

1 Boyle (1666/1999, V, p. 308). 2 *Ibid.* V, pp. 308–310, 316. 3 *Ibid.* V, p. 314.
4 *Ibid.* V, pp. 314, 310–311. 5 *Ibid.* V, p. 317.

the problem of how the disposition of the colored body and the disposition of its perceiver were related to each other. A critic might object, he noted:

> Whereas we explicate Colours, Odours, and the like sensible Qualities by a *relation to our Senses*, it seems evident, that they have an *absolute* Being irrelative to *Us*; for, Snow . . . would be white, and a glowing Coal would be hot, though there were no Man or any other / Animal in the World: and 'tis plain, that Bodies do not onely by their Qualities work upon *Our senses*, but upon *other*, and those, Inanimate, *Bodies*.[6]

He acknowledged that this would be a serious objection to his dispositional explanation if he could not reconcile two conflicting claims: (a) the claim that sensible qualities do really exist because a body still possesses its sensible qualities "though all the Men and sensitive Beings in the World were annihilated," and (b) the claim that there are in a body no real qualities beyond the "Size, Shape, and Motion, or Rest of its component Particles, together with that Texture of the whole."[7]

Boyle's solution was to stipulate a crucial distinction:

> If there were no Sensitive Beings, those Bodies that are now the Objects of our Senses, would be but *dispositively* . . . endow'd with Colours, Tasts, and the like; and *actually* but onely with those more Catholick Affections of Bodies, Figure, Motion, Texture, &c.[8]

In other words, a dispositional property, such as the color of a body, exists even when it is not manifested to a perceiver because the body always possesses the categorical properties that are the causal basis of the disposition, namely, the size, shape, and motion of the body's atoms and their overall composition, or texture. By "texture," Boyle here referred to the posture and order of the several atoms that together compose a distinct middle-sized body. The body's color – if considered independently of any human perceiver – can be said only to exist *dispositively* because the color cannot be manifested unless the body interacts with a perceiver, whereas the atoms and their combined texture can be said to exist *actually* because they always exist in the body and they constitute the causal basis of its color disposition regardless of whether a perceiver interacts with it.

Having stipulated this crucial distinction between the *dispositive* existence of the body's color, when it is not being perceived, as contrasted with the *actual* existence of the body's categorical properties that constitute the causal basis of its color disposition, Boyle wrongly believed that he had thereby addressed the problem of how the disposition of a colored body and the disposition of its perceiver are related to each other. But his attempted solution to this problem

6 *Ibid.* V, p. 317. Italics are Boyle's.
7 *Ibid.* V, p. 317.
8 *Ibid.* V, p. 319. Italics are Boyle's.

did not adequately distinguish between two sorts of questions: (1) How do atoms and their combined texture constitute the dispositions of whole, middle-sized bodies? That is, how does the causal interaction of one body's atoms with the atoms of another body operate *to make manifest the respective dispositions* of the two bodies? (2) In cases where one of the bodies is a human perceiver, how does the causal interaction between the observed body's atoms and the human body's atoms produce the human's *idea, or mental representation*, of a sensible quality of the other body?

I will henceforth refer to these two sorts of questions jointly as Boyle's problem concerning the dispositional explanation of sensible qualities. To answer the first sort, he employed his concepts of texture and form, while his answers to the second sort involved his concepts of texture and sensories, or organs of sense. These uses of the concept of texture tended to blur the distinction between the two sorts of questions.

Boyle's reliance on the concept of texture was evident in his account of how a Boylean, non-Aristotelian form sets apart *each kind* of middle-sized body from *all other kinds* of middle-sized bodies in the world. He defined a Boylean form as follows:

> This *Convention of Essential Accidents* being taken . . . *together* for the Specifical Difference that *constitutes* the Body, and *discriminates* it from all other sorts of Bodies, is by one Name, because consider'd as one *collective* thing, call'd its *Forme* (as Beauty, which is made up both of Symmetry of Parts, and Agreeablenesse of Colours,) which is consequently but a certain *Character*, . . . or a *peculiar state of Matter*, . . . or, if I may so name it, an *Essential Modification*.[9]

But he added that such a form is also what enables *each individual body* belonging to a kind to be "consider'd *per modum unius*, as one Entire Corporeal Agent."[10] Its form is the determiner of the identity of each middle-sized body composed of atoms. Its form determines its identity by specifying the function that the middle-sized body performs, as illustrated by the body of a watch whose function is to operate as a time-piece. The parts of the watch, like the atoms of a body, are organized in a combined texture that constitutes its disposition to keep time. The watch's function – its disposition to keep time – defines its identity as an individual body despite its composition from separable parts, such as a spring and wheels, which could otherwise have constituted the disposition of another middle-sized body quite unlike a watch.

Moreover, according to Boyle, the entire world is an organized "Self moving Engine" composed of middle-sized bodies, each of which performs its functional role in relation to the others, and each of whose identities is functionally defined by its role in the world-wide organization of this "Self moving

9 *Ibid.* V, p. 334. Italics are Boyle's. 10 *Ibid.* V, p. 325. Italics are Boyle's.

Engine."[11] Hence it is not surprising that he did not seek a further explanation of how and why a particular group of atoms unites and acquires their combined texture. Their combined texture – i.e., their middle-sized body's disposition to perform its function – is organized by the "Self moving Engine" of the world as this Engine specifies the changing forms of all middle-sized bodies over various periods of time:

> According to our Doctrine, the World we live in is not a Movelesse or Indigested Mass of Matter, but an 'Αυτόματον, or *Self moving Engine*, wherein the greatest part of the common Matter of all Bodies is alwaies . . . in motion . . . The various *manner* of the *Coalition* of several *Corpuscles* into one visible *Body* is enough to give them a peculiar Texture, and thereby fitt them to exhibit diverse sensible Qualities, and to become a Body, sometimes of one Denomination, and sometimes of another.[12]

This extensive reliance on his concept of texture also kept Boyle from adequately explaining the relationship *between* (a) the interacting textures of an observed body and its perceiver's body, *and* (b) the perceiver's ideas, or mental representations, of the sensible qualities of the observed body. Beyond describing the observed body's effect on the texture of the perceiver's sense organs, he had little to say concerning the nature of the perceptions or ideas themselves:

> As when a Pin, being run into my Finger, causeth pain, there is no distinct Quality in the Pin answerable to what I am apt to fancie Pain to be, but the Pin in it self is / onely slender, stiff, and sharp, and by those qualities happens to make a Solution of Continuity in my Organ of Touching, upon which, by reason of the Fabrick of the Body, and the intimate Union of the Soul with it, there ariseth that troublesome kind of Perception which we call Pain, and I shall anon more particularly shew, how much that depends upon the peculiar fabrick of the Body.[13]

In the pin example, Boyle ignored the need to explain the *phenomenon of pain* and its representation in the perceiver's mind. His account of the functions of human sense organs focused on the physical texture of the organs and assumed without further argument that the perceiver also possesses a mind that is both distinct from her sense organs yet united with them. Nonetheless it is the perceiver's mind that must perform those important mental functions which her sense organs' textures are incapable of handling alone, functions such as the perceiving and naming of sensible qualities:

> The body of Man having several of its external parts, as the Eye, the Ear, &c. each of a distinct and peculiar Texture, whereby it is capable to receive Impressions from the Bodies about it, and upon that account it is

11 *Ibid.* V, pp. 318, 331–332. 12 *Ibid.* V, pp. 331–332. Italics are Boyle's.
13 *Ibid.* V, p. 317.

> call'd an Organ of Sense, we must consider . . . that these Sensories may be wrought upon by the Figure, Shape, Motion, and Texture of Bodies without them, after several waies . . . And to these Operations of the Objects on the Sensories, *the Mind of Man, which upon the account of its Union with the Body perceives them*, giveth distinct Names, calling the one Light or Colour, the other Sound, the other Odour, / &c.[14]

In this and other passages of *The Origin of Forms and Qualities*, Boyle seems to refer to a mind–body union when describing the mind's functions of perceiving and naming.[15] Still none of these references are fully spelled out, and hence Boyle does not satisfactorily explain, for instance, the relationship between an observed body's physical color disposition and its perceiver's mental disposition to have color perceptions. Although he repeatedly states that the textures of matter in the observed body and the perceiver's body constitute the causal grounding of both dispositions, he allows the perceiver's mind–body union to elude his dispositional analysis. Consequently, his reference to the mind's ineliminable role in perception serves as a mere ad hoc assumption supplementing his theory of dispositions without solving its problems.

It should be recalled, however, that Boyle was a working scientist as well as a philosophical theorist, and his dispositional analysis of sensible qualities was formulated chiefly as a philosophical theory, not an empirical investigation.[16] As a philosophical theorist, he expressed confidence that his principles concerning the size, shape, and motion of corpuscles of matter and their combined texture would better explain the nature of sensible qualities than could any rival Aristotelian metaphysics of substance. But he was also well acquainted with the practical difficulties of conducting empirical investigations, such as those in his ambitious unfinished project, *The History of Particular Qualities*, where he tried to provide an exhaustive analysis of qualities like fluidity and firmness, colors, heat and cold, and countless others.[17] As a working scientist investigating dispositions, he thus often expressed a notable lack of confidence in his own views – a diffidence that Boyle scholars have variously characterized as his scrupulosity about religion or his epistemological skepticism regarding

14 *Ibid.* V, p. 316. Italics are mine. 15 See also *Ibid.* V, pp. 317–321, 334.

16 My distinction between Boyle as a philosophical theorist and a working scientist focuses only on his theorizing about what are *dispositions*, and not on the general relations between philosophical theory and empirical investigations in his overall chemical science. The latter relations may have been much closer, as Newman (2006, pp. 175–189), has argued. By contrast, however, Clericuzio (2000, pp. 103–148), and Chalmers (2002) have drawn a fundamental distinction between what they regard as these two separate domains of Boyle's work.

17 M. A. Stewart, "Introduction," in Boyle (1991, p. xix).

any scientist's achievement of a complete knowledge of the interacting bodies in an ever-changing natural world.[18]

In Section 13.3, I will examine Newton's radically different approach to defining an important dispositional property of bodies, the property of mass. Newton turned Boyle's philosophical theory of dispositions on its head by showing that mass could be conceived as an exclusively dispositional property of bodies without requiring that mass be causally grounded in the categorical properties of Boyle's matter. Newton's science and philosophy were thus open to the revolutionary possibility that the disposition of mass, when conceived as a natural force acting according to certain mathematical laws, constitutes an existence more fundamental than that of matter itself.

13.3 Newton's account of mass as a dispositional property of bodies

It is fair to say that Newton regarded Boyle's corpuscular hypothesis and, especially, his knowledge of alchemical subjects with genuine interest and respect.[19] Although he did not always agree with Boyle's specific treatments of chemical and optical phenomena,[20] he, too, put a premium on observation and experiment, and sought to distinguish between the merely apparent qualities of bodies and their real qualities without presupposing an Aristotelian metaphysics that defined what should count as real qualities. Newton nevertheless discriminated between apparent and real qualities in ways that were far subtler than many of Boyle's treatments. For example, he discriminated between the apparent motions and true motions of bodies, between the weight and mass of bodies, and between the color and angles of refrangibility of light. His studies of mass and motion scrutinized what for Boyle had been the unquestionable categorical properties of atoms of matter – their size, shape, and motion – properties that Boyle had argued should be the causal basis of scientific explanations and not *the explananda* to be explained. Newton's willingness to investigate the categorical properties of Boyle's matter as themselves *phenomena*, whose fundamental nature required further scientific explanation, separated his natural philosophy from Boyle's as did his mathematical treatment of them.

I begin my analysis of Newton's dispositional account of mass by identifying two versions of the account which were developed at successive stages of his studies of motion in two different works.

(1) Mass was described by him, in his *Philosophiae naturalis principia mathematica* (1687), both in dispositional terms describing a body's inertial force and in terms of the mathematical relations describing a body's changes of

18 See, for instance: Hunter (2000, pp. 132–134, 232–234), Sargent (1995, pp. 205–216), Henry (1994b, pp. 119–138).

19 Westfall (1980, pp. 268, 282–284, 289, 293, 307–309, 371–377, 388–389).

20 *Ibid.* pp. 308–309.

motion, whether instantaneous or continuous over a period of time. A body's manifestation of an inertial force, whenever it reacts to an impressed force, was conceived, in the *Principia*, as its manifesting a single disposition to which Newton's three laws of motion referred. But even when the body's states of motion were described by such mathematical laws, these mathematical propositions only acquired their sense of physical necessity when they were interpreted as referring to the dispositions of one or more bodies. The three laws of motion and their role in Newton's account of mass will be examined in greater detail in Section 13.3 after this initial summary. Later, in Section 13.4, I will discuss how his dispositional account of mass also contained important philosophical resources for addressing Boyle's problem of how the disposition of an observed body and the disposition of its perceiver are related to each other.

(2) Prior to the *Principia*,[21] however, Newton had already explored the question of how dispositions might define the nature of bodies in *De gravitatione et aequipondio fluidorum* (c. 1664–85): "Whether matter could be created in one way only, or whether there are several ways by which different beings similar to bodies could be produced" by the divine power of God.[22] He had argued, "It must be agreed that God, by the sole action of thinking and willing, can prevent a body from penetrating any space defined by certain limits."[23] This impenetrability, imagined here by Newton as the *disposition of a part of space* to become impervious to bodies, was then elaborated by him as the dispositional property that might explain all other properties of bodies, including their shape, tangibility, mobility, and perceptibility:

> If we should suppose that that impenetrability is not always maintained in the same part of space but can be transferred here and there according to certain laws, yet so that the quantity and shape of that impenetrable space are not changed, there will be no property of body which it does not possess. It would have shape, be tangible and mobile, and be capable of reflecting and being reflected, and constitute no less a part of the structure of things than any other corpuscle, and I do not see why it would not equally operate upon our minds and in turn be operated upon . . . For it is certain that God can stimulate our perception by means of his own will, and thence apply such power to the effects of his will.[24]

In these passages of *De gravitatione*, Newton had envisioned how the dispositions of parts of space to become impenetrable might explain the way that imaginary beings similar to bodies could be produced by God's thought and

21 On the problem of dating *De gravitatione et aequipondio fluidorum*, see Andrew Janiak, Introduction, n. 14, p. xviii, and Note on Texts and Translations, p. xxxvii, in Newton (2004). See also Stein (2002, n. 39, pp. 302–303).

22 Newton (2004, p. 27). Newton (1962, Latin text, p. 105, and English translation, p. 138).

23 Newton (2004, p. 27). 24 *Ibid.* p. 28.

action. But he had also analyzed the property of impenetrability as part of a heuristic model that he believed could explain how *real bodies*, not merely imaginary ones, are actually produced in nature. According to the model, there are three dispositions – mobility, impenetrability, and perceptibility – all of which the parts of space must actualize if they are to constitute real bodies. He highlighted these three dispositions in the outline of his model, denoting them as the three conditions with which God endows extension:

> If they are bodies, then we can define bodies as *determined quantities of extension which omnipresent God endows with certain conditions*. . . : (1) that they be mobile, . . . [i.e.,] definite quantities which may be transferred from space to space; (2) that two of this kind cannot coincide anywhere, that is, that they may be impenetrable, and hence that oppositions obstruct their mutual motions and they are reflected in accord with certain laws; (3) that they can excite various perceptions of the senses and the imagination in created minds, and conversely be moved by them.[25]

Such a model, if it were exemplified by real bodies, would have transformed the ontological status of matter as understood by Boyle. The reality of matter would become a dependent reality because matter would no longer have an absolute existence that grounds the dispositional existence of properties of bodies like impenetrability. Rather, according to this model, the dispositional existence of properties like impenetrability, which are thought and willed by God, would instead ground the reality of matter.

I want next to consider how exactly Newton's three laws of motion defined his dispositional account of mass in the *Principia* and how that account, like the one given in *De gravitatione*, posed a significant philosophical challenge to Boyle's concept of matter. Later, in Sections 13.4 and 13.5, I will discuss other reasons why the *Principia*'s account of mass can be profitably interpreted in terms of Newton's heuristic model of dispositional properties in *De gravitatione*.

How exactly should mass be defined? Newton's multiple treatments of the concept of mass were a pivotal development in the science of matter because his concept of mass helped to define a single set of principles that described the dynamics of moving bodies on Earth, in the solar system, and beyond. These principles also unified certain of the existing well-confirmed principles of astronomy, mechanics, and fluid dynamics into an empirically adequate system of mathematical laws. Nevertheless, stating what exactly is mass was a philosophical task of the first order. How did Newton pull this rabbit out of the hat? He did so by employing interchangeably these terms: "quantity of matter," "body," "mass," "inherent force of matter," and "inertia, or force of inertia."[26] Some of these terms were already in use among astronomers and

25 *Ibid.* pp. 28–29. Italics are Newton's.
26 Newton (1999, Definitions, pp. 403–408; Axioms, or the Laws of Motion, pp. 416–417).

natural philosophers, but Newton revised their prior meanings by identifying them with each other and re-defining them in terms of his own laws of motion. He also employed ordinary language descriptions of certain observable phenomena such as the density, volume, and weight of middle-sized bodies. But he nonetheless distinguished carefully between the terms "mass" and "weight," and he restricted the use of the ordinary language term "weight" to statements about measuring the masses of different bodies relative to each other.[27]

Underlying these re-definitions of his predecessors' terms was his treatment of mass as a dispositional property of bodies in three distinct senses. The disposition of mass could be manifested by a body in three types of *behaviors*: the body's *resisting* a change in its prior state of motion or rest; the body's *changing* its prior state; and the body's *reciprocally altering* another body's state. The fact that Newton, in the *Principia*, described the motions of all physical objects – including non-living as well as living bodies – in the language of dispositional behaviors was significant. Indeed even his formulation of the laws of motion made reference to the dispositional behaviors of animals and the intentional actions of human beings. Consider his statements of the first and third laws:

> Law 1. Every body perseveres in its state of being at rest or of moving uniformly straight forward, except insofar as it is compelled to change its state by forces impressed.
>
> Law 3. To any action there is always an opposite and equal reaction; in other words, the actions of two bodies upon each other are always equal and always opposite in direction.
>
> ... If anyone presses a stone with a finger, the finger is also pressed by the stone. If a horse draws a stone tied to a rope, the horse will (so to speak) also be drawn back equally toward the stone, for the rope, stretched out at both ends, will urge the horse toward the stone and the stone toward the horse by one and the same endeavor to go slack and will impede the forward motion of the one as much as it promotes the forward motion of the other.[28]

Newton's three laws of motion, expressed by him not only mathematically but also in the language of dispositional behaviors, transformed the kinematic study of moving bodies into a dynamic study of the dispositions of the bodies, focusing on their capacities to maintain or change their states of motion or rest, and to act reciprocally on other bodies that exert impressed forces on them. First, a body was held to possess a disposition called the "force of inertia" or "inherent force of matter," which is "the power of resisting by which every body, so far as it is able, perseveres in its state either of resting or of moving

27 *Ibid.*, Definition 1, pp. 403–404. 28 *Ibid.*, Law 1, p. 416; Law 3, p. 417.

uniformly straight forward."[29] In this sense, mass is a disposition described by his first law of motion. Second, any change in a body's inertial state was held to be necessarily related to various impressed forces that could act on the body, including percussion, pressure, and centripetal forces such as gravity.[30] Here mass was conceived by Newton as the disposition that is exhibited whenever an impressed force causes a body to change its prior state of motion or rest. By exhibiting such a disposition to change, the body conforms to a law-like relation between the impressed force and a change in its own prior state of motion or rest. In this sense, mass is a disposition described by his second law of motion.[31] Finally, according to his third law of motion, whenever a body changes its inertial state owing to the action of an impressed force, the body will act reciprocally to alter the state of whatever other body is responsible for the impressed force.[32] In other words, the body will exhibit its disposition to exert an inertial force not only by resisting a change in its own state but also by functioning simultaneously as an agent that alters the state of another body.

Furthermore the manifestations of the types of dispositional behaviors described by his three laws of motion have a double aspect. These manifestations count as the *appearances of motions* to observers, but they also count as the *absolute, or true, motions* of bodies caused by forces. Before Newton, they might have been described kinematically as appearances of a body's merely apparent motions before and after a given change. But Newton chose to interpret such appearances differently, in terms of the body's exhibiting the relevant dispositional behavior of mass, so that he could specify the relationship between how the body appears to an observer and its true motions. The merely apparent motions could be analyzed as the differences between two or more true motions, and, subsequently, the forces that cause them could be inferred from these true motions. He remarked about this way of interpreting the appearances in the Scholium to the Definitions at the beginning of the *Principia*:

29 *Ibid.*, Definition 3, p. 404.

30 *Ibid.*, Law 2, p. 416; Definition 4, p. 405.

31 The question of whether Newton interpreted his second law of motion as referring to a body's "change of motion" (its translation from one location to another) or to a body's "change in quantity of motion" (its momentum) is insightfully examined in Pourciau (2006, pp. 160–161, 165). Pourciau shows that Newton intended his second law to refer to a body's translation or displacement, not to its momentum, and that his second law thus describes motive forces that can be either continuous forces or impulses. It should be noted that my analysis of Newton's dispositional account of mass can be applied to both of these interpretations of the second law.

32 *Ibid.*, Law 3, 417. Howard Stein has argued that it is the conjunction of all three laws of motion that should be considered as Newton's principle characterizing the intrinsic force of matter as a natural power. See Stein (2002, p. 289). In my analysis here, I argue that the three laws of motion are best understood as alternative manifestations of a single property, which I call the "dispositional property of mass."

> Relative quantities, therefore, are not the actual quantities whose names they bear but are those sensible measures of them (whether true or erroneous) that are commonly used instead of the quantities being measured. But if the meanings of words are to be defined by usage, then it is these sensible measures which should properly be understood by the terms 'time,' 'space,' 'place,' and 'motion,' and the manner of expression will be out of the ordinary and purely mathematical if the quantities being measured are understood here.
>
> . . . It is certainly very difficult to find out the true motions of individual bodies and actually to differentiate them from apparent motions, because the parts of that immovable space in which the bodies truly move make no impression on the senses. Nevertheless, the case is not utterly hopeless. For it is possible to draw evidence partly from apparent motions, which are the differences between the true motions, and partly from the forces that are the causes and effects of the true motions.[33]

Because the *Principia*'s account of mass combined all these features – (a) the behavioral dispositions of bodies to move and to exert forces, (b) the mathematical description of their various motions, and (c) the relationship between absolute, or true, motion and a human observer's perceptions of relative as well as true motions – Newton's treatment of mass as a dispositional property amounted to a rejection of Boyle's theory of the dispositions of bodies. For Newton's account showed how mass can be conceived as an exclusively dispositional property that does not require a causal grounding in the categorical properties of matter. Ernst Mach famously belittled the *Principia*'s definition of mass as circular: "If density is mass per unit volume, how can mass be defined as jointly proportional to density and volume?"[34] This alleged circularity was one that Bernard Cohen later tried to disprove by attending to the Latin text of the *Principia*'s definition of mass, which he translated as follows: "Quantity of matter is a measure of matter that arises from [*orta est*] its density and volume jointly."[35] Yet this disagreement between Mach and Cohen, over whether or not Newton's definition of mass is viciously circular, seems benign when compared to the issue of whether or not mass can be conceived as an exclusively dispositional property of bodies. It may also take a back seat to certain questions about the role of human and divine observers in a possible Newtonian solution to Boyle's problem. Such a solution will be developed in Sections 13.4 and 13.5, where I will use it to show why the *Principia*'s account of mass should be interpreted in terms of Newton's heuristic model of dispositional properties that he originally outlined in *De gravitatione.*

33 Newton (1999, Scholium to Definitions, pp. 413–414).

34 I. Bernard Cohen's paraphrase of a statement by Mach (1893/1960, chapter 2, section 7), quoted in Cohen (2002, p. 59).

35 Newton (1999, Definition 1, p. 403). See also Cohen (2002, n. 8, p. 82).

13.4 Newtonian and Humean approaches to Boyle's problem

Boyle's problem, as stated in Section 13.2, consisted of two sorts of questions: (1) How does the causal interaction of one body's atoms with the atoms of another body operate *to make manifest the respective dispositions* of the two bodies? (2) In cases where one of the bodies is a human perceiver, how does the causal interaction between the observed body's atoms and the human body's atoms produce the human's *idea, or mental representation*, of a sensible quality of the observed body? These questions had arisen because of Boyle's denial that certain sensible qualities like color are categorical properties of bodies. To explain the phenomenon of color, for example, he had stipulated that color has a merely dispositive, or dispositional, existence that is somehow grounded in the actual existence of the size, shape, and motion of the body's atoms and their combined texture. But he could not coherently explain the workings of such a disposition. He had been especially stymied by the question of how a colored body's disposition to appear colored is related to a color perceiver's disposition to experience the body's color. Is the quality manifested by the body the *same thing* as the observer's perception of the quality? Or are they different? Boyle's concept of a disposition rested on his lock-and-key metaphor, whereby one body's disposition is a propensity to exhibit a quality whose actual occurrence needs to be unlocked, metaphorically speaking, by the key which is its interaction with another body. Yet he himself did not think his lock-and-key metaphor fully captured the complexity of a colored body's interaction with a human perceiver. Hence he had supplemented this analysis with his fundamental distinction between a body's dispositional properties like color and its categorical properties like size and shape – a distinction that Locke would later christen "the primary-secondary quality distinction."[36] Boyle had further stipulated that these categorical properties should be treated as the unquestionable basic entities in terms of which to explain a body's dispositional properties.

When Newton introduced mass as a dispositional property of bodies, treating mass as the body's disposition both to react to an impressed force and to exhibit to an observer a change of motion, he had thereby opened to scrutiny the unquestionable basic entities of Boyle's science. Unlike Boyle, he had also explicitly discussed the ineliminable role of the observer's perception of apparent motions in revealing the differences between true motions and hence the observer's role in determining the existence of the forces that cause true motions. His concern about how to interpret the observer's perceptions was evident in his distinction between relative and true motions in the

36 Locke (1975, Book 2, chapter 8, sections 8–26, pp. 134–143). For an analysis of the differences between Locke's and Newton's respective conceptions of primary qualities, see Stein (1990, pp. 29–31).

Scholium to the *Principia*'s Definitions. For, if a body's mass is understood as its disposition *both* to react to an external impressed force *and* to be perceived by an observer as changing its motion, then what one means by "perceived by an observer" will be crucial in defining mass itself.

What could it mean for mass as a disposition to depend in part on its being perceived by an observer? Why should one entertain this possibility? To appreciate why a Newtonian scientist should be well motivated to do so, it is helpful to consider a post-Newtonian philosopher's skeptical attack on the primary–secondary quality distinction as it had been formulated by Boyle and Locke. Hume's attack on the primary–secondary quality distinction, which he believed to be the chief error of modern philosophy,[37] appeared in his *Treatise of Human Nature* (1739) over a half century after Newton had first published the *Principia.* As Berkeley had already tried to do, Hume was engaged in correcting what he took to be the errors in modern accounts of the relationship between human perceivers and the objects of perception. His philosophical project thus shared with Berkeley's project at least one important goal, that of criticizing the primary–secondary quality distinction, including its use in causal explanations of perception and its use in epistemological analyses of scientific knowledge.[38] However, Hume's own argument against the primary–secondary distinction did not aim to question Boyle's assumption that corpuscular matter exists independently of the mind of the observer. Nor did it comprise part of a general Berkeleyan program of discrediting the prevalence of abstract ideas and materialist principles in Newtonian science.[39] Hume's argument focused instead on an example much like one that Newton had cited when explaining his third law of motion, the man-pressing-a-stone example.[40] In Hume's version of the example, he argues that the primary quality of solidity, which is constituted by the mutual interaction between the man and the stone, would be wholly inconceivable if the secondary quality of touch were eliminated from the interaction:

> The impressions of touch are simple impressions... And from this simplicity I infer, that they neither represent solidity, nor any real object. For let us put two cases, viz. that of a man, who presses a stone, or any solid body, with his hand, and that of two stones, which press each other; 'twill

37 Hume (2000, 1.4.4.3–9, pp. 149–151).

38 Berkeley (1982, part I, sections 9–10, 18–19, 22, 87; pp. 26–27, 30–32, 60). On Hume's familiarity with Berkeley's views, beginning during his student days at Edinburgh University, see Mossner (1980, pp. 48–49, 133).

39 Berkeley's general program of discrediting abstract ideas and materialist principles in Newtonian science was undertaken in several of his works, including Berkeley (1982, part I, sections 96–117, pp. 64–74). See also Berkeley's criticisms of Newton's concept of force and Newton's method of fluxions in mathematics in Berkeley (1992).

40 Newton (1999, Law 3, 417): "If anyone presses a stone with a finger, the finger is also pressed by the stone."

> readily be allow'd, that these two cases are not in every respect alike, but that in the former there is conjoin'd with the solidity, a feeling or sensation, of which there is no appearance in the latter. In order, therefore, to make these two cases alike, 'tis necessary to remove some part of the impression, which the man feels by his hand, or organ of sensation; and that being impossible in a simple impression, obliges us to remove the whole, and proves that this whole impression has no archetype or model in external objects.[41]

This argument is interesting because it probes the nature of a human observer's perception of solidity, a quality that Boyle had classified as a categorical property of matter.[42] Hume here resembles Newton in his willingness to investigate solidity, or impenetrability, as itself a phenomenon whose fundamental nature is open to question. His argument targets what he regards as the conceptual incoherence of the primary–secondary quality distinction by showing that a modern philosopher's commitment to this distinction will ultimately compel the philosopher to give up an even more basic commitment, his commitment to the existence of a causal relationship between the human perceiver and the object of his perception. For if the philosopher claims that the secondary qualities of bodies, unlike primary qualities, have no real existence in a body apart from their being perceived, he will be unable to explain the difference between a causal interaction between two stones and a causal interaction between a man and a stone. This is because there is an ineliminable *phenomenal part*, the secondary quality of touch, in the man–stone interaction. The secondary quality of touch differs both from the solidity of the stone as it acts on the solidity of the man, and from the solidity of the man as it acts on the solidity of the stone.

Of course the modern philosopher might try to eliminate such a phenomenal part from the man–stone interaction by arguing that only the primary quality of solidity – not the secondary quality of touch – can be truly attributed to the stone. But, if he insists on denying that there exists such a phenomenal part of the man–stone interaction, he will then be unable to claim that the stone possesses the quality of solidity at all. Why? Because, by eliminating the phenomenal part of the man–stone interaction, the philosopher will have prevented the man from acquiring any sensation of the stone. But it follows from such a prevention that the causal relationship between the stone and the *man considered as a perceiver* has also been eliminated. Hence the stone will not be able to manifest its disposition of solidity, or impenetrability, by causally interacting with the *man considered as a perceiver*. Therefore the stone will not possess the quality of solidity, or impenetrability, at all.

41 Hume (2000, 1.4.4.14, p. 152). 42 Boyle (1666/1999, V, pp. 325, 333).

What did Hume's version of the man-pressing-a-stone argument reveal about Newton's third law of motion and Newton's own citing of a similar example to explain the third law? Hume did not intend his argument to question the third law. Rather his criticism addressed those theories of matter typified by Boyle's and Locke's causal explanations of sensible qualities. This type of theory had relied on the primary–secondary quality distinction to finesse questions concerning the causal relationship between a material object's disposition to manifest a quality and an observer's perception of that quality. By contrast, Newton's dispositional account of mass, including his third law of motion, did not rely at all on the primary–secondary quality distinction to determine the relationship between the phenomena of motion perceived by an observer and the forces that cause them. For Newton treated mass as a body's disposition *both* to react to an external impressed force *and* to be perceived by an observer as changing its motion. Hence, on his account, it is the body's *dispositional property of mass* that necessitates the relationship between the phenomena of motion perceived by an observer and the inertial and impressed forces that cause these phenomena. If bodies do actually possess such a disposition, then Newton will have succeeded in establishing that there must be a causal relationship between an observer's perception of a quality and a material object's manifestation of the same quality, and furthermore he will have identified the precise dispositional nature of that relationship. In short, he will have answered one of the main questions posed by Boyle's problem concerning sensible qualities.

13.5 Newton's divine phenomenalism

But do bodies actually possess such a disposition? The answer to this question rests largely on whether or not two of the features that define mass as a dispositional property of bodies are in fact necessarily linked by a single disposition. According to Newton's account of mass, a body's disposition to react to an impressed force is *the same as* its disposition to be perceived by an observer as changing its motion. Thus the existence of the body's mass depends not only on its disposition to react to an impressed force but also on its disposition to be perceived by an observer as changing its motion. However, human observers may seriously doubt the necessary linking of these two features in the same disposition. For a human observer can perceive a body's translation from one location to another, and this perception does indeed enable her to experience the body's disposition to be perceived as changing its motion. Yet this perception does not necessarily enable her to experience, in addition, the body's disposition to react to an impressed force since the observer has no direct perceptual access – to the body's exertion of or its reaction to any forces – beyond her perception of the body's translation from one location to another.

Even in cases of circular motion, such as the case of two balls connected by a cord and revolving about a common center of gravity,[43] it is an open question in what sense an observer can have direct perceptual access to the tension of the cord. One might ask, for instance, whether the forces causing the circular motion of the two balls can be *directly perceived* as the tension of the cord, or whether the existence of the tension itself can only be inferred *indirectly* when the observer perceives a displacement in the parts of the cord. Therefore, even in special cases of circular motion, human observers have good reason to inquire what warrants the treatment of these two features – a body's reaction to an impressed force and its being perceived by an observer as changing its motion or displacing its parts – as manifestations of the *same disposition.*

Newton believed that the relevant warrant lies in an analogy of reason between a human being's direct perceptual access to her own body and God's divine access to the bodies and minds comprising the created world. His clearest statement of this analogy was presented in *De gravitatione*:

> Since each man is conscious that he can move his body at will, and believes further that other men enjoy the same power of similarly moving their bodies by thought alone, the free power of moving bodies at will can by no means be denied to God, whose faculty of thought is infinitely greater and more swift. And for the same reason it must be agreed that God, by the sole action of thinking and willing, can prevent a body from penetrating any space defined by certain limits.
>
> If he should exercise this power, and cause some space projecting above the earth, like a mountain or any other body, to be impervious to bodies and thus stop or reflect light and all impinging things, it seems impossible that we should not consider this space really to be a body from the evidence of our senses (which constitute our sole judges in this matter); for it ought to be regarded as tangible on account of its impenetrability, and visible, opaque, and colored on account of the reflection of light, and it will resonate when struck because the adjacent air will be moved by the blow.
>
> . . . If we should suppose that that impenetrability is not always maintained in the same part of space but can be transferred here and there according to certain laws, yet so that the quantity and shape of that impenetrable space are not changed, there will be no property of body which it does not possess. . . . [A]nd I do not see why it would not equally operate upon our minds and in turn be operated upon, because it would be nothing other than the effect of the divine mind produced in a definite quantity of space. For it is certain that God can stimulate our perception by means of his own will, and thence apply such power to the effects of his will.[44]

43 Newton (1999, Scholium to Definitions, pp. 414–415).

44 Newton (2004, pp. 27–28). See also Newton (1962, pp. 105–106, 138–139).

Newton here employed an analogy between the human will and the divine will that other philosophers had previously invoked for a variety of purposes. Yet he put the analogy to a new use by relating God's thinking and willing of the sensible quality of impenetrability to a material object's manifestation of the same quality. Human beings, he noted, are capable of willing their own bodies' movements so that there is no essential difference between their mental actions of willing and their bodies' actual movements – barring external obstacles that might accidentally constrain their intended movements. Their thoughts of intended movements and their bodies' actual movements are linked by their wills in such a way that they perceive their mental actions of willing to move and their movements as equivalent states. God, by analogy, is capable of a comparable sort of willing so that there is, as God perceives it, an equivalence between his willing the impenetrability of a part of space and the actual impenetrability of a material object at that spatial location. Furthermore there is, as God perceives it, an equivalence between his willing the impenetrability of *successive parts* of space and the actual motion of a material object through those spatial locations.

Newton's explanation by analogy of this equivalence between God's mental actions and a material object's properties, such as impenetrability and mass, constitutes what I call "Newton's divine phenomenalism." His phenomenalism was distinctive by virtue of its realist purpose, its attempt to ground the existence of matter in the dispositional properties of bodies. It differed from Berkeley's phenomenalism because, unlike Berkeley, Newton did not claim that phenomenalism necessitates immaterialism.[45] He affirmed instead that the dispositional behaviors of material objects, manifested in their changing relations over time, fully ground the existence of material objects.

However, this grounding depends on *something mental* which defines what it means to be impenetrable or capable of reacting to an impressed force. Without that *something mental* – such as God's willing that impenetrability be transferred to different parts of space according to certain laws – the dispositions of impenetrability and mass would be unintelligible. It is therefore not surprising that, on Newton's view in the *Principia*, human beings have sufficient warrant to believe that a body's disposition to react to an impressed force is *the same as* its disposition to be perceived by a human observer as changing its motion. A human observer is here warranted by an analogy of reason to think, as Newton's God does, about the body's mass as a disposition that necessarily links these two features. Nor is it surprising that the human observer, who perceives only one feature of this disposition, namely, the body's changing motion, can also acquire – by describing these phenomena in terms of the three laws of motion – a warranted belief in the body's disposition to react to an impressed force.

45 Berkeley (1982, part I, sections 7–11, 46–58, 87–96; pp. 26–28, 42–47, 60–64).

For Newton's God likewise thinks of the body's disposition to react to various forces in terms of the phenomena as described by the three laws of motion.

Indeed Newton's God is indispensable to his physics as a whole because, as I have just shown, to separate his theology from his physics is to challenge the credibility of his dispositional account of mass itself. God is required to underwrite his empiricism.[46] Newton's theology is not a compartmentalized part of his work, separable from the other parts, whose main interest lies in its providing a cultural context for his science. Our understanding of the *Principia*'s dispositional account of mass is significantly deepened by interpreting it in terms of his heuristic model of dispositional properties which, in *De gravitatione*, depend on a divine power. What emerges from my integrated analysis of these works in Sections 13.3, 13.4, and 13.5 is Newton's development of a radically new concept of dispositions. His other innovations in various areas of mathematics, physics, and natural philosophy are well known, and his influence on early modern philosophical theories of matter and causation has been the subject of stimulating debate among recent Newton scholars and Hume scholars, who have explored the similarities and differences between Newton's and Hume's views of causal powers.[47] My present contribution to this ongoing discussion is to show how the *very idea of a disposition* itself underwent a major conceptual change between Boyle's theory of the dispositions of bodies and Newton's dispositional treatment of impenetrability in *De gravitatione* and mass in the *Principia*. These properties could now be analyzed without presupposing that they possessed a causal grounding in any more basic property of bodies, including the categorical properties of Boyle's matter. As such, Newton's new concept of a dispositional property was incompatible with Boyle's theory, which required that a body's dispositions be ultimately reducible to the size, shape, motion, and combined texture of its atoms. Moreover, as I have argued in Sections 13.4 and 13.5, Newton's dispositional accounts of impenetrability and mass also contained important philosophical resources for trying to solve Boyle's problem of how the disposition of an observable body and the disposition of its perceiver are related to each other.

13.6 Newton's legacy to Hume, and Hume's legacy to us

Suppose, however, that you are a religious skeptic and yours is an empiricism that, while it continues to accept his physics, nonetheless rejects his analogy of

46 See Joy (2006, pp. 99–105) for my account of Newton's scientific epistemology and the central role that his conception of God performs in this epistemology.

47 See these excellent studies: Schliesser (2007), Schliesser (2008), Janiak (2007), Janiak (2008, pp. 87–129), De Pierris (2002), Beebee (2006, pp. 193–225), Strawson (2000), Winkler (2000).

reason between human and divine observers. How would your acceptance of his physics, if it were divorced from its divine phenomenalism, shape your beliefs about matter and your beliefs about yourself as an observer of phenomena?

Newton's most important legacy to Hume[48] was his showing how mass can be conceived as an exclusively dispositional property of bodies without presupposing its grounding in a more basic categorical property of bodies. Mass, as a dispositional property, did not require any causal grounding in Boyle's matter because, when it was conceived as a natural force acting according to the laws of motion, it constituted an existence more fundamental than that of matter itself. But Hume never accepted divine phenomenalism, and, as a result, the Newtonian elements of his own work sometimes subverted certain claims of the very science which had inspired it. Nowhere was this more evident than in his re-interpretation of Newton's dispositional account of mass as instead a *relational account* of the perceptual states of an observer of the phenomena of motion.

Hume's rejection of Newton's analogy of reason between human and divine observers deprived him of the warrant for believing that a body's disposition to react to an impressed force is the same as its disposition to be perceived by a human observer as changing its motion. He was thus left with an account of the human observer that featured only the observer's perceptions of changes in a body's motion and one that postulated no causal basis of these perceptions in the body's disposition. So Hume had no reason to presuppose that material objects possess dispositions that necessitate a causal relationship between an object's manifesting a particular quality and an observer's perceiving the same quality. Rather his account of the human observer was focused on the relations *between the observer's distinct perceptions* of a body's changing motions because it was these relations between perceptions that now defined Newtonian forces for him.[49]

This focus was notable in two ways. First, by questioning Newton's divine guarantee of the belief that our perceptions of changes of motion are caused by a body exerting or reacting to a force, Hume proceeded to examine the nature of all our beliefs about material objects external to the mind. Yet his examination was not aimed at discovering a correspondence between our beliefs, on the one hand, and the forces and material objects external to the mind, on the other hand. That is, Hume did not try to prove that the impressions and ideas that constitute our beliefs are true representations of material objects. For given his refusal to stipulate any divininely guaranteed, necessary *causal relationship*

48 Several recent studies that differ from my view of the relationship between Newton's natural philosophy and Hume's metaphysics and epistemology are cited in footnotes 47 and 52.

49 Hume (2000, 1.3.14.11–31, pp. 108–115). Hume (1999, sections 5.5–8, 5.20–22, 7.5–8, 7.25 n. 16, 7.28–30; pp. 121–124, 128–130, 136–137, 143–147).

between an object's manifesting a particular quality and an observer's perceiving the same quality, he had no reason to infer, from such a causal relationship, that there exists an additional *cognitive relationship of representation* between an external material object and its observer. His only way to evaluate the reliability of our beliefs was to examine the perceptions that constitute those beliefs and to analyze the variety of relations among them. Hume's account of the nature of our beliefs was therefore directed at discovering and explaining all of the significant relations among our impressions and ideas themselves.[50]

Moreover, this aim was notable for a second reason. Hume, in implementing it, was able to drive a wedge between his own relational interpretation of the laws of motion and Newton's dispositional interpretation of those laws. Apprised of the type of connection between impressions and ideas that constitutes a belief, he then interpreted an observer's perceptions of changes of motion as providing the *phenomenal content* of the observer's belief that a body is exerting or reacting to a force. By thus translating the statement of a belief about a body's mass into a statement describing the relations between the observer's perceptual states, he could now claim that a justificatory connection exists between the observer's perceiving certain impressions and ideas and her forming the relevant belief. Therefore her belief that a body is exerting or reacting to a force could be reduced to and hence warranted by regularities in the relations between the observer's perceptions. And Hume could, accordingly, explain what makes this particular belief *a reliable belief*.[51]

There is of course much in Hume that we might now question, but his relational account of the phenomena of motion in terms of the perceptual states of the observer is still a philosophical position well worth defending. It originated in his acceptance of Newtonian physics divorced from divine phenomenalism and greatly aided his efforts to apply a Newtonian methodology more broadly to his own philosophy of mind. It showed how our beliefs about a body's physical disposition can be reduced to the relations between our perceptual states. Furthermore it anticipated the view that our beliefs about scientific laws of nature can be interpreted as and warranted by relations between the perceived phenomena without reference to the dispositions of material objects and the forces they exert. Unlike Newton, therefore, Hume did not revise and rehabilitate Boyle's lock-and-key concept of a disposition for the purpose of causally relating the phenomena of motion to the forces that cause them. Rather he chose to translate previous generations of empiricists' talk of the dispositions of matter into talk of a *non-dispositional account of bodies* based on his relational account of belief. He could scarcely have done otherwise if he was to remain a Newtonian – yet a Newtonian for whom God could play no part

50 Hume (2000, 1.1.1.7–12; 1.1.4–5; 1.3.1–2; pp. 9–10, 12–16, 50–55). Hume (1999, sections 2.1–5, 3.1–3, 4.1–13, 5.5–8, 5.20–22; pp. 96–97, 101–102, 108–113, 121–124, 128–130).

51 See Hume (2000) and Hume (1999) references in footnote 49.

in his theorizing. It has often been thought that it was Hume's empiricism which compelled him to exclude God from his explanations of both mind and body.[52] However, I want to conclude by suggesting that, at least in so far as his beliefs about material objects were concerned, it was the exclusion of God from Hume's explanations that compelled him to become an empiricist.[53]

52 For example, see: Buckle (2001, pp. 325–327); Fogelin (2003, p. 61).

53 In my *Dispositions and Intentionality in the Humean Tradition* (work-in-progress), I offer an extended discussion of how Hume's atheism shaped both his empiricism and philosophy of mind.

14

Newton and Kant on absolute space

from theology to transcendental philosophy

MICHAEL FRIEDMAN

In my previous work on Newton and Kant I have primarily emphasized methodological issues: why Kant takes the Newtonian Laws of Motion (as well as certain related propositions of what he calls "pure natural science") as synthetic a priori constitutive principles rather than mere empirical laws, and how this point is intimately connected, in turn, with Kant's conception of absolute space as a regulative idea of reason – as the limit point of an empirical constructive procedure rather than a self-subsistent "container" existing prior to and independently of all perceptible matter. I have also argued that these methodological differences explain the circumstance that Kant, unlike Newton, asserts that gravitational attraction *must* be conceived as an "action at a distance through empty space," and even formulates a (rare) criticism of Newton for attempting to leave the question of the "true cause" of gravitational attraction entirely open. In this chapter I emphasize the importance of metaphysical and theological issues – about God, his creation of the material world in space, and the consequences different views of such creation have for the metaphysical foundations of physics. I argue, in particular, that Kant's differences with Newton over these issues constitute an essential part of his radical transformation of the very meaning of metaphysics as practiced by his predecessors. I also suggest that the metaphysical and theological issues in question form an essential part of the intellectual context for the methodological issues I have emphasized previously – especially the issue of action at a distance.

It is now well known that the main target of Newton's rejection of "relationalism" in favor of an "absolutist" metaphysics of space was Descartes, and the *locus classicus* for Newton's own metaphysics of space is his unpublished *De Gravitatione.*[1] What was most important for Newton was decisively to reject

1 This point was first made in Stein (1967), and we can now also cite Stein (2002) for an authoritative account of Newton's metaphysics. *De Gravitatione* first appeared, together with an English translation, in Newton (1962). An improved translation by Christian Johnson, made with the assistance of Andrew Janiak, and consulting an earlier unpublished translation by Stein, appears in Newton (2004): my parenthetical page references to *De Grav.* – and to Newton's writings more generally – are to this edition.

Descartes's identification of matter with extension and to defend, accordingly, the concept of absolute (empty) space existing prior to and independently of matter. Yet Newton, like Descartes before him, also appropriated philosophical ideas from the neo-Platonic tradition,[2] which he incorporated into his own metaphysics. For Newton, the most salient source of such ideas was the Cambridge Platonism represented especially by Henry More, and Newton employs them in his doctrine that absolute space is neither a substance nor an accident, but what he calls "an emanative effect of God and an affection of every kind of being" (*De Grav.*, p. 21).[3] In particular, absolute space or pure extension is even an affection of God himself, since God is omnipresent or everywhere. God can thereby create matter or body (as something quite distinct from pure extension) by endowing certain determined regions of space with the conditions of mobility, impenetrability, and obedience to the laws of motion. God can do this anywhere in space, in virtue of his omnipresence, by his immediate thought and will, just as our souls can move our bodies by our immediate thought and will. It is essentially this doctrine which surfaces in Newton's well-known published statements, in the General Scholium to the *Principia* and the Queries to the *Optics*, that space is the "sensorium" of God.[4]

The sharp differences between Descartes's and Newton's metaphysics of space – their different conceptions of the relationships among space, God, and matter – are of fundamental importance. For Descartes, since space is simply identical with matter, God creates matter by creating space itself, and it is precisely this act of creation of space at successive moments of time that is responsible for the laws of motion. In particular, the conservation of what Descartes called the total "quantity of motion" results from the unity and simplicity of God, whereby God continually recreates the entire Universe (the whole of pure extension, whose various parts may have different instantaneous tendencies to motion at any given time) at each instant while constantly expressing the very

2 For Descartes's appropriation of neo-Platonic metaphysics, as mediated by Augustine, see Menn (1998).

3 Some of the most important writings of the Cambridge Platonists are collected in Patrides (1980). For discussion of the idea that space is an emanative effect of God see the exchange between J. E. McGuire and John Carriero in Bricker and Hughes (1990). See also the very careful discussion in Stein (2002, pp. 266–272). In the course of his discussion Stein is led to claim (p. 269) that "the grounds for thinking that Newton's theory of emanation is neo-Platonic, or 'Cambridge Platonic,' are very weak." Whatever one may think of Stein's particular reasons for this claim, it seems to me very hard to deny, in any case, that Newton is *appropriating* neo-Platonic (and, indeed, 'Cambridge Platonic') ideas for his own purposes here.

4 In Query 31, for example, Newton describes God as (p. 138) "a powerful ever-living agent, who being in all places, is more able by his will to move the bodies within his boundless uniform sensorium, and thereby to form and reform the parts of the universe, than we are by our will to move the parts of our own bodies."

same divine essence.[5] For Newton, by contrast, matter and space have radically different statuses vis-à-vis God's creation. Space is "an emanative effect of God and an affection of every kind of being," including God, while matter is the result of God's creative activity in space, wherein certain determined regions are then endowed with the conditions of mobility, impenetrability, and obedience to the laws of motion. By instituting the laws of motion, in particular, God thereby endows certain regions of space with Newtonian *mass* or quantity of matter (*vis inertiae*), and the presence of this quantity, specifically, clearly distinguishes matter from empty space. This not only leads, following earlier work of Wren, Wallis, and Huygens, to a much more adequate formulation of the laws of impact (whereas Descartes's inadequate formulation had no room for the quantity of mass, and thus no room for momentum or quantity of motion in the Newtonian sense), it eventually leads to the theory of universal gravitation of Book III of the *Principia*. And this theory, in turn, puts the notions of absolute space, time, and motion to real physical work in determining the center of mass of the solar system as the true "center of the world."

Nevertheless, despite these fundamental differences, both Descartes and Newton are using neo-Platonic ideas to support an essentially mathematical approach to physics over the older qualitative approach of Aristotelian physics. For Descartes, the world described by physics is, in its essence, the object of pure geometry. God, in creating this world, not only brings about (what Descartes takes to be) the (mathematical) laws of motion of the new physics, he also, in creating us as mind–body composites located within this world, guarantees that we can use our purely intellectual mathematical knowledge in successively correcting and refining our knowledge of the material world – as we apply pure mathematics, that is, to the initially misleading deliverances of our senses.[6] For Newton, although the world described by physics is not, in its essence, the object of pure geometry, space (which *is* the object of pure geometry) nonetheless constitutes the "frame of the world" – an emanative effect of the divine existence wherein God then creates matter by an immediate act of his will. The bare existence of space suffices for the existence of all the shapes and figures studied in pure geometry (*De Grav.*, p. 22): "there are everywhere all kinds of figures, everywhere spheres, cubes, triangles, straight lines, everywhere circular, elliptical, parabolical, and all other kinds of figures, and those of all shapes and sizes, even though they are not disclosed to sight." And thus pure geometry is *ipso facto* applicable to all material bodies as well (pp. 22–3): "the delineation of any material figure is not a new production of that figure with respect to space, but only a corporeal representation of it, so that what was formerly insensible

5 Article 36 of the *Principles of Philosophy* (1644). See the edition by Miller and Miller (Descartes 1991, pp. 57–58).

6 This, in a nutshell, is how I read the argument for the existence of matter of the Sixth Meditation: see Friedman (1997), (2008).

in space now appears before the senses." Therefore, in virtue of their (differently) neo-Platonic conceptions of a metaphysics of space, neither Descartes nor Newton has any room for a necessary gap (as there was in Plato's original "Platonism") between pure mathematics, on the one side, and the sensible and material world, on the other.

The significance of this point becomes clearer if we contrast the conceptions of both Descartes and Newton with the quite distinct approach of Leibniz, who was explicitly opposed to both Descartes and Newton in correspondingly different ways. Leibniz began, in fact, by reacting to Descartes's failure adequately to formulate the basic laws of impact, which were supposed to govern, according to the then dominant paradigm of the mechanical natural philosophy, all phenomena in the material or corporeal world. Leibniz responded to this problem by emphasizing the importance of a new, essentially dynamical quantity, which he called *vis viva* or living force (in modern terms, mass multiplied by the square of the velocity), where the basic law of motion is now formulated as the conservation of the total quantity of *vis viva*. Leibniz also strongly emphasized that living force is not purely geometrical or mechanical, so that, in particular, this quantity (unlike Descartes's purely mechanical "quantity of motion") reintroduces an element of Aristotelian teleology into the mechanical philosophy. For *vis viva*, on Leibniz's view, is the counterpart of the Aristotelian notion of *entelechy*: namely, that internal (non-spatial) principle by which an ultimate simple substance or monad determines (by a kind of "appetition") the entire future development of its own internal state. Moreover, in accordance with this same renewed emphasis on Aristotelian teleology, Leibniz then articulated a doctrine of divine creation in terms of God's choice of the best among all merely logically possible worlds. The distinction between what is logically possible and what is actual – between all merely thinkable worlds available to the divine intellect and the best and most perfect of these worlds as determined by the divine will – thereby corresponds to the distinction between principles of pure mathematics (including geometry) and principles of natural science or physics (the laws of motion). In particular, the laws of motion, unlike the principles of pure mathematics, precisely express the divine wisdom in actualizing or creating the best and most perfect of all possible worlds.[7]

Leibniz thereby breaks decisively with Descartes's metaphysics of space, for the actual world of material substances results from a special act of the divine will which introduces additional non-spatial, and essentially teleological, elements into the mechanical laws of motion. Indeed, Leibniz's break with

7 Leibniz first articulated his criticism of Descartes concerning *vis viva* in "A Brief Demonstration of a Notable Error of Descartes and Others Concerning a Natural Law" (1686), and he developed the wider metaphysical implications of *vis viva* in his *Discourse on Metaphysics*, written in the same year. Both of these, together with a very wide selection of Leibniz's works, are translated in Leibniz (1976).

Descartes on this issue is deeper still, for, on Leibniz's view, the entire mechanical physical world (including the space in which bodies move) is a secondary appearance or phenomenon (a "well-founded phenomenon" like the rainbow) of an underlying metaphysical reality of mind-like simple substances or monads – substances which, at this level, are not spatial at all, but rather have only purely internal properties and no external relations. This point, in turn, is closely connected with a fundamental disagreement with Descartes about the nature of the intellect: whereas Descartes entirely rejects traditional Aristotelian logic and takes purely intellectual knowledge to be exemplified by the procedure of his new analytic geometry instead, Leibniz self-consciously returns to the idea that purely intellectual knowledge is essentially logical.[8] And, although Leibniz appears to have envisioned some sort of extension of Aristotelian logic capable of embracing the new algebraic methods of his calculus, there is no doubt that the traditional subject-predicate structure of this logic pervades his monadic metaphysics: it is precisely because ultimate metaphysical reality is essentially intellectual in the logical sense that the entire mechanical world, including space, is a merely secondary reality or phenomenon. Thus, although Leibniz, like everyone else in the period, holds that there are exact mathematical laws governing the sensible and material world, he reintroduces a new kind of necessary gap between reality as known by the intellect and this sensible world.

For Newton, by contrast, space – the very space in which bodies exist and move – is metaphysically fundamental, for, as we have seen, it is "an affection of every kind of being," including God himself. Indeed, Newton puts the point even more strongly several pages later (*De Grav.*, p. 25): "Space is an affection of a being just as a being. No being exists or can exist which is not related to space in some way." In particular, God, through his omnipresence, creates matter in space by endowing certain determined regions with mass (*vis inertiae*), and God thereby institutes the (Newtonian) Laws of Motion by singling out momentum (mass multiplied by velocity) as the fundamental dynamical quantity governing all changes of motion of matter. For Newton, moreover, impressed force (*vis impressa*) is a further dynamical quantity involved in such changes – where this refers to any action on the body in question by which a change of momentum is produced. Impressed force, in the Newtonian sense, is an external action on a body by something else, not an internal principle of change like Leibnizean *vis viva*, and, what is more, the changes it effects are not intrinsically limited to the condition of contact. On the contrary, the principal instantiation of this concept, in the *Principia*, is precisely the force of universal gravitation, whereby one body exchanges momentum with another body immediately and at a distance; and it is the theory of universal gravitation, as we have said, which

8 See the classic discussion of Descartes's and Leibniz's very different conceptions of the relationship between mathematics and logic in Hacking (1980).

then puts the notions of absolute space, time, and motion to real physical work in determining the true "center of the world."

It is by no means surprising, therefore, that Newton also rejects the traditional Aristotelian notion of substance, and replaces it, in effect, with space itself – or, more precisely, with space plus God (*De Grav.*, p. 29):

> For the existence of these beings [bodies] it is not necessary that we suppose some unintelligible substance to exist in which as subject there may be an inherent substantial form; extension and an act of the divine will are enough. Extension takes the place of the substantial subject in which the form of the body is conserved by the divine will; and that product of the divine will is the form or formal reason of the body denoting every dimension of space in which the body is to be produced.

For Leibniz, by contrast, space, as we have seen, is a mere "well-founded phenomenon," and pure intellectual knowledge is explicitly modelled on Aristotelian subject-predicate logic: (a modified version of) the Aristotelian concept of substance *must* be metaphysically fundamental.

Newton's struggles with the problem of action at a distance result in significant complications here. Although later Newtonians (including Kant) were happy to conceive gravitation as an immediate action of one body on another body across empty space, Newton himself was seriously troubled. He appeared deliberately to leave it open in the first (1687) edition of the *Principia* that gravity may ultimately be explained by mechanical impact; and he also speculated in the *Optics* about an interplanetary aetherial medium as the cause of gravity.[9] Moreover, Newton famously declared that the idea of action at a distance is an "absurdity" in his well-known letter to Bentley of February 5, 1693 (pp. 102–103):

> It is inconceivable that inanimate brute matter should, without the mediation of something else, which is not material, operate upon and affect other matter without mutual contact, as it must be, if gravitation in the sense of Epicurus, be essential and inherent in it. And this is one reason why I desired you would not ascribe innate gravity to me. That gravity should be

9 Thus, for example, in the Scholium to Section 11 of Book I of the *Principia*, after discussing the three-body problem at some length, Newton says (p. 86, my emphasis): "I use the word 'attraction' here in a general sense for any endeavor whatever of bodies to approach one another, whether that endeavor occurs as a result of the action of the bodies either drawn toward one another or acting on one another by means of spirits emitted or whether it arises from the action of the aether or air or of any medium whatsoever – whether corporeal or incorporeal – *in any way* impelling toward one another the bodies floating therein." However, as explained in note 21 below (which also discusses the aetherial medium proposed in the *Optics*), Newton definitely appears to exclude mechanical impact from the possible candidates in the second (1713) edition of the *Principia*.

> innate, inherent, and essential to matter, so that one body may act upon another at a distance through a vacuum without the mediation of anything else, by and through which their action and force may be conveyed from one to another, is to me so great an absurdity, that I believe that no man who has in philosophical matters a competent faculty of thinking can ever fall into it. Gravity must be caused by an agent acting constantly according to certain laws; but whether this agent be material or immaterial, I have left to the consideration of my readers.

And what is most striking, from our present point of view, is the suggestion that the true cause of gravity may be an *immaterial* agent – perhaps even God himself.

It is natural, in the first place, that the mediating agent between distantly gravitating bodies be immaterial, for it is essential to Newton's argument for universal gravitation in Book III of the *Principia* that such mutually attracting bodies – Jupiter and Saturn, for example – directly and immediately exchange momentum with one another, entirely independently of any other matter that may be located in between. Whatever is playing this mediating role must therefore experience negligible exchanges of momentum with the two attracting bodies themselves, and the most natural way to achieve this, in general, is to conceive the mediating agent as massless or immaterial. Moreover, in the second place, since God exists or is omnipresent everywhere in space, and he thereby creates matter and its fundamental laws by an immediate act of the divine will, it is natural to suppose that the ubiquitous immaterial agent ultimately responsible for gravitational attraction is either God himself or an ubiquitous immaterial spirit directly resulting from God's own ubiquity.[10] Finally, in the third place, God is described in the General Scholium added to the second edition of the *Principia* in 1713 as an omnipresent acting substance (p. 91): "God is one and the same God always and everywhere. He is omnipresent not

10 In Query 31 to the *Optics* (unlike *De Grav.*), Newton suggests that God's act of creating matter in space is responsible not only for impenetrability and mass (in accordance with the three "passive" laws of motion), but also for specific forces or "active principles," including gravity (pp. 136–137): "[I]t seems probable to me, that God in the beginning formed matter in solid, massy, hard, impenetrable, moveable particles, of such sizes and figures, and with such other properties, and in such proportion to space, as most conduced to the end for which he formed them; . . . It seems to me farther, that these particles have not only a *vis inertiae*, accompanied with such passive laws of motion as naturally result from that force, but also that they are moved by certain active principles, such as that of gravity, and that which causes fermentation, and the cohesion of bodies. These principles I consider, not as occult qualities, supposed to result from the specific forms of things, but as general laws of nature, by which the things themselves are formed; their truth appearing to us by phenomena, though their causes be not yet discovered." Compare also the passage quoted in note 4 above, which can easily be taken to suggest that God himself "moves" the bodies interacting in accordance with universal gravitation.

only *virtually* but also *substantially*; for action requires substance."[11] Therefore, Newton does not so much entirely reject the traditional notions of substance and active agency, but reinterprets them in light of his metaphysics of space. He continues to conceive of efficient causality as the (local) action of one substance on another, and God, in particular, is the ultimate substantial agent underlying all causal action in the material world. His true opposition to Descartes concerns the notion of specifically *material* substance, and he uses his neo-Platonic (Cambridge Platonic) metaphysics of space to craft a further argument against Descartes's metaphysics from the (apparent) phenomena of gravitational attraction at a distance.

The importance of Newton's metaphysics of space in underwriting his principled rejection of the mechanical philosophy has not, I believe, been sufficiently appreciated. For, from a post-Newtonian perspective, the requirement that all causal interaction in the material world be limited to the communication of motion by impact may appear as an entirely arbitrary restriction on the basic principles governing the exchange of momentum, and there is then no reason, from this point of view, that a direct (equal and opposite) exchange of momentum at a distance via universal gravitation may not be viewed as a perfectly legitimate example of causal interaction.[12] At the time when Newton was first formulating this theory, however, everyone took it for granted that one substance could act on another by efficient causality only if the one is locally present to the other: this principle was shared by contemporary Aristotelians, by mechanical philosophers, and (as we have just seen) by Newton himself.

11 Immediately following this passage Newton adds (*ibid.*): "In him all things are contained and moved, but he does not act on them nor they on him." And, at the very end of the General Scholium, after pointing out that he has "not yet assigned a cause to gravity," and that, nonetheless, it is not to be reckoned among the "occult qualities," but is rather derived by induction from the phenomena, Newton continues (p. 93): "A few things could now be added concerning a certain very subtle spirit pervading gross bodies and lying hidden in them; by its force and actions, the particles of bodies attract one another at very small distances and cohere when they become contiguous; and electrical bodies act at greater distances, repelling as well as attracting neighboring corpuscles; . . . " If, in accordance with the above passage from Query 31 of the *Optics*, we suppose that this "very subtle spirit" is also the cause of gravitational attraction, it would follow that it is this (presumably immaterial or massless) "spirit" which mediates gravitational action in line with the letter to Bentley. God's own active agency would then be confined to creating both matter and the spirit in question, which then *interact* with one another to produce the phenomena of gravitational attraction.

12 This, for example, is how Robert DiSalle views the matter in his excellent recent philosophical history of space-time physics. In particular, according to DiSalle (2006, p. 42): "[I]n the Newtonian view, any interaction is physically intelligible as long as, and just to the extent that, it conforms to the laws of motion." This, however, was not the view of Newton himself; rather, it is a (certainly very natural) conception arising in a post-Newtonian context where Newton's physics itself is then taken as a reliable guide to metaphysics – for example, as we shall soon see, in Kant.

Everyone also took it for granted that the clearest and most fundamental example of causal agency is the creative activity of God. Newton's metaphysics of space then made it possible for him to maintain that universal gravitation involves an immediate exchange of momentum across empty space (as his physics requires) while, at the same time, preserving the more traditional ideas of causality and agency he shared with his contemporaries.[13]

From Newton's own point of view, his conception of the creation of matter by God makes maximal room for divine creative activity, and thereby avoids the threat of atheism opened up by the Cartesian conception of material substance. Nevertheless, this conception of divine agency is of course highly unorthodox, and, from a more traditional point of view, one would certainly not constrain God's creative activity by the requirement of local presence governing the interactions of material substances. From this point of view, it is Newton who opens up the threat of atheism (or rather pantheism) by seeming to materialize God. However, although Leibniz, for example, thus stands on much firmer theological ground than Newton, he does not have a competing metaphysics adequate for natural philosophy and physics. Kant, as we shall see, simply takes Newtonian physics (including action at a distance) for granted, while simultaneously rejecting Newton's theologically and metaphysically objectionable notion of absolute space.[14] Kant's problem, accordingly, is to construct a radically new approach to metaphysics along broadly Leibnizean lines, while also doing full justice to Newtonian physics.

In the pre-critical period, Kant attempts to fashion a direct unification of Leibnizean and Newtonian ideas, by starting with a Leibnizean metaphysics of monads and then building a Newtonian metaphysics of space, as it were, on top of this monadic metaphysics. The primary reality remains a non-spatial realm of ultimate simple substances, but these substances, for Kant, now have *both* purely internal, intrinsic properties *and* external or extrinsic relations.

13 My conception of the relationship of the importance of Newton's metaphysics to his physics (contrary to views like those of DiSalle) has much in common with the recent very extensive treatment in Janiak (2008): compare Janiak (2008, p. 174, note 19).

14 Kant famously rejects Newtonian absolute space and time (along with the opposing Leibnizean conception) in the transcendental aesthetic of the *Critique of Pure Reason.* He begins by forcefully rejecting the conception of "mathematical natural scientists," who "must assume two eternal and infinite self-subsistent non-entities (space and time), which are there (yet without there being something actual), only in order to comprehend everything actual within themselves" (A39/B56). He then adds (A40/B57): "The [mathematical natural scientists] succeed in so far as they keep the field of appearances free for mathematical assertions. On the other hand, they confuse themselves very much by precisely these conditions when the understanding pretends to go beyond this field." Compare note 16 below. (All translations from Kant's writings are my own, and I cite all writings – except for the first *Critique*, which is cited by the "A" and "B" pagination of the first (1781) and second (1787) editions – by volume and page numbers of the standard Akademie edition of *Kant's gesammelte Schriften.*)

Such external relations among the monads are not necessary for them to be the simple substances which they are, but they are necessary for them to exist – or, more precisely, to co-exist – together in a common world. In this way, God's creative activity has two distinguishable aspects: one act by which the simple substances themselves are created in the first place, and a second by which a number of such simple substances are joined together into a single world. This second act occurs in conformity with what Kant calls a "schema of the divine intellect," and it is in virtue of just such a schema, in the end, that what we know as the laws of nature then arise. More precisely, what we know as the fundamental forces of matter (attraction and repulsion) – together with the laws that govern them – are a direct expression of the divinely instituted external relations (of *co*-existence) between monads; and what we know as space is then the phenomenal expression of this same system of divinely instituted relations. Space is thus a secondary reality, derivative from the monads and their external relations, but, since the external relations between monads, for Kant, are just as real as their internal properties, it is a reality nonetheless – and not, as in Leibniz, a merely ideal "well-founded phenomenon." Indeed, since the fundamental force of attraction, for Kant, is explicitly modelled on Newtonian universal gravitation (as an immediate action at a distance through empty space), Kant explicitly links his pre-critical conception of space with the Newtonian conception of divine omnipresence.[15]

It is in the *Inaugural Dissertation* of 1770 that Kant makes a fundamental break with the Leibnizean philosophy – and, in a somewhat different fashion, with the Newtonian philosophy as well. Kant here first articulates his characteristic distinction between two independent rational faculties of the human mind – the pure understanding or pure intellect, on the one side, and pure sensibility or pure intuition, on the other. The former embodies the traditional categories and concepts of rational (Leibnizean) metaphysics, but it is the latter, for Kant, which now embodies the concepts and principles of pure mathematics. In particular, Kant now holds that mathematical knowledge is in no way purely intellectual, but is rather essentially intuitive or sensible, requiring the forms of pure sensibility, space and time. The world as we know it therefore bifurcates into two: the intellectual world described by traditional metaphysics

15 Kant makes this connection in the *New Exposition of the First Principles of Metaphysical Cognition* and the *Universal Natural History and Theory of the Heavens*, both appearing in 1755. For discussion, and references, see Friedman (1992, pp. 5–14). As I point out there, an echo of the Newtonian doctrine of divine omnipresence occurs as late as the Scholium to §22 of the *Inaugural Dissertation* (1770). (Kant of course had no knowledge of Newton's unpublished *De Gravitatione*, but, as observed above, essentially the same metaphysics of space surfaces in such well-known published writings as the General Scholium to the *Principia* and the Queries to the *Optics*.) For further recent discussions of Kant's pre-critical metaphysics see Laywine (1993), Schönfeld (2000), Watkins (2005). A recent volume of translations is Kant (1992).

(the Leibnizean metaphysics of ultimate simple substances as modified by the earlier Kant), and the sensible world as described by mathematics and mathematical physics in space and time. Although something like Newtonian space therefore remains as the foundation of this sensible world, space can no longer be conceived, as in Newton, as the sensorium of God – it is rather, as it were, the form of *our* sensorium, the form of our pure sensibility. Yet it is an unresolved problem, in the *Inaugural Dissertation*, how these two worlds are now supposed to be related, and, in particular, how the world described by mathematics and mathematical physics (the world as it appears to us) is related to the ultimate metaphysical reality of the intellectual world.

It is precisely this problem which finally gives birth to the critical philosophy in 1781. Kant now declares that purely intellectual, metaphysical knowledge – whether of immaterial things like God and the soul or of the ultimate simple substances which (according to both Leibniz and the pre-critical Kant) underlie the material world – is completely impossible, at least from a theoretical point of view. The pure intellect, considered entirely on its own and independently of any possible relation to sensibility, can issue only in the empty logical forms of Aristotelian syllogistic: in what Kant calls the "logical forms of judgement." And, while it is true that these forms then yield, in what Kant calls the "metaphysical deduction," the pure concepts or categories of the understanding (substance, causality, community, possibility, actuality, necessity, and so on), such pure concepts of the understanding are themselves entirely empty and without any "relation to an object" (again from a purely theoretical point of view) considered independently of our particular (human) forms of sensibility – space and time.[16] In short, it is only in virtue of spatio-temporal

16 Kant takes particular pains, in the second edition of the *Critique*, to emphasize that his conception of space and time as pure forms of sensibility is the only real alternative to the theologically disastrous Newtonian view (B71–72): "In natural theology, where one thinks an object that is not only no object of sensible intuition for us, but cannot even be an object of sensible intuition for itself, one takes care to remove the conditions of space and time from all of its intuition (for all of it cognition must be intuition and not *thought*, which is always a manifestation of limitations). But with what right can one do this, if one has previously made both into forms of things in themselves – and, indeed, into forms which, as a priori conditions of the existence of things, even remain when one has annihilated the things themselves? (For, as conditions of all existence in general, they must also be conditions for the existence of God.) There is therefore no alternative, if one does not pretend to make them into objective forms of all things, except to make them into subjective forms of our outer and inner mode of intuition. [This kind of intuition] is called sensible, because it is *not original* – i.e., it is not such that the existence of objects of intuition is itself given through it (which, as far as we can comprehend, can only pertain to the primordial being), but it depends on the existence of the objects, and is thus only possible in so far as the representative faculty of the subject is affected by them." This passage clarifies what Kant had in mind earlier in the aesthetic when he criticized the Newtonians for extending their conception of space and time *beyond* the "field of the appearances" (see note 14 above).

"schemata" produced by our pure intellect that rational knowledge of the phenomenal world is possible, and the task of showing how the pure intellect thereby injects itself into pure sensibility (space and time) so as to apply the pure categories of the understanding to sensible experience then becomes the problem of the transcendental deduction.[17] Such an injection of *our* pure intellect into *our* pure forms of sensibility now takes the place, as it were, of Kant's pre-critical doctrine that a schema of the divine intellect, by an analog of Newtonian divine omnipresence, is ultimately responsible for the order we perceive in the physical world.[18]

Pure metaphysical concepts – pure concepts of the understanding – can now be used for genuine (theoretical) knowledge only when applied to spatio-temporal "appearances," and thus only when "schematized" in terms of space and time: substance in terms of permanence, causality in terms of succession, and so on. When we do this, moreover, we find that specifically outer or spatial

17 Since, for Kant, the pure mathematician inscribes figures in space – in the process of Euclidean construction – by this same activity of the understanding, we thereby obtain, at the same time, an explanation of why all empirical objects in the phenomenal world (appearances) are necessarily subject to pure mathematics. This explanation essentially involves the categories of quantity and, in particular, the Axioms of Intuition (A165–166/B206): "The synthesis of spaces and times, as the essential form of all intuition, is that which also makes possible the apprehension of the appearance, and thus all outer experience, and therefore all cognition of the objects of experience; and what mathematics in its pure use demonstrates of the former [the essential form of all intuition], it is also necessarily valid for the latter [all outer experience, etc.]." And it is in precisely this way, too, that Kant demonstrates the necessary applicability of mathematics to sensible experience (and forestalls any possible Platonic gap between the two) which Newton secured by his metaphysics of space: compare the paragraph to which note 6 above is appended.

18 See note 15 above, together with the paragraph to which it is appended. As I observed, there is an echo of the pre-critical theory of divine omnipresence even in the *Inaugural Dissertation*, where Kant has already drawn a fundamental distinction between understanding and sensibility. The question Kant raises there (in the Scholium to §22) concerns precisely the *causes* of our sensible intuitions, and, in particular, the relationship between our sensible intuitions and the assumed ultimate substances constituting the intelligible world. The answer Kant (tentatively) suggests is that, since both our mind and these "external things" are sustained by a single infinite being, space, as the "sensibly cognized universal and necessary condition for the co-presence of all things" can thus be characterized as (God's) *phenomenal omnipresence.* In light of §22 itself, it appears that Kant is thereby invoking a pre-established harmony (instituted by God) between the purely intellectual reality of ultimate substances and our spatio-temporal sensibility to explain the necessary connection between this reality as it is in itself and as it appears to us. In §27 of the second edition *Transcendental Deduction*, Kant explicitly rejects such an explanation of the agreement between experience and its objects (which he calls a "**preformation-system** of pure reason") in favor of his new, critical explanation (which he calls an "**epigenesis** of pure reason") – where, as I understand it, the understanding rather creates the a priori order of sensible experience by injecting *itself* into the pure forms of sensibility.

intuitions are also necessarily required, so that, in particular, "in order to give something *permanent* in intuition corresponding to the concept of substance (and thereby to verify the objective reality of this concept), we require an intuition *in space* (of matter), because space alone is determined as permanent, but time, and thus everything in inner sense, is continually flowing" (B291). There is no longer any room (among the objects of theoretical knowledge) for mind-like or spiritual substances in the traditional sense, and there is no such room, therefore, for Leibnizean simple substances having only purely internal properties:

> Only that is internal in an object of pure understanding which has no relation at all (with respect to its existence) to anything different from itself. By contrast, the internal determinations of a *substantia phaenomenon* in space are nothing but relations, and it itself is nothing but a totality of mere relations. We are only acquainted with substance in space through forces that are active in space, either driving others into [this space] (attraction) or stopping their penetration into it (repulsion and impenetrability). We are acquainted with no other properties constituting the concept of a substance which appears in space and which we call matter. As object of the pure understanding, on the other hand, every substance must have internal determinations and powers, which pertain to [its] internal reality. However, what can I entertain as internal accidents except those which my inner sense presents to me – namely, that which is either itself a *thought* or is analogous to it? Therefore, Leibniz, after he had taken away everything that may signify an external relation, and therefore also *composition*, made of all substances, because he represented them as noumena, even the constituents of matter, simple substances with powers of representation – in a word, **monads**.
>
> (A265–266/B321–322)

The entire conception of the Leibnizean monadology – along with the more traditional conception of purely mental or spiritual substances – is now seen to rest on a fundamental mistake: neglecting the necessary spatio-temporal schematization of the pure concepts of the understanding.

But it now follows, similarly, that our basic concepts of action and efficient causality – by which one substance effects a change in another – must also be limited to the necessary conditions of specifically outer or spatial intuition (B66–7): "[E]verything belonging to intuition in our cognition (and thus excluding the feeling of pleasure and displeasure, and the will, which are certainly not cognitions) contains nothing but mere relations – [relations] of position in an intuition (extension), change of position (motion), and laws in accordance with which this change is determined (moving forces). But what may be present in the position, or what may be active in the thing itself aside from the change of position, is not thereby given." Aside from the intuitively presented laws governing the spatio-temporal changes of phenomenal substances, in other words, we have absolutely no conception of inter-substantial efficient causality at all – at least, once again, from a purely theoretical point of view.

It is in the *Metaphysical Foundations of Natural Science* of 1786 (appearing between the first and second editions of the first *Critique*) that Kant develops the "special metaphysics of corporeal nature" governing matter or material substance.[19] In particular, in the second or Mechanics chapter, the three Analogies of Experience governing the pure categories of substance, causality, and community are here specifically instantiated or realized by what Kant calls the three "laws of mechanics" – the conservation of the total quantity of matter, the law of inertia, and the equality of action and reaction – which Kant takes to be very close to (although not completely identical with) the three Newtonian Laws of Motion. In the case of matter or material substance, therefore, its possible changes and interactions are entirely delimited by these laws, in the sense that what it now *means* for one (material) substance to exert a causal action on another (so as, in this case, to effect a change of motion in it) is simply for a well-defined exchange of momentum to take place between the two. Thus, if two bodies exchange momentum at a distance across empty space (as, in Newton's theory of universal gravitation, they must), then they do in fact causally interact with one another at a distance, and there are absolutely no remaining grounds for raising metaphysical or theological objections.[20]

The second or Dynamics chapter introduces the two fundamental forces of repulsion and attraction – the one responsible for impenetrability, the other for gravitation. Proposition 7 states (Ak. IV, p. 512): "The *attraction essential to all matter* is an immediate action of matter on other matter through empty space." And, in the first remark to this proposition, Kant argues that to confine the activity of matter by the condition of contact would be an entirely arbitrary restriction (Ak. IV, 513):

> [T]o say that matters cannot act immediately on one another at a distance, would amount to saying that they cannot act immediately on one another except through the forces of impenetrability. But this would be as much as to say that repulsive forces are the only ones whereby matter can be

19 All translations from this work are taken from Kant (2004).

20 Leibniz's main theological objection to the Newtonian force of gravity, it will be recalled, was that it would be a "perpetual miracle" if a body could persist in orbital motion (without flying off along the tangent in accordance with the law of inertia) unless the material in a celestial vortex acted upon it by impact or pressure to maintain this orbital motion. Since Newton himself shared the widespread rejection of action at a distance at the time, he could not give the straightforward rejoinder later available to Kant: the Sun itself causes the planets to persist in their orbits, by precisely its immediate attraction across empty space. Compare notes 12 and 14 above, together with the paragraphs to which they are appended.

> active, or that they are at least the necessary conditions under which alone matters can act on one another, which would declare attractive force to be either completely impossible or always dependent on the action of repulsive forces. But these are both groundless assertions.

Once we conceive both impenetrability and gravitation as impressed forces in the Newtonian sense, governed solely by the Newtonian laws of motion, then there is no longer any reason to take one to be more intrinsically intelligible than the other.

In the second remark to the same proposition, however, Kant goes on to make a much stronger claim – that, in explicit opposition to Newton, gravitational attraction *must* be conceived as an essential active power of matter, operating immediately at a distance through empty space (Ak. IV, p. 515):

> [O]ne cannot adduce this great founder of the theory of attraction as one's predecessor, if one takes the liberty of substituting an apparent attraction for the true attraction he did assert, and assumes the *necessity* of an impulsion through *impact* to explain the phenomenon of [gravitational] approach. He rightly abstracted from all hypotheses purporting to answer the question of the cause of the universal attraction of matter, for this question is physical or metaphysical, but not mathematical. And, even though he says in the advertisement to the second edition of his *Optics*, "to show that I do not take *gravity* for an *essential* property of bodies, I have added one question concerning its cause," it is clear that the offense taken by his contemporaries, and perhaps even by Newton himself, at the concept of an original attraction set him at variance with himself. For he could not say that the attractive forces of two planets, those of Jupiter and Saturn, for example, manifested at equal distances of their satellites (whose mass is unknown), are proportional to the quantity of matter of these heavenly bodies, if he did not assume that they attracted other matter merely as matter, and thus according to a universal property of matter.[21]

21 The "one question concerning its cause" added to the second edition of the *Optics* is of course Query 21, where Newton famously speculates that a universal "Aetherial Medium" growing denser at greater distances from the heavenly bodies might explain the gravitational interactions between these bodies. However, this aether does not act by impact (as in the vortex theory favored by the mechanical philosophers), but is rather governed by short-range repulsive forces (between the particles of aether) responsible for its pressure (and thus density). So far, therefore, this particular speculation about a possible cause for gravity is consistent with Newton's remarks in the General Scholium to the second edition of the *Principia* (compare note 9 above), where he denies that such a cause can be mechanical (p. 92): "[T]his force arises from some cause that penetrates as far as the centers of the sun and the planets without any diminution of its power to act, and that acts not in proportion to the quantity of the *surfaces* of the particles on which it acts (as mechanical causes are wont to do) but in proportion to the quantity of *solid* matter, and whose action is extended everywhere to immense distances, always decreasing as the squares of the distances."

Kant's point here, specifically, is that Newton cannot leave the question of the "true cause" of universal gravitation entirely open, without fatally compromising the fundamental property of this interaction that the mutual accelerations in question are directly proportional to the masses or quantities of matter of the two interacting bodies.

I have considered Kant's argument in detail elsewhere,[22] so let me simply state it briefly here. Consider the system consisting of Jupiter, Saturn, and two of their respective moons. Newton's argument in Book III of the *Principia* crucially involves the idea that one can determine the masses of the primary bodies in question by the gravitational accelerations produced in their satellites. Newton assumes, in order to make this determination, that there are also gravitational accelerations of Saturn produced by Jupiter and vice versa. Then, in the most important step, Newton applies the equality of action and reaction directly to these two accelerations, so that the acceleration of Jupiter towards Saturn, multiplied by the mass of Jupiter, is equal and opposite to the acceleration of Saturn towards Jupiter, multiplied by the mass of Saturn. Newton assumes, in other words, that we can apply the conservation of momentum directly to this particular exchange, entirely independently of what other matter may or may not be found in between.[23] For Kant, this amounts, from a methodological point of view, to assuming, in effect, that no other matter is in fact involved, and that conservation of momentum within such an exchange is both necessary and sufficient for true causal action. So it is at precisely this point, therefore, that any metaphysical conception of cause pretending to compete with the conservation of momentum must now most definitely fall away.[24]

The importance of this argument is underscored, for Kant, by the circumstance that Newton's own inductive inference to the law of universal gravitation crucially involves such direct applications of conservation of momentum to gravitational interactions at a distance (in showing, for example, that Saturn's gravitational acceleration towards Jupiter is proportional to the mass of Jupiter and vice versa).[25] And it is further underscored, in particular, by the fact that

22 See Friedman (1992, pp. 153–159), and compare Friedman (1990).

23 As is well known, Roger Cotes objected to Newton on this score in their correspondence, and argued that Newton himself must therefore assume that gravitational attraction – as an immediate action at a distance – is in fact essential to matter. See Koyré (1968, chapter 7), and also Stein (1967).

24 By contrast, for Newton himself, as we have seen, this particular problem is solved by taking the ultimate causal agent here to be immaterial and, indeed, divine (compare again notes 12 and 14 above, together with the paragraph to which they are appended).

25 Thus, Newton's adherence to a neo-Platonic (Cambridge Platonic) metaphysics of space is not simply an additional (and arbitrary) assumption on his part, one which could easily be dropped. On the contrary, his own inductive argument for universal gravitation, in the context of the prevailing ideas about efficient causation and ultimate (divine) agency, more or less uniquely single out this metaphysics among the available alternatives.

the resulting determinations of the masses of the primary bodies in the solar system play a central role in Kant's parallel constructive procedure, articulated in the fourth chapter or Phenomenology, for arriving at the true motions of bodies from their apparent motions. We begin with our parochial perspective here on Earth, from which we can record both the observable phenomena governed by Galileo's law of fall and the observable relative motions of a variety of satellites in the solar system with respect to their primary bodies (the Moon relative to the Earth, the planets relative to the Sun, the moons of Jupiter and Saturn relative to their planets). The latter are just the phenomena expressed in Kepler's laws, and what we now find is that we can first determine the true state of rotation of the Earth (using small deviations from the law of fall manifesting what we now call Coriolis forces), and we can then determine the masses of all the primary bodies in the solar system (at least those actually having satellites) – with the result (as Newton shows) that the center of mass of the solar system is always very close to the center of the Sun.

Hence we can empirically determine, from the observable phenomena themselves, the true center of motion of the solar system, and this thereby counts as an approximation, for Kant, of Newtonian absolute space. However, since it is also true, for Kant, that the solar system itself rotates around the center of the Milky Way galaxy, this galaxy rotates around the center of a larger system of such galaxies, and so on ad infinitum, absolute space (the true center of motion of the entire Universe) is in the end what he calls an "idea of reason" – a forever unreachable regulative ideal we can only successively approximate in experience but never fully attain.[26]

Finally, since Newtonian absolute space is thus viewed as a regulative idea of reason, there is also an associated reconfiguration, for the critical Kant, of the relationships among space, the interactions of matter, and the idea of God. For the idea of God, too, is a regulative idea of reason. Indeed, there is an important sense in which it is the ultimate such regulative idea, since all human activity, together with the whole of nature, is ultimately subject to the idea of the Highest Good – the idea of a perfect community of all rational beings in a moral realm of ends, for which our only ground even to hope this could actually be achieved in nature (or, more precisely, successively approximated) is the idea of God (or, more precisely, divine providence). Moreover, Kant saw a deep analogy between the community of all rational beings in a moral realm

26 This constructive procedure for approximating absolute space in experience is analogous, in important respects, to the constructive method of Euclidean geometry (compare note 17 above). But the circumstance that the former can never be completed marks an essential difference between the two, closely related to Kant's view that the mathematical principles of pure understanding are *constitutive* with respect to intuition while the dynamical principles are merely *regulative* with respect to intuition (but constitutive with respect to experience): for further discussion see Friedman (1992, pp. 159–164).

of ends and the thoroughgoing community effected among all material bodies in the Universe by universal gravitation, and this is the basis, in fact, for his late (and very striking) re-interpretation of the Newtonian doctrine of divine omnipresence in a footnote appended to the General Remark to the Third Part of *Religion Within the Limits of Reason Alone* (1793):

> When Newton represents [the universal gravitation of all matter in the world] as, so to speak, divine universal presence in the appearance (*omnipraesentia phenomenon*), this is not an attempt to explain it (for the existence of God in space contains a contradiction), but rather a sublime analogy, in which it is viewed merely as the unification of corporeal beings into a world-whole, in so far as we base this upon an incorporeal cause. The same would happen in the attempt to comprehend the self-sufficient principle of the unification of the rational beings in the world into an ethical state and to explain the latter from the former. We know only the duty that draws us towards this; the possibility of the intended effect, even when we obey this [duty], lies entirely beyond the limits of all our insight.
>
> (Ak. VI, pp. 138–139)

For the critical Kant, in other words, the only possible meaning the idea of divine omnipresence (and divine providence) can now have is a purely *practical* meaning, in so far as we unconditionally obey the command of morality to strive to realize the realm of ends here on Earth, and, accordingly, we take the whole of that material nature of which we are a part to be in principle *capable* of such a realization (or, more precisely, its successive approximation). Kant thereby brings the characteristic mode of metaphysical investigation into the relationships among space, God, and matter practiced by his predecessors to a close, and transforms it – without remainder – into transcendental philosophy.

15

How Newton's *Principia* changed physics

GEORGE E. SMITH

Newton expressly intended his *Principia* to produce three revolutionary changes in the way physics and astronomy were being conducted:

1. Theorizing in physics should center on identifying fundamental *forces* of nature and characterizing them as quantities related by laws to other measurable quantities.
2. Astronomy should abandon the 1500 year tradition of trying to describe complex orbital motions directly from observations and instead derive them from the forces acting on the orbiting bodies.
3. Physics and astronomy should demand of themselves a much higher standard of evidence in theorizing than just success in deriving observed phenomena from speculative hypotheses.

The *Principia* did indeed ultimately effect all three of these revolutions – the first two obvious to anyone familiar with the subsequent history of physics and orbital astronomy, but the third less obvious. Here accordingly we shall focus on the third, though explaining how the book changed the standards of evidence in physics will involve us with the first two as well. While those two emerged in Newton's thinking only with the *Principia*, the third he had set as a goal more than a decade and a half earlier with the remark, "But truly with the help of philosophical geometers and geometrical philosophers, instead of conjectures and probabilities that are being blazoned everywhere, we shall finally achieve a natural science supported by the greatest evidence" (Newton 1984, p. 87).

One reason why the revolution in evidence is less obvious has been a longstanding, but nonetheless ill-informed misconstrual of the evidential reasoning not only in the *Principia*, but in subsequent research in orbital mechanics as well. The first two sections of the chapter will contrast that construal of the reasoning with the evidence problem Newton saw himself as facing when he started writing the *Principia*. Another reason for the revolution in evidence being less obvious has been the complexity of the *Principia*'s approach to marshalling evidence, involving as it does several distinct elements that are usually discussed, when at all, in isolation from one another. Sections 15.3 through 15.6, forming the main body of the chapter, will lay out those revolutionary elements one by one, indicating how each reflects this evidence problem. A final section will then consider the whole formed by those elements, asking how clearly Newton saw it as a response to the principal worries he had about potential shortcomings in his evidence.

15.1 Introduction: the question

Prima facie, the evidence put forward on issues about orbital and other kinds of motion at the time of Laplace's *Celestial Mechanics* (1799–1805) was of much higher quality than the evidence on those issues at the time of Copernicus or, for that matter, Kepler. Furthermore, of the several advances that were made between Copernicus and Laplace that enabled more decisive evidence to be developed, nothing appears to have been more important than Newton's *Principia.* The central question of this chapter is, How did Newton's *Principia* change the way in which evidence was marshalled in orbital research, and thereby in physics generally?

Prompting that question is a view that empirical science is first and foremost a process of *turning data into evidence.* Evidence is a two-place relation between data and claims that reach beyond them.[1] Data, in and of themselves, are not evidence for one claim more than another; something beyond data is always needed for them to become evidence for anything. In experimental research novel data are often generated, sometimes with masterful artifice, precisely because their likely value as evidence is clear beforehand. Often, however, data are abundantly available in nature, and the problem is one of figuring out what they show about the world. Linguistics provides a clear example of this, for data on the syntax of our native languages are immediately at hand, but we still do not have a fully adequate account of the syntax of any natural language.[2]

In orbital astronomy, too, data have always been accessible in the form of nightly observations of relative positions of objects on the celestial sphere, and efforts to turn those data into evidence go back at least as far as the Babylonians. The introduction of the telescope at the beginning of the seventeenth century provided access to new data, but almost all of the evidence bearing on orbital astronomy until the middle of the eighteenth century came from instrument-aided naked-eye observation. New ways of turning those data into evidence concerning celestial physics emerged between Copernicus and Laplace. Those

1 In deference to Charles Saunders Peirce, who surely would have insisted that evidence involves a third place as well as the two cited, perhaps I should say "two- (or more) place relation."

2 In conversation a few years ago Noam Chomsky and I were unable to figure out which of us first began speaking of science as an endeavor to turn data into evidence, followed immediately by the remark that evidence is a relation and being a datum is not. Regardless of who did, the thought was originally no less his than mine.

new ways are the central concern of this chapter. How precisely did Newton's *Principia* contribute to them?

Seen from that perspective, the most frequently quoted portion of the Preface to the first edition of the *Principia* indicates that Newton saw it as illustrating a new way of turning data into evidence:

> our present work sets forth mathematical principles of natural philosophy. For the whole difficulty of philosophy seems to be to discover the forces of nature from the phenomena of motions and then to demonstrate the other phenomena from these forces. It is to these ends that the general propositions in books 1 and 2 are directed, while in book 3 our explanation of the system of the world illustrates these propositions. For in book 3, by means of propositions demonstrated mathematically in books 1 and 2, we derive from celestial phenomena the gravitational forces by which bodies tend toward the sun and toward the individual planets. Then the motions of the planets, the comets, the moon, and the sea are deduced from these forces by propositions that are also mathematical. If only we could derive the other phenomena of nature from mechanical principles by the same kind of reasoning! For many things lead me to have a suspicion that all phenomena may depend on certain forces by which the particles of bodies, by causes not yet known, either are impelled toward one another and cohere in regular figures, or are repelled from one another and recede. Since these forces are unknown, philosophers have hitherto made trial of nature in vain. But I hope the principles set down here will shed some light on either this mode of philosophizing or some truer one.
>
> (Newton 1999, p. 382)

Newton's approach to turning data from astronomical observations into evidence about forces governing orbital motions and then about those motions themselves was more multi-faceted than is generally appreciated. This chapter aims to lay out his approach and the rationale behind it and then to indicate ways it altered orbital astronomy and physics generally.

There is a commonplace answer to the question of how Newton's *Principia* resulted in exceptionally high quality evidence, an answer that can be extracted from undergraduate textbooks in physics, if not explicitly found in them: What Newton did in the *Principia* was to put forward the law of gravity, together with his three laws of motion, by way of explaining Kepler's so-called laws; and the resulting theory then turned out to explain ever so much more, including the respects in which actual planetary motions deviate from Kepler's laws. In other words, until a small residual discrepancy in the precession of the perihelion of Mercury emerged in the second half of the nineteenth century, Newton's theory turned out to be consistent with all observations, and in that sense passed every test to which it was put. On this view, the high quality of the evidence coming out of the *Principia* lay in the *range* of observations with which the laws it proposed turned out to be in agreement and the *precision* of that agreement.

That view of the *Principia* offers a conception of the enterprise of science in sharp contrast with my "process of turning data into evidence." Science is instead first and foremost a process of coming up with basically correct theories.[3] Once such a theory is in hand, the evidence for it will largely just fall into place as tests of it emerge and it survives them. The task of marshalling evidence itself presents no special challenge save for an occasional need for ingenuity in devising new, more telling tests. Granted this is a stick-figure summary. Even in this form, however, it explains why textbooks in science include so little discussion of details of the evidence.

As a preliminary step toward motivating the view of the evidence for Newton's theory to be presented below, let me offer two objections to the view I have just sketched. First, it distorts history. For example, Kepler's rules for calculating orbits were far from established at the time Newton began drafting the *Principia*. Indeed, they appear never to have been called "laws" before the *Principia*.[4] Furthermore, a number of so-called tests of Newton's theory were not expressly offered as tests of it at the time. A blatant example of this is Cavendish's experiment, which in physics textbooks is usually presented as a decisive test of Newton's law of gravity even though Cavendish himself said that what he was doing was to measure the (mean) density of the Earth.[5]

A second objection lies in Newton's own outspoken dismissal of hypothetico-deductive evidence. As quoted above, Newton claimed in the first edition of the *Principia* to have derived the law of gravity from phenomena of orbital motion; and at the end of the second edition he added that "hypotheses, whether metaphysical or physical, or based on occult qualities, or mechanical, have no place in experimental philosophy" (Newton 1999, p. 943). Important to note here is Newton's lifelong reason for dismissing hypothetico-deductive evidence: "For if the possibility of hypotheses is to be the test of the truth and reality of things, I see not how certainty can be obtained in any science; since numerous hypotheses may be devised, which shall seem to overcome new difficulties" (Newton 1978, p. 106). That Newton was responding to such worries alone gives reason for examining whether he had an alternative approach.

This chapter does not aim to argue against deductivist accounts of evidence. It lays out an alternative account of the logic of the evidential reasoning in gravitation research and the way in which this logic derives from the *Principia*.

3 Through most of this chapter *theory* designates bodies of lawlike relations among quantities. Here, however, it designates more a way of conceptualizing a range of phenomena. Sylvain Bromberger has labeled those two senses of "theory" "theory$_1$" and "theory$_2$" in Bromberger (1992).

4 Curtis Wilson (2000, p. 225) has noted that Kepler's orbital rules appear never to have been called "laws" in print before Leibniz did so shortly after publication of the first edition of Newton's *Principia*.

5 Cavendish (1798) simply assumed the law of gravity throughout. Had he been trying to test it, he would have at least varied the masses of the spheres in his trials.

This alternative, I claim, is more accurate historically and more consistent with Newton. Most of all, however, I want to claim that, on its very face, it is a more tenable account of why the *Principia* had the effects it did on how evidence is developed in physics.

15.2 Complexity and parochialism: the evidential problem

At the time Newton began drafting the *Principia* in 1685, there were several competing approaches to calculating planetary orbits, at least seven of which were known to him. Kepler's approach employed the ellipse and his area rule – *planets sweep out equal areas with respect to the Sun in equal times* – but he did not use his 3/2 power rule – *the semi-major axes of the ellipses vary as the 2/3 power of the orbital periods* – to infer the lengths of the semi-major axes directly from the periods. Instead, he inferred these lengths from observations. Jeremiah Horrocks found that he could improve on Kepler's *Rudolphine Tables* by inferring the semi-major axes directly from the periods, which were known to very high precision (Wilson 1978). The other five approaches employed some alternative to Kepler's area rule for determining where each planet is on its orbit at any given time. Ismaël Boulliau (1657, pp. 29–31) used a geometric construction involving the empty focus. Thomas Streete (1661, pp. 53f. and 39f.) used this same geometric construction, but followed Horrocks in inferring the semi-major axes from the periods. Vincent Wing (1651, p. 44ff.) initially used an oscillating equant – that is, a center of equiangular motion oscillating about the empty focus – and later (1669, pp. 130, 144, 151, 170, 176) switched to his own geometric construction. And Nicolaus Mercator (1676, pp. 163–171) used a still different geometric construction.[6]

Of these different approaches, Kepler's was computationally the most complicated. None of them gave predictions that were consistently within the accuracy of pre-telescopic observations (Wilson 1989). The errors were more or less comparable in all seven – around a third of the apparent width of the Moon. The only thing common to all of them was the ellipse, which is striking because the orbits are actually so near to being circular; the most elliptical of the orbits then known, Mercury's, has a minor axis only two percent shorter than its major axis. Equally striking, the ellipse itself was something that Newton did not consider appropriate to use as evidence for his law of gravity (Smith 2002b).

A question accordingly at the forefront of orbital astronomy in 1679 when Hooke first put the matter to Newton, and still in 1684 when Halley did the same, was which of the different ways of calculating orbital trajectories was to be preferred. The brief tract Newton had registered with the Royal Society in December 1684, "De Motum Corporum in Gyrum," gives an answer to

6 Mercator precedes his account of his new hypothesis with an extensive review of Kepler's area rule and alternatives to it.

the question: "The major planets orbit, therefore, in ellipses having a focus at the center of the Sun, and with their *radii* drawn to the Sun describe areas proportional to the times, exactly as Kepler supposed" (Newton 1967–1981, VI, p. 49). A plausible construal of Newton's reasoning here is that Kepler's (and Horrocks's) calculation rules admit of a physical explanation – they result from an inverse-square centripetal force – while the competing rules of calculation appear unlikely to do so. On this construal, the evidential problem to which Newton was offering a solution was one that had been posed by Kepler: given the imprecision of observation and measurement, any number of distinct curves can fit observations to any given level of precision; only physical considerations can pick out the true curve from among these. That, however, is not the evidential problem Newton addressed in the *Principia*, for by the time he began drafting it a few weeks later, he had come to see ways in which orbital motion poses a far more ramified challenge.

Specifically, Newton had come upon a deep reason why none of the ways of calculating the orbits were yielding results within observational accuracy. In the registered version of "De Motu" he had concluded that there are what we would now call inverse-square centripetal acceleration fields not merely around the Sun, but also around the Earth, Jupiter, and Saturn.[7] He saw no reason why the inverse-square accelerative tendency toward, for instance, Jupiter exhibited by its four known satellites would not extend all the way to the Sun, so that Jupiter and the Sun would be interacting with one another. If they do interact, then this interaction should not produce any change in the motion of the center of gravity of the system. (This is just the principle of inertia applied to a system of interacting bodies.) From this, Newton reached an extraordinary conclusion in an augmented version of the "De Motu" tract that did not become public until 1893:

> By reason of the deviation of the Sun from the center of gravity, the centripetal force does not always tend to that immobile center, and hence the planets neither move exactly in ellipses nor revolve twice in the same orbit. There are as many orbits of a planet as it has revolutions, as in the motion of the Moon, and the orbit of any one planet depends on the combined motions of all the planets, not to mention the action of all these on each other. But to consider simultaneously all these causes of motion and to define these motions by exact laws admitting of easy calculation exceeds, if I am not mistaken, the force of any human mind.
>
> (Newton 1962, pp. 256 and 281)[8]

7 While Newton never employed the term "field," my use of it is not so anachronistic as it might at first seem, for he did speak of centripetal motive forces being "propagated through the surrounding regions" (Def. 8).

8 I have altered the translation along lines derived from Curtis Wilson. This passage did not become public until Ball (1893).

In other words, no calculation scheme like Kepler's or any of the others was ever going to yield exact predictions, not for the comparatively uninteresting reason that observation itself is always imprecise, but for the far more important reason that the true motions are too complex to allow exact computation.

Newton was not the first to decry the complexity of true motions in the world. Both Galileo and Descartes had concluded that motion under air resistance forces (the other topic of Newton's *Principia*) is too complex to allow a science (Galileo 1974, 224; Descartes 1985–1991, III, p. 9f). Newton knew that Descartes had said much the same of the motions of the planets, adding that their trajectories are sure to change from one epoch to another:

> Finally, we must not think that all the centers of the Planets are always situated exactly on the same plane, or that the circles they describe are absolutely perfect; let us instead judge that, as we see occurring in all natural things, they are only approximately so, and also that they are continuously changed by the passing of the ages.
>
> (Descartes 1991, p. 98)

This raised a second worry: not only might Keplerian motion be but one of several comparably accurate approximations to the true trajectories, whose complexity defies exact description; but also, Kepler's and all the other approximations might be mere epochal parochialisms, projected from a few decades of observations that were assumed to be representative, but instead were systematically misleading historical accidents.

As noted earlier, Newton saw his *Principia* as illustrating a new way of doing science. I contend that Newton's new "experimental philosophy" – as he came to call it – was in response to the complexity of the real world and the risk that our straightforward empirical access to it is parochial. That is, it is an approach to developing evidence in the face of, first, a complexity that leaves room for many competing descriptions of observed regularities and, second, a lack of any immediate means of obviating respects in which the observed regularities we invoke as evidence might be misleadingly parochial. In forming this new approach, Newton introduced a number of changes in approach that have persisted at least in the subsequent history of gravitational research, if not in physics generally. It is to these changes that we now turn.

15.3 The Newtonian conception of theory

One change came out of Newton's realization that physics – or, more specifically, mechanics – cannot help but include, within its scope, its own theory of measurement. Newton hints at this in the Preface to the first edition of the *Principia* when he concludes that "*geometry* is founded on mechanical practice and is nothing other than that part of *universal mechanics* which reduces the art of measuring to exact propositions and demonstrations" (Newton 1999,

p. 382). The point emerges more forcefully in the opening section of the book, "Definitions." This section might just as well have been called "Critical Reflections on Measurement." The definitions of *quantity of matter* or *mass*, *quantity of motion* (our *momentum*), and *force*, besides indicating how the terms are going to be used in the *Principia*, emphasize their measures. Indeed, the definitions of *quantity of matter* and *quantity of motion* expressly identify each as "a measure." The discussions following their definitions make clear that both *mass* and *force* are what we now call "theoretical quantities." That is, values for them must be inferred from other measurements and the inferences in question presuppose theoretical claims within mechanics. In the case of *mass* Newton even invokes a pendulum experiment to justify inferring values from *weight* (Newton 1999, pp. 404 and 807).

Following the explicit definitions is the famous Scholium on space and time, the central concern of which is the distinction between "absolute, true, or mathematical space, time, and motion" and "relative, apparent, or common space, time, and motion." The space, time, and motion that we observe fall into the relative, apparent, or common category. Values in the absolute, true, or mathematical category have to be inferred from them. In the paragraph ending the Scholium, Newton remarks:

> It is certainly very difficult to find out the true motions of individual bodies and actually to differentiate them from apparent motions, because the parts of that immovable space in which the bodies truly move make no impression on the senses. Nevertheless, the case is not utterly hopeless. For it is possible to draw evidence partly from apparent motions, which are the differences between true motions, and partly from the forces that are the causes and effects of the true motions... But in what follows, a fuller explanation will be given of how to determine true motions from their causes, effects, and apparent differences, and conversely, of how to determine from motions, whether true or apparent, their causes and effects. For this was the purpose for which I composed the following treatise.
>
> (Newton 1999, p. 414f)

Viewed from the perspective of the rest of the treatise, the natural way to interpret what Newton is saying here is that true motions are ones for which all theory-mediated measurements of the relevant forces yield the same values. But then, not just values of *mass*, *force*, and *quantity of motion*, are theory-mediated; so too are values of *velocity*.

Newton may not have been the first to realize that physics must include its own theory of measurement. In one respect the point is obvious, for measurement is itself a physical process and measurements in mechanics involve mechanical processes. Still, Newton does appear to have been the first to appreciate two of its implications. One is that any method of measurement is provisional, subject to replacement by a method that is deemed preferable at some later point. Newton expressly calls attention to this in the case of *time*:

> In astronomy, absolute time is distinguished from relative time by the equation of common time. For natural days, which are commonly considered equal for the purpose of measuring time, are actually unequal. Astronomers correct this inequality in order to measure celestial motions on the basis of a truer time. It is possible that there is no uniform motion by which time may have an exact measure. All motions can be accelerated and retarded, but the flow of absolute time cannot be changed. The duration or perseverance of the existence of things is the same, whether their motions are rapid or slow or null; accordingly, duration is rightly distinguished from its sensible measures and is gathered from them by means of an astronomical equation. Moreover, the need for using this equation in determining when phenomena occur is proved by experience with a pendulum clock and also by eclipses of the satellites of Jupiter.
>
> (Newton 1999, p. 410)

Newton's defense of sidereal time by appealing to the pendulum clock and the eclipses of the satellites of Jupiter is striking because both of these presuppose theories that were first published in the 1670s – the latter including a theoretical redetermination of *simultaneity* in astronomy.[9] The evidence for both the regularity of pendulum clocks and the eclipses of Jupiter's satellites in turn invokes sidereal time. In other words, a confluence of theoretical considerations lies behind the choice of sidereal time. But then, if preferred methods of measurement are subject to change as new theoretical considerations emerge, any lawlike relationship between measured quantities must also be provisional, subject to change as science advances.

The second implication of physics having to include its own theory of measurement that Newton appears to have been the first to appreciate is less spectacular, but no less important. It concerns how theory-mediated measurement can enter into evidence. Insofar as all measurement presupposes theoretical considerations of one sort or another, there is no reason to insist that a theory be firmly established first, before new methods of measurement are derived from it. Huygens in 1659 had used his theoretical laws for the cycloidal and conical pendulums to measure the strength of surface gravity in two different ways, obtaining the same value to four significant figures (Yoder 1988). Huygens, however, seems never to have viewed the stability, convergence, and precision of his measurements as evidence for the theory of uniform gravity from which he derived his pendulum laws. Newton saw not only this, but also that in the long run such stability, convergence, and precision of measurement

9 The former presupposed the theory of the pendulum in Huygens's *Horologium Oscillatorium* of 1673, and the latter, Olaus Römer's determination of the finite speed of light in 1676.

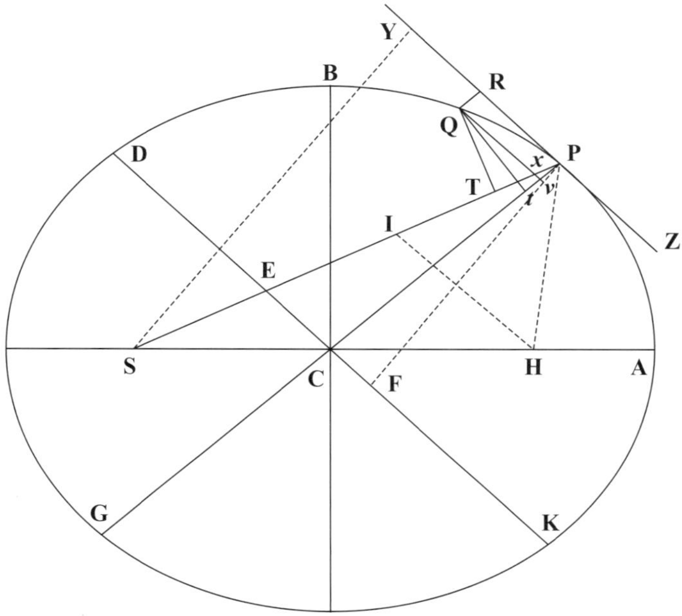

Figure 15.1 The figure accompanying both Propositons 10 and 11 on elliptical orbits in the First Edition

cannot help but be a primary form of evidence for any theory in mechanics. But then, success in theory-mediated measurement should be regarded as evidence for a theory right from the outset, even before any other evidence for it is available.

Freeing quantities like *force* and *time* from any specific way of measuring them allowed them to be considered in the abstract, as mere mathematical quantities separate from any question about physical mechanism. This in turn allowed Newton to introduce a new way of employing mathematical theory in physics. Galileo and Huygens had used mathematics to derive testable consequences from their theories of motion. Newton, by contrast, developed a generic theory of motion under centripetal forces, deriving results not only for inverse-square forces, but also for forces that vary linearly with distance, that vary as the inverse-cube of distance, and finally that vary as any function whatever of distance. Two consecutive propositions of Book I of the *Principia* illustrate this new form of mathematical theory (Figure 15.1):

> *Proposition 10*: Let a body P revolve in an ellipse; it is required to find the law of centripetal force tending toward the center [C] of the ellipse.
>
> *Solution*: The force varies as CP directly.

> *Proposition 11*: Let a body P revolve in an ellipse; it is required to find the law of centripetal force tending toward a focus [S] of the ellipse.
>
> *Solution*: The force varies inversely as the square of SP.

(Keep these two in mind, for I will be making a further point about them in the next section.) In both of these, questions about how such centripetal forces might be physically effected are irrelevant. Book I of the *Principia* consists of more than ninety "if-then" propositions linking motions to forces, forces to motions, and macrophysical forces to microphysical forces composing them; throughout, *force* is treated as a quantity, independent of what brings it about.

Book I is best described as giving a generic mathematical theory of motion under forces directed toward a center *with no regard to how such forces might be physically realized.* Mathematical theory of this sort is a second way in which Newton changed physics.

Late in Book I Newton indicates why he wants a generic mathematical theory in which forces are treated without regard to the question of the physical mechanisms producing them:

> Mathematics requires an investigation of those quantities of forces and their proportions that follow from any conditions that may be supposed. Then, coming down to physics, these proportions must be compared with the phenomena, so that it may be found out which conditions of forces apply to each kind of attracting bodies. And then, finally, it will be possible to argue more securely concerning the physical species, physical causes, and physical proportions of these forces.
>
> (Newton 1999, p. 588)

The passage brings out two points. First, generic mathematical theory for Newton is an instrument for turning data into evidence, more specifically for enabling phenomena to answer theoretical questions about physical forces and processes. The idea is to have generic mathematical theory and phenomena together dictate physical theory. Second, Newton is prepared to leave questions within physical theory open when he can't find phenomena to answer them. In particular, he answered questions about the physical species of celestial orbital forces – they are one in kind with terrestrial gravity – and questions about their physical proportions – the law of gravity; but he found no way of addressing their physical causes. His theory of gravity was rejected by many of the leading figures of his time precisely because it left the question of the cause of gravity open, and therefore offered no explanation of how gravitational forces can act over vast distances. A third way in which Newton's *Principia* ended up changing physics was by making limited physical theory – theory without mechanistic explanations – respectable.

The phrase, *limited physical theory*, describes a methodological innovation. I would be remiss not to point out how revolutionary that innovation also was from a substantive standpoint. Newton was the first to propose that physical theory focus on fundamental kinds of force. In doing so he introduced an intermediate level of theory, between mere description of observed regularities in the manner of Galileo's *Two New Sciences*, on the one hand, and laying out full mechanisms in the manner of Descartes' *Principia*, on the other. Newton's *Principia* showed that this intermediate level of theory, with laws of force but no mechanism, is still sufficient to answer a whole host of questions about observed regularities – especially questions about whether observed regularities, as described, are suitable for playing a role in evidence. This, I take to be the point Newton was making in his remark about his predecessors having "hitherto made trial of nature in vain" in the passage from the Preface to the first edition I quoted in Section 15.1. It is to this intermediate level of theory that we now must turn.

15.4 Evidential reasoning in Newton's *Principia*

The features of Newton's approach identified so far give Newtonian theory what Pierre Duhem called an "abstract, symbolic" character. But as Duhem himself showed, that character need not be peculiar to generic theories. To appreciate the advantage Newton found in insisting on a generic theory of motion under centripetal forces, we need to look in detail at how he reasoned from orbital phenomena to physical proportions of force.

Newton inferred that the force acting on the planets is centripetal from Kepler's area rule, and he inferred that it is inverse-square first from Kepler's 3/2 power rule, and then more strictly from the absence of precession of the orbits. The following three theorems from his generic mathematical theory are the "inference-tickets" licensing those inferences:

> *From Propositions 1–3*: A body sweeps out equal areas in equal times with respect to a second body if and only if the net force on the body is compounded of a centripetal force directed toward the second body and the whole accelerative force acting on the second body.
>
> (Newton 1999, pp. 444–448)
>
> *From Corollaries to Proposition 4*: The periodic times of bodies moving uniformly in circular orbits about a central body vary as the 3/2 power of their distance from the central body if and only if the centripetal forces acting on the orbiting bodies vary inversely as the square of the distances.
>
> (Newton 1999, p. 451)
>
> *From Proposition 45 and its Examples*: The centripetal force acting on a body moving in a nearly circular orbit is inverse-square if and only if the perihelion (or perigee) of its trajectory does not precess.
>
> (Newton 1999, pp. 539–545)

We will discuss below why it was important to Newton to establish not merely the conditionals licensing his inferences from phenomena, but the bi-conditionals as well.

Each of these enabling theorems is richer than it first appears to be. In corollaries to the first Newton adds that an increasing areal velocity with respect to a point entails that the net force is directed forward of that point, and vice versa. In a corollary added in the second edition to the second, Newton points out that the period varies as the distance to the power n (where n need not be an integer) if and only if the centripetal force varies inversely as distance to the power $2n - 1$. And his precession theorem actually gives an algebraic formula tying the rate of precession of a nearly circular orbit to the exponent in the force rule:

> Let θ be the angle at the force center from aphelion to perihelion in a very nearly circular orbit; then the centripetal force varies as $R^{(n-3)}$, where $n = (180/\theta)^2$.

In other words, the enabling theorems show that a real acceleration is a theory-mediated measure of the direction of the force on an orbiting body; the exponent in the power rule relating periods to distances of a collection of bodies moving uniformly in circular orbits is a theory-mediated measure of what we now call the strength of the acceleration field around the central body; and the rate of orbital precession is a theory-mediated measure of the exponent of distance from the center in the centripetal force rule for any one orbiting body.

Still more important, if either clause in any of the three enabling bi-conditionals holds only approximately – Newton's phrase is *quam proxime*, very nearly – then the other clause still holds *quam proxime*. This follows trivially from the algebraic relations in the second and third cases, and Newton expressly points it out in a corollary to the first:

> *From Proposition 3, Corollaries 2 and 3*: The areas with respect to the central body are as the times *quam proxime* if and only if the force retaining the moving body in an orbit around the central body tends toward the central body *quam proxime*.

Thus, for every "if-then" statement that Newton uses to reason from orbital phenomena to conclusions about forces, he takes the trouble to show that the consequent still holds *quam proxime* so long as the antecedent holds *quam proxime*! In effect, then, the logical form of the propositions that serve to license Newton's reasoning from orbital phenomena is not really "if-then", but rather "if *quam proxime*, then *quam proxime*," and hence the premises describing the orbital phenomena in question are required to hold only *quam proxime*. Consequently, Newton is not begging any questions about whether the area rule or some other rule is the proper one for locating planets in their orbits as a function of time, for he knew that the area rule agrees with all the other rules at least *quam proxime.*[10]

That Newton was consciously engaged in such a form of approximative reasoning explains why he did not use Proposition 11, given above, to infer the inverse-square proportion from the Keplerian ellipse. For, he had shown that, when the eccentricity is small, as it is in the case of several of the planets, the contrast between Propositions 11 and 10 – that is, between the proportion implied by ellipses with equal areas about a focus and about the center – becomes a problem. It is demonstrably not true that, if the trajectory is a Keplerian ellipse *quam proxime*, then the exponent in the force rule is -2 *quam proxime.* Current textbooks typically do present Newton as reasoning from the Keplerian ellipse to the inverse-square proportion, and this inference is certainly inviting on its face. Newton, however, was more careful than these textbooks. Nowhere, not even in the original "De Motu" tract, did he make this move. Instead he always relied on the 3/2 power rule and the absence of orbital precession to infer the inverse-square (Smith 2002b).

Newton's not inferring the inverse-square from the Keplerian ellipse, together with his taking the trouble to show that the "if-then" statements he did employ hold in *quam proxime* form, provides the strongest evidence that he was self-consciously engaged in approximative reasoning. There are several other signs of it as well. The phrase, "*quam proxime*," occurs 139 times in the *Principia.* The numerical summaries of the observed relations between periods and mean distances at the beginning of Book III all display some deviation from an exact 3/2 power relation, and Newton openly acknowledges that the Moon is not in perfect accord with the area rule and that its orbit is not stationary. Moreover, Newton had decided before he began writing the *Principia* that the area rule does not hold exactly for the planets, and he had concluded while writing the *Principia*, if not before, that their orbits are not perfectly stationary. Granting that he was engaged in approximative reasoning is thus a way of absolving him of accusations of rank hypocrisy (Lakatos 1978). Finally, it undercuts the complaint made by Duhem and others that the law of gravity cannot be deduced from Keplerian phenomena, taken as premises, because it entails these premises are false: the seeming self-contradictory element of Newton's "deduction" disappears once the reasoning is construed as approximative.

Of course, what this means is that, strictly speaking, the evidence Newton offers for his law of gravity shows that it is true of the motions of the planets and their satellites, but only *quam proxime*, only to high approximation. Newton is perfectly aware that the orbital evidence does not show that the law holds exactly. Nevertheless, he *takes* the law to hold exactly – or, what in practice amounts to the same thing, to hold unqualifiedly within the limits of

10 See Mercator (1676).

observational accuracy. In the third edition of the *Principia*, he gives a rule of reasoning to authorize this leap from approximate to exact:

> Rule 4: *In experimental philosophy, propositions gathered from phenomena by induction should be taken to be exactly or very nearly true notwithstanding any contrary hypotheses, until yet other phenomena make such propositions either more exact or liable to exceptions.*
>
> (Newton 1999, p. 796)[11]

Note the phrase here, *should be taken to be*. The leap from approximate to exact amounts to a research strategy.[12]

This is a fourth way in which Newton's *Principia* changed physics. On the one hand, because the scope and precision of observation are limited and the real world is complex, evidence in physics can at most show that theoretical claims hold to certain levels of precision over a limited range of observations. On the other hand, when appropriate requirements are met – as expressed in the phrase, *gathered from phenomena* – physicists should nevertheless proceed as if these theoretical claims hold exactly. What we need to do now is to see how this research strategy works.

Newton appears to have required that two conditions be met before he was willing to take the law of gravity as exact. He expressly states in Proposition 8 of Book III that he did not conclude that the inverse-square proportion holds exactly until he had established that it holds exactly around a sphere of uniform (or spherically symmetric) density if it holds with respect to all the particles of matter forming that sphere (Newton 1999, p. 811). In other words, he required there to be *some* configuration for which the macroscopic forces around a body, composed out of forces toward its parts, would accord exactly with the law. Second, he also appears to have required that there be *some* identifiable circumstances in which the phenomena from which the law was inferred *would* hold exactly. As the following quote indicates, the subjunctive here is Newton's, not mine:

> if the Sun were at rest and the remaining planets did not act upon one another, their orbits would be elliptical, having the Sun in their common focus, and they would describe areas proportional to the times.
>
> (Newton 1999, p. 817f.)

11 Newton adds by way of explanation for why the rule is needed, "This rule should be followed so that arguments based on induction may not be nullified by hypotheses."

12 Similarly, the inference that Newton can strictly speaking draw from orbital phenomena has to be restricted as holding only over the period of time for which observations were available, primarily from Tycho Brahe forward. Newton's third rule of reasoning licenses the inference *to be taken* to hold universally into the past and future, just as his fourth rule licenses the inference *to be taken* to hold exactly, or at least to high approximation. For reasons of space I have chosen not to go into his third rule and its strategic role in ongoing research here.

Newton's theory of gravity did not show that the planets revolve in stationary Keplerian orbits, but instead that they decidedly do not. Nevertheless, Kepler's area and 3/2 power rules, and also the absence of orbital precession, are special in one crucial respect. According to the theory, each of these phenomena *would* hold exactly were it not for small gravitational forces directed toward the individual planets. (The deductions of these phenomena are where the converse parts of the enabling bi-conditionals come into play, though here in their exact, not their *quam proxime* form.) Because the actual motions of the planets are exceedingly complex, they can be approximated to any given level of accuracy in an indefinite number of ways. Newton is requiring the approximations from which physical theory is "deduced" to be ones, that according to the theory, *would* hold exactly in specific identifiable conditions. This is a key element of Newton's way of marshalling evidence in the face of complexity.

What Newton has done here is to single out a particular kind of idealization in science: *an approximation that, according to theory, would hold exactly in certain specifiable circumstances.* For want of a better word, I am going to call idealizations of this type "Newtonian" because of the special role they play in his approach to evidence. They include not only the phenomena from which his theory was inferred, but also further phenomena inferred from his theory, such as Kepler's ellipse. Science contains idealizations of all sorts of kinds – mathematical simplifications, schematics of experiments and apparatus, explanatory models, etc. They were commonplace before Newton. Both Galileo and Huygens, for example, had taken the curved surface of the Earth as flat in their treatment of projectile and pendular motion. My point is that Newton singled out and placed great emphasis on one particular kind of idealization: *approximations to the actual world that are deduced from his theory of gravity as holding exactly in specified circumstances.* Idealizations of this sort are not simplifications made in the process of arriving at physical theory; they are offspring of physical theory. Idealizations of this kind and the use to which Newton put them are a fifth way in which his *Principia* changed physics.

As defined here, a Newtonian idealization requires an overarching theory from which the claim of exactness in specified circumstances is inferred. Thus, Newton's law of universal gravity was not itself a Newtonian idealization, for there was no overarching theory which entailed its exactness in specified circumstances. Equally, Galileo's uniform acceleration in vertical fall in the absence of air resistance was not a Newtonian idealization; Galileo claimed it holds exactly in this circumstance, but he did not infer this claim from an overarching theory. As noted above, the theory of gravity in the *Principia* forms an intermediate level of physical theory between mere description of phenomenal regularities in the manner of Galileo's *Two New Sciences*, on the one hand, and laying out full mechanisms in the manner of Descartes's *Principia*, on the other. Intermediate though it may be, this level of theory is nevertheless sufficient to assign certain phenomenal regularities the status of a Newtonian idealization.

In general, what Newtonian idealizations do is to shift the focus of ongoing research to deviations from the ideal – that is, to discrepancies between theoretically deduced, idealized approximations to the world and the world itself. This shifting of the focus in research to discrepancies between theory and observation is a sixth way in which Newton's *Principia* changed physics – perhaps, the most conspicuous way. It was with Newton that the phrase "exact science" took on its current meaning. Instead of explaining away such discrepancies, as for example Galileo had invariably done, they became a source of continuing evidence that had promise of becoming increasingly discriminating. The complexity of the actual motions thus became not an impediment to high-quality evidence, but historically the source of it! And, having discrepancies between theory and observation became not a negative, but a positive.

Newton generally left research into such discrepancies and what they tell us about the world to future generations. The one exception was the non-Keplerian motion of our Moon. In his solution for the systematic deviation from the area rule known as Tycho's variation, Newton starts from the idealization of the Moon in a circular orbit with the Earth at the center and first determines how the gravitational force of the Sun would distort this orbit, elongating it in a direction perpendicular to the line from Earth to Sun. He then calculates the deviation from the area rule, obtaining seven-eighths of Tycho's value, and he ends by pointing out that including the effects of orbital eccentricity would make the calculation still more accurate. Notice what is happening here: one idealization, a simple circular orbit, is being replaced by another idealization that gives a better approximation, the idealization now known as the "variational" orbit produced by the perturbing effect of the Sun on the simple circular orbit. This process can continue, yielding a sequence of successive idealizations that should achieve increasingly closer agreement with observation. They are nonetheless all idealizations.

Newton gets an even more impressive result for the mean motion of the line of nodes, the 18-year cycle in lunar and solar eclipses known since the Babylonians, obtaining a result within three-tenths of one percent without considering eccentricity. The Moon's orbit is exceedingly complex; no attempt just to describe it geometrically had ever come close to the level of accuracy Kepler and others had achieved for the planets. Newton's announced purpose in making his calculations was to show that the best hope for genuine progress lay not in conventional observational astronomy, but instead in his theory of gravity and a sequence of deduced successive approximations.

Systematic deviations from Keplerian motion and other Newtonian idealizations can be thought of as a kind of phenomena in their own right. Only no one can observe them. They arise from the residual discrepancies between observation and idealizations deduced from theory – that is, from the difference that remains after Newtonian idealizations are subtracted from observation. I prefer to call them "second-order phenomena" for just this reason. They presuppose specific theory, and they cease having any meaning – they cease to exist – without that theory. As remarked above, on Newton's approach the focus in ongoing research shifts from primary phenomena and pursuit of a theory covering them to residual discrepancies between that theory and observation. With this shift, the goal in research becomes one of identifying second-order phenomena and determining what they are telling us about the world. If indeed the actual planetary motions are complex to a degree that exceeds exact mathematical description, then residual discrepancies will always remain. The requirement put on ongoing research is that the increasingly refined, and hence more complicated, idealizations result in continually smaller discrepancies.

Notice that, when the focus shifts in this way, further research is being predicated on the theory, and hence the theory has become an instrument entering constitutively into ongoing research. Earlier I said that Newton's generic mathematical theory was a tool for turning data into evidence – more specifically for turning phenomena into evidence for physical theory. Now we see how his physical theory was no less a tool for turning data into evidence – this time, second-order phenomena into evidence about such things as what other forces are contributing to the complex motions of the planets and their satellites. This is a seventh way in which Newton and his *Principia* changed physics. Before him the primary role of physical theory was to explain observed phenomena. With him, that role became subsidiary, superseded by the role physical theory is to play in ongoing research. This was the change that such contemporaries as Leibniz and Huygens had the most trouble seeing and appreciating.

15.5 Beyond the *Principia*: the logic of the continuing evidence

The *Principia* ends up spotlighting a number of potential second-order phenomena beyond the specific lunar inequalities for which Newton obtained results. There were also the past irregularities in the supposed 75-year return of what we call Halley's comet; the thoroughly confusing, not-yet-characterized departures of Saturn and Jupiter from Keplerian motion; and the as-not-yet-confirmed precession of the perihelia of the planets entailed by Newton's theory of gravity. The discrepancy between Newtonian theory and observation that became historically most important, however, was the precession of the apogee (or perigee) of the Moon. The apogee of the Moon shifts in very complicated ways from one orbit to the next. This precession nevertheless has a well-behaved mean value: on average the apogee moves forward slightly more than 3 degrees per revolution. Newton had used his precession theorem to calculate the effect of the gravitational force of the Sun, obtaining a mean precession of 1 degree, 31 minutes, 28 seconds – essentially half the observed value. A question he never managed to answer was why the Sun's gravity could readily account for

90 percent of the Moon's mean departure from the area rule and more than 99 percent of the mean motion of the nodes, yet only 50 percent of the mean motion of its apogee.

Newton was less than candid about this discrepancy in the *Principia.* In an appendix to the first English translation in 1729, however, John Machin, the orbital astronomer who was closest to Newton during the 1720s, made the problem clear to all:

> But the [mean] motion of the apogee, according to this method, will be found to be no more than 1°37′22″ in the revolution of the moon from apogee to apogee, which (according to observations) ought to be 3°4′7.5″.
>
> So that it seems there is more force necessary to account for the motion of the moon's apogee than what arises from the variation of the moon's gravity to the sun, in its revolution about the earth.
>
> But if the cause of this motion be supposed to arise from the variation of the Moon's gravity to the Earth, as it revolves round in the elliptic epicycle, this difference of force, which is nearly double the former, will be found to be sufficient to account for the motion, but not with the exactness as ought to be expected. Neither is there any method that I have ever yet met with, upon the commonly received principles, which is perfectly sufficient to explain the motion of the moon's apogee.
>
> (Machin 1729, p. 30f.)

Machin went on to concede that, so far as he could see, it is impossible to derive the motion of the apogee and the alteration of the eccentricity "from the laws of centripetal forces."

During the 1740s Euler, Clairaut, and d'Alembert took up the problem, each concluding that solar gravity could account for only half the observed precession. Clairaut went the furthest, for he took the eccentricity of the lunar orbit into consideration. Specifically, he adapted a method of Euler's to derive all the terms for the Sun's effect in which eccentricity occurs to the first power, and still found Newtonian theory giving only half of the observed motion. Based on this, and some discrepancies between recent geodetic measurements and Newtonian theory, he registered a paper with the Royal Academy of Paris proposing that Newton's law of gravity has to be amended with a $1/r^4$ term. This provoked quite a debate within the Royal Academy. As this debate continued, Clairaut decided to derive the terms for the Sun's effect to the next highest order, obtaining terms in eccentricity squared and cubed. When he calculated their effect, he discovered to his surprise that, even though the lunar eccentricity is less than 0.06, these higher-order terms are not negligibly smaller in magnitude than the first-order eccentricity terms. Together with those terms they yield a nearly exact value for the mean motion of the apogee out of the inverse-square effect of the Sun's gravity (Waff 1976). D'Alembert, ever the querulous one, went on to calculate the contribution from terms of the next highest order, confirming that they do not mess up Clairaut's result.

Two points should be made about Clairaut's efforts before turning to the historical importance of his result. First, Clairaut's reasoning when he proposed adding an inverse r^4 term was that the data from planetary orbits which Newton had used in deriving the law of gravity involved distances that were too large to expose the need for the additional term; in other words, Clairaut was saying that the data on which Newton had relied were parochial. Second, the perturbational approach used in Clairaut's calculations – and in virtually all subsequent calculations in celestial mechanics – introduces another layer of complication in the logic of the evidence: Clairaut was deriving not a Newtonian idealization for a specific case of the three-body problem of the Sun, Earth, and Moon, but instead a computational *approximation* to such an idealization. The intractability of the mathematics stood in the way of a rigorous derivation of the exact solution called for in my definition of a Newtonian idealization. Given any remaining discrepancy, then, a question arises about the extent to which it reflects imprecision in the method of calculation versus the need for some refinement, like an unaccounted for force, in the idealized model presupposed in the calculation.

Clairaut's result was much heralded. In a private letter to him, Euler remarked,

> the more I consider this happy discovery, the more important it seems to me, and in my opinion it is the greatest discovery in the Theory of Astronomy, without which it would be absolutely impossible ever to succeed in knowing the perturbations that the planets cause in each other's motions. For it is very certain that it is only since this discovery that one can regard the law of attraction reciprocally proportional to the squares of the distances as solidly established; and on this depends the entire theory of astronomy.[13]

A year later Euler made the same point in print, arguing that they could now be certain that there are inverse-square forces between Jupiter and Saturn causing the confusing irregularities in their motions that had been observed:

> since M. Clairaut has made the important discovery that the movement of the apogee of the Moon is perfectly in accord with the Newtonian hypothesis . . . , there no longer remains the least doubt about this proportion . . . And if the calculations that one claims to have drawn from this theory are not found to be in good agreement with observations, one will be always justified in doubting the correctness of the calculations, rather than the truth of the theory.
>
> (Euler 1752, p. 4f.)

13 Letter from Euler to Clairaut, 29 June 1751, in Bigourdan (1929, p. 38f.); translation from Wilson (1980, p. 143).

The hyperbole in these pronouncements is not so extreme as it may at first appear. As Euler explained in his *Theory of the Moon's Motion* of 1753,[14] the mean motion of the apogee provides an exceptionally sensitive measure of the exponent in the rule of centripetal force. The exponent is exactly −2 if and only if the orbit is perfectly stationary *in the absence of any forces beyond the centripetal force holding it in orbit.* The trouble, of course, is that the Moon's orbit is not stationary, but precesses on average 3 plus degrees per revolution. Even so, as Newton showed in the *Principia*, one can still conclude that the centripetal force on the Moon is inverse-square, *quam proxime.* The question whether it is exactly inverse-square can then be addressed by identifying forces beyond the Earth's gravity and seeing whether any discrepancies remain once the effects of these forces are taken into account. How is one to identify such further forces? By first taking into account the gravitational forces from the Sun and the planets, and then seeing what discrepancies, if any, remain.

This brings me to an eighth way in which Newton's *Principia* changed physics. His approach opened the way to a new form of evidence – *evidence indirectly accruing to a theory from the success of research predicated on it.* The original evidence for Newton's law of gravity showed at most that it holds to high approximation, yet he took it to hold exactly and deduced idealizations from it. This strategy leaves open the question, how exact is the law? We now see that ongoing research on deviations from these idealizations can continue to bring evidence to bear on the law in general and on this question in particular – albeit indirect evidence. Clairaut's result, together with the observed lunar precession, provided direct evidence that, whatever other forces are perturbing the lunar orbit, they are much smaller than the perturbing force from the Sun's gravity. Indirectly, however, it provided evidence that Newton's law of gravity holds to a still higher level of approximation than his original evidence implied. More generally, focusing research on deviations from Newtonian idealizations and demanding progressively smaller discrepancies between observation and the idealized model of the world is a strategy for exposing limitations in this law. Correlatively, because the idealizations are deduced from the law, taken as exact, evidence accrues to the law from continuing success in pinning down robust physical sources of still remaining deviations.

I claim that Newton's new approach to marshalling evidence was a response to the complexity of the world. Those before Newton despaired of any such complexity, concluding it would always limit the quality of evidence that can be achieved in science. What Newton did was to find a way to turn the very complexity into a source of increasingly more telling evidence. This, to me, was the ultimate genius of the *Principia.*

14 Euler (1753, pp. 71–72); translation in Wilson (1980, p. 144).

The logic of this evidence needs to be made clear, for the Clairaut example can be misleading. At first glance, one might think that Clairaut deduced the theoretical mean motion of the lunar apogee, and its close agreement with observation therefore provided hypothetico-deductive evidence for the law of gravity. That is a mistake. For Clairaut to have deduced the motion, he would first have had to assume that no other forces are at work beyond the perturbing force of the Sun – a question that was surely still open. (Newton himself at one point intimated that the missing one and a half degrees in the mean precession of the lunar orbit might be coming from the Earth's magnetic field (Newton 1999, p. 880).) Rather, Clairaut was only deducing the effect of a specific perturbing force entailed by Newton's theory of gravity. More generally, all such calculations of orbital motions in celestial mechanics are merely deducing the effects of the forces specified. When the result of any such calculation matches observation very closely, the appropriate conclusion is that any further perturbing forces either do not exist or are of much lesser consequence. A failure to match observation leaves open the possibility that some other force is making a significant contribution. A less outspoken version of Euler's statements about the strong evidence Clairaut had provided for the inverse-square would still have been valid even if it had turned out that the missing one and a half degrees was from the Earth's magnetic field.

To put the matter differently, the test to which Newton's theory is put by the deviations from his idealizations is more subtle than a simple hypothetico-deductive construal suggests. Newtonian idealizations are by definition ones that, according to the theory of gravity, would hold exactly in specified circumstances. But then any deviation from them must result from some physical departure from those circumstances, an additional celestial force not yet taken into consideration. The implication, in other words, is that any deviation from an idealization must be *physically significant* within the context of the theory – this in contrast, for instance, to being merely a reflection of the mathematical scheme that happened to have been chosen in curve fitting. The test to which Newtonian theory is put in ongoing research centers on the question, *Is every deviation from a Newtonian idealization physically significant?* The evidence that accrues to Newtonian theory comes from pinning down robust physical sources of deviations – a continuing process that ought to result in ever smaller discrepancies between observation and the idealized representation of the world. Whenever all residual discrepancies drop below a then-current level of accuracy of observation, the appropriate conclusion must have a somewhat Popperian flavor: at least for the moment, observation has ceased providing any basis for identifying either further complications in the world or respects in which theory is inadequate.

The conclusion, *any other perturbing forces are of much lesser consequence*, is a variant of a problematic auxiliary assumption, *all forces acting on the planets other than the designated gravitational forces have very small effects*, required in hypothetico-deductive construals of the evidence. Both are variants of Karl Hempel's *the constituent bodies of the system are subject to no forces other than their mutual gravitational attraction* – his paradigmatic example of a "proviso" in his paper, "Provisos: A Problem Concerning the Inferential Function of Scientific Theories" (Hempel 1988). On the account of the logic of the evidential reasoning I am offering, these are not assumptions in the deductions of celestial mechanics at all. The deductions are spelling out the (idealized) consequences of a set of specified forces. The point of the resulting Newtonian idealizations is not as such to test the theory of gravity by making predictions with it, but rather to address the question, are any forces beyond those specified of consequence? Hempel's provisos, instead of being assumptions in the deductions, are conclusions that emerge when the answer to this question is no – that is, when the discrepancies between the calculated and the observed motions are sufficiently small.

Our sense that celestial mechanics over a period of centuries generated extraordinary support for Newton's law of gravity stems not from its having continually yielded predicted motions within observed accuracy (which, in fact, it never really did), but from the success in pinning down – that is, identifying and further confirming – the physical sources of forces responsible for ever more subtle complexities in the observed motions. The extent to which orbital motions are dominated by gravitational forces has been among the most remarkable findings of celestial mechanics.

A long tradition of carelessly talking about evidence in celestial mechanics as if it were straightforwardly hypothetico-deductive has obscured the extent to which the focus of ongoing research has been on questions about further forces. In saying that Newton's theory of gravity has been an instrument in post-*Principia* research in celestial mechanics, I mean more than just that this theory has been presupposed in instance after instance of evidential reasoning throughout that research. Because the overall pattern has been one of successive approximations, the evidence for the physical sources of the increasingly smaller deviations from the current ideal presupposes not only Newtonian gravity, but also the previously identified sources of the larger deviations from the earlier ideals. In other words, the ongoing evidential reasoning has presupposed the theory of gravity in an increasingly ramified fashion. To question the law of gravity is to throw into question a huge collection of facts (or, if you prefer, quasi-facts) about the world that post-*Principia* research has established. The burden of proof required to discard the law of gravity thus became increasingly large – which is the same thing as saying that the law became increasingly entrenched.

At the beginning of this chapter I posed a question about the much higher quality of evidence after the *Principia* than before it. The continuing evidence in gravitation research, and not Newton's original evidence, is the high-quality evidence in question. Listing all the evidence of this sort that has unfolded over the last 300 years would be a Herculean task. Still, it is instructive to list a few highlights: (1) Clairaut's prediction of the month of return of Halley's comet in 1759 after taking the gravitational forces of Jupiter and Saturn into account; (2) Laplace's 1785 discovery of the 890-year "Great Inequality" in the motions of Jupiter and Saturn; (3) Leverrier and Adams deducing the existence of an eighth planet, Neptune, from residual anomalies in the motion of Uranus (1846); (4) the Hill–Brown theory of the Moon (1919), involving more than 1400 physically significant terms, which finally brought lunar theory to the level of accuracy of the planets and revealed as well that the Earth's rotation is not uniform, and hence that sidereal time is not an exact measure of time.[15]

The key point, however, is that the process of research is continuing, for there will always be discrepancies. The difference now is that they are at levels of significant figures of which Euler and Clairaut, much less Newton, scarcely ever dreamed.

15.6 Parochialism and the continuity of evidence

The glaring omission in my list of highlights is the precession of the perihelion of Mercury, the discrepancy that finally falsified, so to speak, Newton's law of gravity. Newton already knew that most of the apparent precession of Mercury's perihelion is just that – apparent, stemming from the precession of the equinoxes, the 26,000-year wobble of the Earth. He had no way to calculate the true precession implied by his theory of gravity, in part because he had no way to determine the mass of Venus. By the end of the nineteenth century it became clear that Newton's theory was 8 percent slow for the true 225,000-year precession of Mercury's orbit. This 43 arc-seconds per century residual proved recalcitrant: Newtonian theory was unable to provide any physical source for it, and hence it appeared not to make physical sense within the context of that theory. Later, of course, it turned out to be evidence for the new theory of gravity of Einstein's general relativity.

This residual discrepancy in the very slow motion of Mercury's perihelion shows how Newton's response to the complexity of orbital motion was, at the same time, a response to the risk that our observations are somehow parochial. What better way was there to expose any such parochialism than to push his theory for all it is worth until some subtle discrepancy emerges that might shed light on just how it is parochial? The obvious alternative, *contra* Newton's fourth Rule of Reasoning, was to try to obviate parochialism from the outset by proposing a wide range of competing theories compatible with the available data and then identifying cross-roads experiments to choose among them or to falsify them one-by-one. One problem with this alternative was the degree to

15 These four as well as other contributions from continuing research in orbital mechanics are discussed in Smith (forthcoming b).

which the complexity of the world would have limited the quality of evidence in trying to decide early among the competing theories. The more serious problem, however, was the absence of any way of assuring that the list of proposed alternatives would cover respects in which the available data were indeed parochial. The specific respect in which Newton's data are now known to have been parochial was not something anyone imagined at the time.

The residual in the precession of Mercury's perihelion also brings a distinction into sharper focus that was alluded to in the preceding section, the distinction between Newtonian idealizations and curve-fits. At the end of the nineteenth century Simon Newcomb prepared a new set of planetary tables that, together with the theories of the orbits underlying them, remained the basis for orbital predictions until the switch to direct numerical integration of the equations of motion on high-speed computers in the 1980s. Newcomb's tables were based on Newtonian gravity plus an added term in the calculation of the secular precession of the perihelia of the four inner planets. This term, which turned out to amount to a fudge factor, consisted of a constant times the mean motion of the planet in its orbit.[16] It added 43.37 arc-seconds per century in the case of Mercury, and less for the other planets. With this term included, the calculated orbits involved an element of curve-fitting, and hence they were no longer, strictly speaking, Newtonian idealizations. No longer could any systematic discrepancy between the calculated and observed precessions be automatically taken to be symptomatic of some physical source not yet taken into account. For, the discrepancy might instead represent some physically arbitrary feature in the curve-fit. In particular, suppose a new systematic discrepancy were to emerge in the case of Mercury much smaller than the prior 43 arc-seconds, say a discrepancy around 0.4 arc-seconds. Why should that discrepancy automatically be taken as a sign of some yet-to-be-noted physical effect when it could just as well be attributed to the choice of mean speed as the curve-fitting parameter or to the decision not to include terms in mean-speed squared?

Both Newtonian idealizations and curve-fits can be carried out in sequences of successive approximations in response to a complex world. Curve-fits aim at prediction, with the mathematical scheme chosen to reflect a trade-off between accuracy of fit and calculational ease. Least-squares curve-fits have the virtue of minimizing the expected value of the square of the error in prediction, relative to the adopted mathematical scheme, and errors in prediction are expected to be Gaussian. In general, whether the curve-fitting criterion is least-squares or otherwise, the goal is for errors in prediction to have a random character; and, in that regard, curve-fits attempt not to highlight discrepancies, but to achieve prediction within a certain level of precision, in the process sweeping lesser

16 Specifically 0.0000000806 times the centennial mean motion; see Newcomb (1898, p. 12).

discrepancies under the rug. Discrepancies between Newtonian idealizations and observation, by contrast, are not expected (or even desired) to have a random character, for the driving research question is, what further physical factors, if any, need to be taken into consideration?

The distinction between Newtonian idealizations and curve-fits is especially important when worried about the possibility that available data are somehow misleadingly parochial. With each successive approximation in curve-fitting, physical sources of features in the data become progressively more submerged in a welter of choices embedded in the mathematical scheme. As a consequence, there are multiple potential sources for a recalcitrant discrepancy besides some respect in which the accessible data are physically parochial. By contrast, the further a sequence of successive approximations progresses with Newtonian idealizations, the stronger the grounds are for attributing any recalcitrant discrepancy to some physical parochialism. For, the alternative, that the theory has been (by its own standards) radically wrong all along, is countered by the record of success so far in pinning down robust physical sources of discrepancies.

Within two decades of Newcomb's new orbital tables, Einstein put forward his theory of general relativity, and Newcomb's curve-fitting response to the residual in Mercury's perihelion ceased to matter. Einstein's relativity produced a conceptual revolution in physics, but not really a revolution in evidence. For, Newtonian gravity holds as an asymptotic limit of Einstein's, specifically the static weak-field limit. This had two important consequences. First, save for qualifications about levels of precision, all the evidence for Newtonian gravity carried over immediately to Einsteinian gravity. Physics did not have to go back to an earlier time and begin reconstituting evidence. The data that had been evidence for Newtonian gravity were guaranteed to be evidence for Einsteinian gravity as well insofar as, under the conditions of our solar system, Newtonian gravity amounts to an approximate special case of Einsteinian, and evidence for a special case counts as evidence, though often of reduced strength, for the more general theory of which it is a special case. Second, Einstein's theory did not out-and-out nullify the evidential reasoning supporting Newton's theory. That is, it did not entail that the evidence supporting the prior theory was merely illusory, and never truly evidence at all. For, the steps in the original reasoning can still be justified, though the justifications themselves have to be amended to include qualifications – for example, the qualification that a Euclidean metric provides a good approximation so long as the gravitational field is weak. If the reasoning had not remained so justified, then the 43 arc-second residual, taken in itself, could not have provided evidence for Einstein's theory in the manner in which it did, for that residual is a Newtonian second-order phenomenon that presupposes Newtonian gravity; were the evidence for Newtonian gravity illusory, the specific value of 43 arc-seconds would be nothing but an artifact of an illusion.

This point can be put in another way. The transition from Newtonian to Einsteinian gravity certainly did entail a revolutionary discontinuity in conceptual structure. Nevertheless, because Newtonian gravity holds in the static weak-field limit, the transition did not entail any discontinuity in evidence. The residual 43 arc-seconds per century in the precession of the perihelion of mercury is a Newtonian second-order phenomenon that turns out to be physically significant, but only within the context of the less parochial theory. From the point of view of Einstein's theory, Newton's is a limited special case reflecting a systematic bias in the data to which we have ready access, in particular orbital data from within our solar system. From the point of view of Newton's theory, on the other hand, Einstein's is a more general theory – one among an indefinite number of possible more general theories – that a strictly Newtonian phenomenon helped single out and substantiate. The transition from Newton to Einstein yielded the discovery that one specific respect in which the readily available gravitational data within our solar system are parochial is that the fields are so weak and so nearly static.

What we have here is another kind of idealization in physics: a theory that, even though it would never hold exactly in any realizable circumstance, nevertheless holds in a mathematical limit with respect to a more general theory. Let me call these "limit-case" idealizations. They have a different role in the development of evidence from the Newtonian idealizations I have been emphasizing so far. Their most important contribution is to allow evidence to remain continuous and hence cumulative across theory change, especially across theory change involving removal of parochialisms.

Although it has gone largely unnoticed, continuity of evidence is itself a form of evidence. Of course, the continuity of evidence from Newton to Einstein cannot be evidence for Newton's theory itself, or even Einstein's. It is evidence for something more basic that is common to both. In taking the huge inductive leap from inverse-square gravity and the orbits of six planets to universal gravity, Newton was making two tacit, but nonetheless indispensable taxonomic assumptions: (1) gravity marks a distinct natural kind or, to use Newton's phrase, a physical species; and (2) orbital motions of our planets and their satellites represent a pure enough example to typify this species as a whole. Both of these assumptions, at the time Newton made them, could not help but fall largely in the category of wishful thinking.

The research predicated on Newton's law of gravity over the next two centuries succeeded spectacularly in reducing the gap between theory and observation; and this success provided support for these two taxonomic assumptions. All of this success nevertheless came from phenomena within our planetary system over a very short period of astronomical time. Consequently, none of it spoke directly to the possibility that gravity is an accidental feature of our solar system, in much the way that many geological phenomena are mere artifacts of the Earth's history, and not symptomatic of deep physical laws.

Besides revealing the weak-field parochialism of our planetary system, general relativity has enabled data from the universe at large to become evidence bearing on gravitation theory. A prerequisite for continuity of evidence in theory change is that the taxonomy underlying the old theory remain essentially intact within the new theory. The fact that evidence remained continuous from Newtonian to Einsteinian gravity has accordingly provided much stronger support than ever before for the claim that gravity marks a distinct physical species.

Surprising though it may be, limit-case idealizations are something else that Newton introduced in the *Principia* and hence a ninth way in which this book changed physics. Newton, of course, had concluded that the data supporting the theory of uniform gravity acting along parallel lines developed by Galileo and Huygens were parochial, coming as they all did from the narrowly confined region near the surface of the Earth. This theory and the evidence for it were nonetheless important to Newton for a series of reasons: (1) he expressly invokes results by Galileo confirming that the acceleration of gravity is independent of weight as evidence that mass is proportional to weight (Newton 1999, p. 806); (2) he similarly invokes Galileo's vertical fall and parabolic projection and Huygens's pendulum results as evidence for his first two laws of motion, and indeed Huygens's pendulum measurements of surface gravity offered the best evidence for those laws (Newton 1999, p. 424); and (3) Huygens's measured value of surface gravity, which presupposed uniform gravity acting along parallel lines toward a flat surface, provided crucial evidence that terrestrial gravity extends to the Moon (Newton 1999, pp. 803–805).

The way in which Newton chose to treat uniform gravity as a limit-case of universal gravity will surprise anyone not thoroughly familiar with the *Principia*. Newton does not argue that Galilean gravity is simply an approximation to inverse-square gravity over small distances – that is, distances over which the variation in the acceleration of gravity is too small to matter. Instead, he treats it as a limit-case of gravity that varies linearly with distance from the center of a spherical Earth, specifically the limit at the surface as the Earth's curvature approaches zero. The statement of this limit-case idealization occurs in Section 10 of Book I of the *Principia*, in a corollary to one of the propositions on hypocycloidal pendulums – that is, pendulums the arc of which is defined by the trajectory of a circle rolling not on a flat plane, but on the underside of a spherical surface (Figure 15.2):

> *Prop. 52, Cor. 2.* Hence also follows what Wren and Huygens discovered about the common cycloid. For if the diameter of the globe is increased indefinitely, its spherical surface will be changed into a plane, and the centripetal force will act uniformly along lines perpendicular to this plane, and our cycloid will turn into the common cycloid. But in that case the length of the arc of the cycloid between that plane and the describing point will come out equal to four times the versed sine of half of the arc

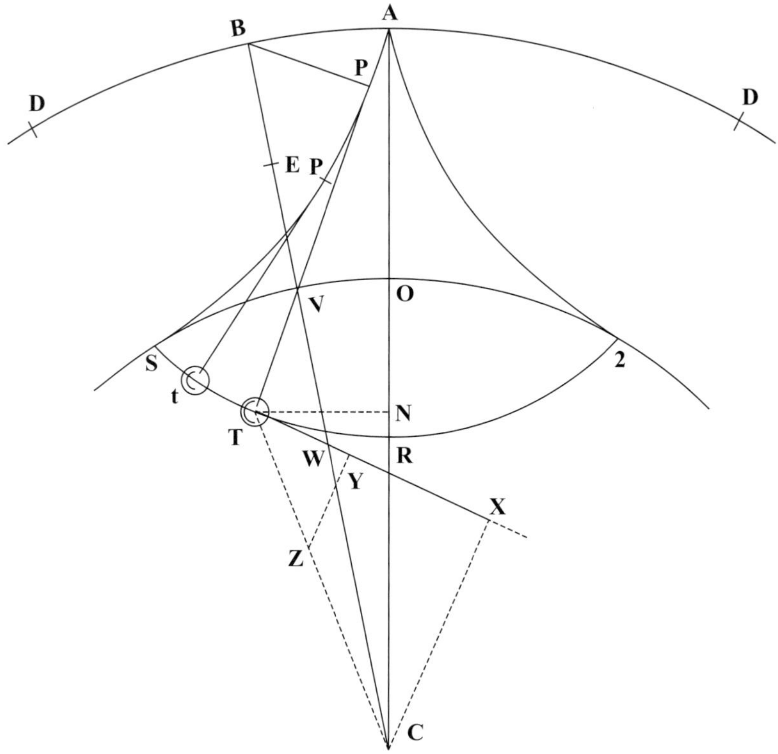

Figure 15.2 The figure accompanying Propositions 51 and 52 on isochronous motion of pendulums

of the wheel between that same plane and the describing point, as Wren discovered; *and a pendulum between two cycloids of this sort will oscillate in a similar and equal cycloid in equal times, as Huygens demonstrated. But also the descent of heavy bodies during the time of one oscillation will be the descent Huygens indicated.*

Moreover, the propositions that we have demonstrated fit the true constitution of the earth, insofar as wheels, moving in the earth's great circles, describe cycloids outside this globe by the motions of nails fastened in their perimeters; and pendulums suspended lower down in mines and caverns must oscillate in cycloids within globes in order that all their oscillations may be isochronous. For gravity (as will be shown in Book 3) decreases in going upward from the surface of the earth as the square of the distance from the earth's center, and in going downward from the surface is as the distance from that center.

(Newton 1999, p. 555f., emphasis added)

As Newton shows later in Book I, his universal gravity entails that, below and up to the surface of a uniformly dense sphere, the net gravitational force varies linearly with distance, while above the surface it varies as the inverse-square (Newton 1999, pp. 593–597 and 617f.).

Galileo's and Huygens's results can be shown to hold to high approximation in inverse-square gravity so long as the vertical distances are small.[17] What then does Newton gain with his limit-case idealization? In Huygens's measurement, the strength of surface gravity is inferred, via his law of the cycloidal pendulum, from the measured period, which he had shown does not vary with the length of the arc of the bob. This isochronism was crucial to Huygens's measurement beyond its being explicit in the law. Thanks to isochronism, no attention needed to be given to the length of the bob's arc and whether it was varying during the measurement of the period. Isochronism was accordingly a key factor in the claimed precision of Huygens's measurement, a precision important to Newton. Now, hypocycloidal pendulums are isochronous under gravity that varies linearly with distance from the center (Prop. 51), but not under inverse-square gravity! Therefore, what Newton's specific limit-case idealization enabled him to show was that the logic underlying Huygens's measurement is not nullified when uniform gravity acting along lines parallel to one another is replaced by his universal gravity. (Notice that this is precisely what Newton said in the portion of the quotation I italicized above.) Remarkably, the *Principia* thus actually goes to the trouble of confirming continuity of evidence in the transition from Galilean to Newtonian gravity.

15.7 Newton or Newtonian?

Employing limit-case idealizations to maintain continuity of evidence across theory change is the ninth and last of my ways in which the *Principia* changed how evidence is developed in physics. Table 15.1 recapitulates the nine ways for the convenience of the reader.

I see these not as nine distinct ways, but as nine aspects of a single change: *a new approach in which theory is first and foremost an instrument for developing evidence, and evidence of increasingly telling quality is then brought to bear on it indirectly through the research predicated on it.* More important than this summary description, however, is the degree to which these nine elements mesh with one another to form a coherent whole. I claim that they gain this unity from their being a response to Newton's conclusion that the true motions of the planets are hopelessly complex and his worry that the data to which we have ready access may be misleadingly parochial. I have trouble imagining a more reasonable response to the complexity of the true motions and the likely

17 Doing so amounts to treating uniform gravity acting along parallel lines as a mere curve-fit approximation to Newton's universal gravity.

Table 15.1 *Nine aspects of how Newton's* Principia *changed physics*

1.	Physics has to include its own theory of measurement
2.	Develop generic mathematical theory to provide "inference-tickets"
3.	Restrict physical theory to principles that phenomena dictate
4.	Leap from approximative evidential reasoning to exact theory
5.	Idealizations that would hold exactly in specified circumstances
6.	Shift focus of ongoing research to deviations from such idealizations
7.	Physical theory becomes an instrument for turning data into evidence
8.	Evidence accrues to a theory from success of research predicated on it
9.	Limit-case idealizations enable continuity of evidence across theory change

parochialism of our observational situation than this one. One can scarcely say of those who have traced the path initiated by the *Principia* that they have "made trial of nature in vain."

Two and a half years after the *Principia* was first published, Huygens published a response to Newton's theory of gravity, *Discourse on the Cause of Gravity*, bound together with his *Treatise on Light*. In the Preface to the latter he offers a wonderfully succinct statement of the then prevailing view about how evidence is to be developed in empirical science:

> One finds in this subject a kind of demonstration which does not carry with it so high a degree of certainty as that employed in geometry; and which differs distinctly from the method employed by geometers in that they prove their propositions by well-established and incontrovertible principles, while here principles are tested by the inferences which are derivable from them. The nature of the subject permits of no other treatment. It is possible, however, in this way to establish a probability which is little short of certainty. This is the case when the consequences of the assumed principles are in perfect accord with the observed phenomena, and especially when these verifications are numerous; but above all when one employs the hypothesis to predict new phenomena and finds his expectations realized.
>
> (Huygens 1888–1950, XIX, p. 454)[18]

Newton's famous pronouncement in the General Scholium that he added at the end of the second edition of the *Principia* twenty-three years later was presumably, at least in part, a response to this statement by Huygens:

> I have not as yet been able to deduce from phenomena the reason for these properties of gravity, and I do not feign hypotheses. For whatever is not deduced from the phenomena must be called a hypothesis; and

18 The English translation is from Matthews (1989, p. 126). The hypothesis which Huygens had most in mind was the longitudinal wave theory of light.

> hypotheses, whether metaphysical or physical, or based on occult qualities, or mechanical, have no place in experimental philosophy. In this experimental philosophy, propositions are deduced from the phenomena and are made general by induction.
>
> (Newton 1999, p. 943)

Newton, however, says nothing more about his approach and why it may be better. In particular, nowhere in the *Principia* does he invoke the complexity of the orbital motions to argue either that too many disparate hypotheses can meet Huygens's requirements or that a hypothetico-deductive approach offers less promise of bringing to light the physical sources of small discrepancies between theory and observation. Indeed, nowhere in the *Principia* does he even intimate that his alternative approach involves remotely the logical intricacy that I have attributed to it. A natural question, then, is whether the approach I have laid out is better called Newtonian rather than Newton's. How much of it did Newton himself see?

While this question is clearly of historical interest, especially to Newton scholars, it is not of central importance to this chapter. The goal of this chapter has been to lay out a picture of the general logical structure of the evidence across the history of research in Newtonian gravity and to trace key constituents of this logical structure to Newton's *Principia*. The data entering into this evidence extends from Tycho Brahe's efforts a century earlier well into the twentieth century in the case of orbital motion. As crucial to the history of research in Newtonian gravity as Newton and his *Principia* were, this research was carried out by a large community that stretched across many generations. The individuals forming that community focused far more heavily on specific, narrow questions in evidence, and not on the general logic of the evidential reasoning across the entire history. Consequently, although what those individuals said and thought is relevant, it is of limited weight when judging the adequacy of the picture of the logic presented in this chapter. One should think of this chapter as emulating the perspective of a review article, unusual only in the scope of time covered and the limited attention given to specific items of evidence. The decisive issue in judging the picture of the overall logic presented here should be its coherence.

That said, let me return to the question of how much of my proposed "Newtonian" approach to evidence Newton himself saw. My guess is, all of its key constituents and, at least on occasion, their *potential* for coming together to form a whole. Much of my reason for saying this is "autobiographical" and hence not of much moment for others: I came to see this logic from repeatedly working through the *Principia* while teaching it cover-to-cover. On a less personal note, each of the items listed in Table 15.1 has been tied in this chapter to specific passages in the *Principia*. I chose Clairaut's work to illustrate continuing indirect evidence from success in using the theory as a tool in ongoing research, but I could almost as well have used Newton's quantitative results on other lunar inequalities. Those results, moreover, are not the only place where Newton proposes to proceed by successive approximation to increasingly refined idealizations – what I. B. Cohen (1980, pp. 3–154) called the "Newtonian style."

Arguing that the *Principia* provides explicit basis for each of the nine items in my list is one thing; arguing that Newton saw much or all of the logical structure I claim they form is another. The argument for that has to come from more subtle features of his text, such as his precise phrasing of his Rules of Reasoning and his careful use of the subjunctive when discussing whether the orbits of the planets are Keplerian and stationary. There are also features of the *Principia* that make totally good sense if he was paying attention to nuances in the logic I have proposed, but are difficult to explain otherwise. The most notable of these are his refusal to infer the inverse-square variation from the Keplerian ellipse and his treatment of Galilean uniform gravity as a limiting case of gravity that varies linearly with distance. Finally, in a similar spirit, my picture of the logic of evidence in the *Principia* absolves Newton of stupidity (or dishonesty) in claiming to have derived the law of gravity from phenomena.

Saying that Newton saw much of the logic I have described does not mean that there was some moment when he had a clear, comprehensive vision of the whole picture and thereafter consciously fashioned the *Principia* accordingly. He appears to have been fully aware from early on that his inferences from phenomena involve "if *quam proxime*, then *quam proxime*" reasoning. Propositions establishing this *quam proxime* form of relevant "if-then" statements occur in the very first draft of Book I, and even the registered version of the "De Motu" tract shows signs of his knowing that the *quam proxime* form of "if a Keplerian ellipse, then inverse-square" does not hold (Smith 2002b, p. 40f.). Not so clear is when, and how fully, Newton saw that deviations from what I have called Newtonian idealizations can provide an evidential basis for a sequence of successive approximations in ongoing research. When he first calls attention to the intractable complexity of planetary motion, in the augmented version of "De Motu," he presents the vagaries as an obstacle in determining the proper Keplerian orbits and seems resigned to never being able to do anything constructive with the deviations from Keplerian motion (Newton 1962, p. 281). In the initial draft of what became Book III of the *Principia* the inequalities in the lunar orbit are treated only qualitatively, and the intent seems merely to be to eliminate the apparent counterexample the Moon offers to Kepler's rules (Newton 1934, p. 577). My suggestion, then, is that Newton saw the possibility of using the deviations as the basis for successive approximations when his quantitative results on the lunar inequalities emerged, between the first draft of Book III in 1685 and the final draft in late 1686 or early 1687.

Newton had good reasons to be cautious about putting too much of the evidential burden for universal gravity on success in pinning down the physical sources of deviations from Keplerian motion. In contrast to the limited quantitative results he achieved on the restricted form of the three-body problem involving the Sun, Earth, and Moon, he obtained no quantitative results at all on the three-body problem posed by the Sun, Jupiter, and Saturn. Worse, the factor of two error in his derivation of the precession of the lunar apogee raised the distinct possibility that the Earth's magnetism was contributing to this effect. Newton tells us that the magnetic force, "in receding from the magnet, decreases not as the square but almost as the cube of the distance, as far as I have been able to tell from rough observations" (Newton 1999, p. 810);[19] and he knew that a superposed inverse-cube centripetal force is precisely what is needed to make an orbit precess (Newton 1999, pp. 535–539). If, however, non-gravitational forces have any significant effect on the motions of the planets or their satellites, the prospects for developing continuing evidence for universal gravity out of the vagaries of the motions is not so straightforward. For, laws of these non-gravitational forces would first have to be established, independently of those vagaries, and even with those laws in place, problems would potentially remain in specifying conditions for their applicability to specific celestial motions – for example, what is the fractional iron content of the Moon?

From his work on the tides Newton knew how much more difficult quantitative analysis becomes when non-gravitational forces are involved. If they are not virtually negligible in orbital motions, then the process of pinning down physical sources of deviations would likely be long and maybe tortuous, and the evidence accruing to universal gravity would be of reduced strength.

Furthermore, we should not lose sight of the limits of observational accuracy in astronomy during Newton's lifetime. The need to correct observations for the effects of solar parallax and atmospheric refraction had long been recognized, but the precise magnitudes of those corrections remained under dispute throughout Newton's lifetime, and no consensus had been reached on the need for a further speed-of-light correction at the time the *Principia* was published.[20] The need for still further corrections for the aberration of light and the nutation of the Earth emerged shortly after Newton died (Bradley 1728 and 1748). Thus, the prospects for increasingly precise observation of the sort needed to support successive approximations beyond the first level of refinements became much clearer only after Newton.

A prominent physicist responded to the account of the evidence for Newtonian gravity given above by remarking, "Newton was lucky."[21] That is surely correct on two counts. He was lucky that a relationship as mathematically

19 Newton was not in error here, for the dipole effect of a magnet gives rise to an inverse-cube variation.

20 Cassini still insisted that the irregularity in the timing of the eclipse of Io came from an inequality in its orbit; see Halley (1694).

21 Kenneth G. Wilson, in conversation.

simple as his law of gravity remained intact across two centuries of pursuit of ever greater precision.[22] And he was lucky in the degree to which gravity dominates celestial motions, making the task of marshalling evidence out of those motions far easier than it would otherwise have been.[23] Newton had no basis for expecting either of these eventualities to work out remotely as well as they did. Gravitation research has been successful, however, not merely because the empirical world happened to cooperate, but also because it has followed an approach that enabled continuing evidence to be brought to bear from increasingly subtle complexities in the motions. Its following that approach was not a matter of luck. In his research in optics Newton conducted experiment after experiment, with only slight variations, in order to address possible loopholes in the experiments that he ultimately published.[24] In the *Principia* Newton shows a similar constant concern for evidential loopholes that

22 The algebraic simplicity of the law as Newton formulated it was an automatic consequence of his inferring the law from phenomena by means of approximative reasoning. For, this blocked him from incorporating any feature into the law unless the phenomena dictated it, and insofar as the original phenomena amounted to first-order approximations to the real motions, nothing in them was going to dictate further complications. But that gave all the more reason to expect that a need for complications might well emerge as research went beyond those first-order approximations. For example, the law does not include time as a variable – something that might at least have raised questions early on.

23 A letter Newton wrote to Leibniz in 1693 shows that he anticipated this possibility:

> For since celestial motions are more regular than if they arose from vortices and observe other laws, so much so that vortices contribute not to the regulation but to the disturbance of the motions of planets and comets; *and since all phenomena of the heavens and of the sea follow precisely, so far as I am aware, from nothing but gravity acting in accordance with the laws described by me*; and since nature is very simple, I have myself concluded that all other causes are to be rejected and that the heavens are to be stripped as far as may be of all matter, lest the motions of planets and comets be hindered or rendered irregular.
>
> (Newton 1959–1977, III, p. 287; emphasis added)

Perhaps Newton is here being disingenuous with Leibniz, who had published his own vortex theory of Keplerian motion four years earlier, but he knew perfectly well that tidal phenomena do not all follow precisely from the laws described by him, and his suggestion that celestial phenomena follow *precisely* was at best wishful thinking. (Two years before this letter Newton had asked Flamsteed for observations of Jupiter and Saturn over a fifteen-year period, presumably because he wanted to answer the question of how precisely their motions follow from the law of gravity.) The evidence that gravity is the overwhelmingly dominant force in celestial motions was incomparably stronger a century after this letter to Leibniz, when Laplace was setting to work on his *Celestial Mechanics*. The extraordinary quality of evidence achieved in gravitation research over the two centuries following the *Principia* would have been far more difficult to attain if non-gravitational forces were more prominent in celestial motions.

24 To quote Alan Shapiro (2002, p. 230), "Sometimes, as in the *Optical Lectures*, the large number of experiments with slight variations to establish various points may seem tedious, but Newton attempted to leave no room for objections."

might arise from the gap between complex motions of the actual world and mathematical representations of them. Each of the items in my list of ways in which the *Principia* changed physics surfaces in a context in which explicit attention is given to this gap. So, regardless of how clearly Newton ever saw the total package formed by the items listed in Table 15.1, the mutual coherence they acquire from their forming a response to a specific evidential challenge truly is owing to him.

A second prominent physicist offered a different response to my account: "[Smith] makes very clear that Newton's celestial mechanics was something truly novel, namely that it displays the currently used method of doing mathematical physics."[25] No comment on my efforts on Newton has ever pleased me more.

Acknowledgements

This is the essay form of a talk prepared in 2001 in response to invitations from the Physics Departments of Tufts University, the University of Colorado, and the University of California at Santa Cruz. The talk was also given at the Department for History and Philosophy of Science of Indiana University and the Dibner Institute for the History of Science and Technology at MIT in 2001; as a History of Science and Technology Colloquium at the University of Minnesota and before an audience of physicists and philosophers at the University of Western Ontario in 2002; and at the New York Academy of Sciences in 2004. I received too many helpful questions and suggestions on those occasions to cite them all here.

Overlap with my subsequent much longer and philosophically more ambitious essay, "Closing the Loop: Testing Newtonian Gravity, Then and Now," had originally convinced me not to publish the essay version of the talk, especially after the editors of the present volume requested the long essay. When, at the last moment, Cambridge University Press decided that the requested essay was too long for the volume, I substituted this one.

25 Markus Fierz, in a letter to Silvan S. Schweber.

REFERENCES

Works by Isaac Newton

(1687) *Philosophiae naturalis principia mathematica.* London: Royal Society.

(1711) W. Jones, ed. *Analysis per Quantitatum, Series, Fluxiones, ac differentias cum Enumeratio Linearum Tertii Ordinis.* London: ex Officina Paersoniana.

(1713) R. Cotes, ed. *Philosophiae naturalis principia mathematica.* Second edition. London.

(1729) A. Motte, trans. *The Mathematical Principles of Natural Philosophy.* London: Benjamin Motte.

(1736) J. Colson, trans. *The Method of Fluxions and Infinite series. . . .* Printed by H. Woodfall and sold by J. Nourse. Cambridge.

(1850/1969) J. Edleston, ed. *Correspondence of Sir Isaac Newton and Professor Cotes including letters of other Eminent Men.* London: Cass.

(1934) F. Cajori, ed. and trans. *Sir Isaac Newton's Mathematical Principles of Natural Philosophy and his System of the World, Motte's Translation Revised.* Berkeley and Los Angeles: University of California Press.

(1959–1977) H. W. Turnbull *et al.*, ed. *The Correspondence of Isaac Newton.* Cambridge University Press.

(1962) A. R. Hall and M. B. Hall, eds. and trans. *Unpublished scientific papers of Isaac Newton: a selection from the Portsmouth collection in the University Library, Cambridge.* Cambridge University Press.

(1967–1981) D. T. Whiteside, ed. *The Mathematical Papers of Isaac Newton.* Cambridge University Press.

(1972) A. Koyré and I. B. Cohen, eds. *Isaac Newton's Philosophiae Naturalis Principia Mathematica.* The third edition with variant readings. Cambridge, MA: Harvard University Press.

(1978) I. B. Cohen and R. Schofield, eds. *Isaac Newton's Papers and Letters on Natural Philosophy.* Revised edition. Cambridge, MA: Harvard University Press.

(1979) *Opticks,: or a Treatise of the Reflections, Refractions, Inflections and Colours of Light.* With a preface by I. Bernard Cohen. New York: Dover Publications.

(1983) J. E. McGuire and M. Tamny, eds. *Certain Philosophical Questions: Newton's Trinity Notebook.* Cambridge University Press.

(1984) A. Shapiro, ed. *The Optical Papers of Isaac Newton*, vol. I: *The Optical Lectures, 1670–1672.* Cambridge University Press.

(1999) I. B. Cohen and A. Whitman, eds. and trans. *The Principia: Mathematical Principles of Natural Philosophy.* With "A Guide to Newton's Principia" by Cohen *et al.* Berkeley: University of California Press.
(2004) A. Janiak, ed. *Isaac Newton: Philosophical Writings.* Cambridge University Press.

Works cited

Ahnert, T. (2004) "Newtonianism in early Enlightenment Germany, c.1720–1750." *Studies in the History and Philosophy of Science* 35: 471–491.
Aiton, E. J. (1972) *The Vortex Theory of Planetary Motion.* New York: American Elsevier.
d'Alembert, J. L. R. (1743/1967) *Traité de dynamique.* Paris: David.
Alexander, H. G., ed. (1956) *The Leibniz-Clarke Correspondence.* Manchester University Press.
Alexander, P. (1985) *Ideas, Qualities and Corpuscles: Locke and Boyle on the External World.* Cambridge University Press.
al Khwārizmī, M. ibn Musa. (2007) *Le commencement de l'algébre.* R. Rashed, trans. Paris: Blanchard.
Anstey, P. R. (2002) "Locke, Bacon and Natural History." *Early Science and Medicine* 7: 65–92.
(2003) "Locke on Method in Natural Philosophy." In P. R. Anstey, ed. *The Philosophy of John Locke: New Perspectives.* London: Routledge.
(2005) "Experimental versus Speculative Natural Philosophy." In P. R. Anstey and J. A. Schuster, eds. *The Science of Nature in the Seventeenth Century.* Dordrecht: Kluwer.
Aristotle. (1984) *The Complete Works of Aristotle.* J. Barnes, ed. Princeton University Press.
Arthur, R. (1994) "Space and Relativity in Newton and Leibniz." *British Journal for the Philosophy of Science* 45: 219–240.
Auger, L. (1962) *Gilles Personne de Roberval (1602–1675).* Paris: Blanchard.
Ayers, M. (1977) "The Ideas of Power and Substance in Locke's Philosophy." In I. C. Tipton, ed. *Locke on Human Understanding.* Oxford University Press.
(1991) *Locke: Epistemology and Ontology.* London and New York: Routledge.
Bacon, F. (1999) *Selected Philosophical Works.* R.-M. Sargent, ed. Indianapolis: Hackett Publishing Company.
Ball, W. W. R. (1893) *An Essay on Newton's "Principia."* London: Macmillan.
Barber, K. F. and J. J. E. Gracia (1994) *Individuation and Identity in Early Modern Philosophy.* Albany, NY: State University of New York Press.
Barbour, J. B. (1989) "Absolute or Relative Motion? A study from a Machian point of view of the discovery and the structure of dynamical theories." In *The Discovery of Dynamics*, Volume I. Cambridge University Press.
Barrow, I. (1670) *Lectiones Geometrica....* London: Typi G. Godbid, & prostant venales apud J. Dunmore, & O. Pulleyn Juniorem.

Bechler, Z., ed. (1982) *Contemporary Newtonian Research.* Dordrecht: Reidel.

Beebee, H. (2006) *Hume on Causation.* London: Routledge.

Beeley, P. (1995) "Les sens dissimulants. Phénomènes et réalité dans l'Hypothesis physica nova." In M. d. Gaudemar, ed. *La notion de nature chez Leibniz.* Stuttgart: Steiner.

Beeson, D. (1992) *Maupertuis: An Intellectual Biography. Studies on Voltaire and the Eighteenth Century.* Volume 299. Oxford: The Voltaire Foundation.

Belkind, O. (2007) "Newton's Conceptual Argument for Absolute Space." *International Studies in the Philosophy of Science* 21: 271–293.

Bennett, J. and P. Remnant (1978) "How Matter Might at first Be Made." Canadian Journal of Philosophy, supplementory volume 4: 1–11.

Bentley, R. (1838/1996) *The Works of Richard Bentley, D.D.* A. Dyce, ed. London: Francis MacPherson/New York: AMS Press.

Bernstein, H. R. (1981) "Passivity and Inertia in Leibniz's *Dynamics.*" *Studia Leibnitiana* 18: 97–113.

Berkeley, G. (1948–1957) *The Works of George Berkeley, Bishop of Cloyne.* A. A. Luce and T. E. Jessop, eds. London: Nelson.

(1982) *A Treatise Concerning the Principles of Human Knowledge.* K. Winkler, ed. Indianapolis: Hackett.

(1992) *De Motu and The Analyst.* D. M. Jesseph ed. and trans. Dordrecht: Kluwer.

Bertoloni Meli, D. (1993) *Equivalence and Priority: Newton versus Leibniz.* Oxford University Press.

(2006a) "Inherent and Centrifugal Forces in Newton." *Archive for History of Exact Sciences* 60: 319–335.

(2006b). *Thinking with Objects: The Transformation of Mechanics in the Seventeenth Century.* Baltimore: Johns Hopkins University Press.

Bigourdan, G. (1929) "Lettres inédites d'Euler à Clairaut." *Comptes rendus du Congrès des sociétés savantes de Paris et des Départments tenu à Lille en 1928, Section des Sciences.* Paris: Imprimerie Nationale.

Blackwell, R. J., trans. (1977) "Christiaan Huygens' *The Motion of Colliding Bodies.*" *Isis* 68: 574–597.

Blay, M. (1995) *Les 'Principia' de Newton.* Paris: Presses Universitaires de France.

Bloch, L. (1908) *La philosophie de Newton.* Paris.

Boas, M. (1952) "The Establishment of the Mechanical philosophy." *Osiris* 10: 412–541.

Boulliau, I. (1657) *Astronomiae Philolaicae Fundamenta clarius explicata, & asserta.* Paris: Sebastian Cramoisy.

Boyle, R. (1772) *The Works of Robert Boyle.* T. Birch, ed. London.

(1666/1999) *The Origin of Forms and Qualities.* In M. Hunter and E. B. Davis, eds. *The Works of Robert Boyle.* London: Pickering and Chatto.

(1991) *Selected Philosophical Papers of Robert Boyle.* M. A. Stewart, ed. Indianapolis: Hackett.

Brackenridge, J. and M. Nauenberg. (2002) "Curvature in Newton's Dynamics." In Cohen and Smith, eds.

Brading, K. (2011) "On Composite Systems: Descartes, Newton, and the Law-constitutive Approach." In D. Jalobeanu and P. R. Anstey, eds. *Vanishing Matter and the Laws of Motion.* New York and London: Routledge.

Brading, K. and A. Skiles. (forthcoming) "Underdetermination as a Path to Ontic Structural Realism." In E. Landry and D. Rickles, eds. *Structure, Object, and Causality: Proceedings of the Banff Workshop on Structural Realism.* University of Western Ontario Series in Philosophy of Science. Springer.

Bradley, J. (1728) "A Letter from the Reverend James Bradley Avilian Professor of Astronomy at Oxford, and F. R. S. to Dr. Edmond Halley Astronom. Reg. etc. giving an Account of a new discovered Motion of the Fix'd Stars." *Philosophical Transactions of the Royal Society* 35: 637–661.

(1748) "Letter to the Rt. Hon. George, Earl of Macclesfield, concerning an Apparent Motion of Some Fixed Stars." *Philosophical Transactions of the Royal Society* 45: 1–43.

Bricker, P. and R. I. G. Hughes, eds. (1990) *Philosophical Perspectives on Newtonian Science.* Cambridge, MA: MIT Press.

Bromberger, S. (1992) "Science and the Forms of Ignorance." In *On What We Know We Don't Know.* University of Chicago Press.

Brown, H. (2005) *Physical Relativity: Space-Time Structure from a Dynamical Perspective.* Oxford University Press.

Brunet, P. (1931) *L'introduction des theories de Newton en France au XVIIIe siècle.* Paris: Blanchard.

Buchwald, J. Z. and I. B. Cohen, eds. (2001) *Isaac Newton's Natural Philosophy.* Cambridge, MA: MIT Press.

Buchwald, J. Z. and G. E. Smith. (2002) "Incommensurability and Discontinuity of Evidence." *Perspectives on Science* 9: 463–498.

Buckle, S. (2001) *Hume's Enlightenment Tract.* Oxford University Press.

Burtt, E. A. (1924/1932/1959) *The Metaphysical Foundations of Modern Physical Science.* London: Routledge & Kegan.

Caroti, S. (2004) "Some Remarks on Buridan's Discussion on Intension and Remission." *Vivarium* 42: 58–85.

Cassirer, E. (1951) *The Philosophy of the Enlightenment.* F. C. A. Koelln and J. P. Pettegrove, trans. Princeton University Press.

Cavendish, H. (1798) "Experiments to determine the Density of the Earth." *Philosophical Transactions of the Royal Society* 88: 468–526.

Chalmers, A. (2002) "Experimental versus Mechanical Philosophy in the Work of Robert Boyle: A Reply to Anstey and Pyle." *Studies in History and Philosophy of Science* 33: 191–197.

Chandrasekhar, S. (1995) *Newton's* Principia *for the Common Reader.* Oxford: Clarendon Press.

Child, J. M., ed. (1916) *The Geometrical Lectures of Isaac Barrow.* Chicago and London: Open Court.

Christensen D. (1983) "Glymour on evidential relevance." *Philosophy of Science* 50: 471–481.

(1990) "The irrelevance of bootstrapping." *Philosophy of Science* 57: 644–662.

Ciufolini, I. and J. A. Wheeler. (1995) *Gravitation and Inertia.* Princeton University Press.

Clarke, S. (1738) "A Discourse Concerning the Unalterable Obligations of Natural Religion." In *The Works of Samuel Clarke*, Vol. II. London: Knapton.

Clericuzio, A. (2000) *Elements, Principles, and Corpuscles: A Study of Atomism and Chemistry in the Seventeenth Century.* Dordrecht: Kluwer.

Cohen, I. B. (1956) *Franklin and Newton.* Philadelphia: American Philosophical Society.

(1971) *Introduction to Newton's* Principia. Cambridge, MA: Harvard University Press.

(1980) *The Newtonian Revolution: With Illustrations of the Transformation of Scientific Ideas.* Cambridge University Press.

(1982) "The *Principia*, Universal Gravitation, and the 'Newtonian Style', in Relation to the Revolution in Science." In Bechler, ed.

(2002) "Newton's Concepts of Force and Mass, with Notes on the Laws of Motion." In Cohen and Smith, eds.

Cohen, I. B. and G. E. Smith, eds. (2002) *The Cambridge Companion to Newton.* Cambridge University Press.

Condorcet, Marquise de, J.-A.-N. (1955/1979) *Sketch for a Historical Picture of the Progress of the Human Mind.* J. Barraclough, trans. Westport: Hyperion Press.

Cozzoli, D. (2007) "Alessandro Piccolomini and the Certitude of Mathematics." *History and Philosophy of Logic* 28: 151–171.

Darwin, G. H. (1907) *Scientific Papers.* Cambridge University Press.

de Gandt, F. (1995) *Force and Geometry in Newton's Principia.* C. Wilson, trans. Princeton University Press.

(1999) "Newton's Nature: Does Newton's Science Disclose Actual Knowledge of Nature?" *St. John's Review* 45: 7–19.

De Pierris, G. (2001) "Hume's Pyrrhonian Skepticism and the Belief in Causal Laws." *Journal of the History of Philosophy* 39: 351–383.

(2002) "Causation as a Philosophical Relation in Hume." *Philosophy and Phenomenological Research* 64: 499–545.

(2006) "Hume and Locke on Scientific Methodology: The Newtonian Legacy." *Hume Studies* 32: 277–329.

Dear, P. (1995) *Discipline and Experience: The Mathematical Way in the Scientific Revolution.* University of Chicago Press.

Densmore, D. (1996) *Newton's* Principia*: The Central Argument.* With translations by W. Donahue. Santa Fe: Green Lion Press.

Desaguliers, J. T. (1734) *A Course of Experimental Philosophy.* Volume I. London: John Senex.

Descartes, R. (1637) *Discours de la méthode.... Plus la Dioptrique, Les Météores, et la Géométrie qui sont des essaies de cette Méthode.* Leiden: I. Maire.

(1644) *Principia Philosophiae.* Amsterdam: L. Elzevirium.

(1647) *Les Principes de la Philosophie.* Paris: H. LeGras.

(1659–1661) *Geometria, à Renato des Cartes Anno 1637 Gallicè edita... Operâatque studio Francisci à Schooten....* Amsterdam: L. & D. Elzevirios.

(1897–1910) *Oeuvre de Descartes.* C. Adam and P. Tannery, eds. Paris: Vrin.

(1954) *The Geometry of René Descartes.* D. E. Smith and M. L. Latham, eds. New York: Dover.

(1985–1991) *The Philosophical Writings of Descartes.* J. Cottingham, R. Stoothoff, D. Murdoch, and A. Kenny, trans. Cambridge University Press.

(1991) *Principles of Philosophy.* V. R. Miller and R. P. Miller, trans. Dordrecht: Kluwer.

Dijksterhuis, E. J. (1961) *The Mechanization of the World Picture.* C. Dikshoorn, trans. Oxford University Press.

DiSalle, R. (1991) "Conventionalism and the Origin of the Inertial Frame Concept." In A. Fine, M. Forbes and L. Wessels, eds. *Proceedings of the 1990 Biennial Meeting of the Philosophy of Science Association.*

(1994) "On Dynamics, Indistinguishability and Spacetime Ontology." *British Journal for the Philosophy of Science* 45: 265–287.

(2002) "Newton's philosophical analysis of space and time." In Cohen and Smith, eds.

(2006) *Understanding Space-Time.* Cambridge University Press.

Dobbs, B. J. T. (1975) *The Foundations of Newton's Alchemy.* Cambridge University Press.

(1988) "Newton's Rejection of the Mechanical Aether: Empirical Difficulties and Grounding Assumptions." In A. Donovan, L. Laudan, and R. Laudan, eds. *Scrutinizing Science: Empirical Studies of Scientific Change.* Dodrecht: Kluwer.

(1991) *The Janus Faces of Genius: The Role of Alchemy in Newton's Thought.* Cambridge University Press.

Domski, M. (2003) "The Constructible and the Intelligible in Newton's Philosophy of Geometry." *Philosophy of Science* 70: 1114–1124.

(2010) "Newton as Historically-Minded Philosopher." In M. Dickson and M. Domski, eds. *Discourse on a New Method: reinvigorating the marriage of history and philosophy of science.* LaSalle: Open Court Press.

Dorling, J. (1973) "Demonstrative induction: Its significant role in the history of physics." *Philosophy of Science* 40: 360–372.

(1990) "Reasoning from phenomena: Lessons from Newton." *Proceedings of the 1990 Biennial Philosophy of Science Association* 2: 197–208.

Downing, L. (1997) "Locke's Newtonianism and Lockean Newtonianism." *Perspectives on Science* 5: 285–310.

(1998) "The Status of Mechanism in Locke's *Essay*." *The Philosophical Review* 107: 381–414.

(2007) "Locke's Ontology." In L. Newman, ed. *The Cambridge Companion to Locke's* Essay. Cambridge University Press.

(2008) "The 'sensible object' and the 'uncertain philosophical cause.'" In D. Garber and B. Longuenesse, eds. *Kant and the Early Moderns.* Princeton University Press.

Duchesneau, F. (1994) *La dynamique de Leibniz.* Paris: J. Vrin.

Ducheyne, S. (2005) "Mathematical models in Newton's Principia: A new view of the 'Newtonian style.'" *International Studies in the Philosophy of Science* 19: 1–19.

Duhem, P. (1906) *La théorie physique son objet et sa structure.* Paris.

(1982) *The Aim and Structure of Physical Theory.* P. P. Weiner, trans. Princeton University Press.

Earman, J. *et al.*, eds. (1977) *Foundations of Space-Time Theories. Minnesota Studies in the Philosophy of Science.* Volume VIII. Minneapolis: University of Minnesota Press.

(1989) *World Enough and Space-Time: Absolute Versus Relational Theories of Space and Time.* Cambridge, MA: MIT Press.

Euler, L. (1748a) *Introductio in analysin infinitorum.* Lausanne: M.-M. Bousquet & Soc.

(1748b) "Reflexions sur l'espace et le temps." *Histoire de l'Académie Royale des sciences et belles lettres* 4, 324–33. Reprinted in L. Euleri (1942) *Opera Omnia*, Series III, Volume 2.

(1752) "Recherches sur irrégularities du mouvement de Jupiter et de Saturne." In *Recueil des pièces qui ont remporté les prix de l'Académie des Sciences*, vol. 7. Paris.

(1753) *Theoria Motus Lunae, Exhibens Omnes Eius Inaequalitates.* Reprinted in *Opera Omnia*, Series II, vol. 23, Basil, 1969.

(1911) *Opera Omnia.* E. Hoppe *et al.*, eds. Lipsiae et Berolini: Typis et in aedibus B.G. Teubneri.

Feingold, M. (1988) "Partnership in Glory: Newton and Locke Through the Enlightenment and Beyond." In P. B. Scheurer and G. Debrock, eds. *Newton's Scientific and Philosophical Legacy.* Dordrecht: Kluwer.

(2000) "Mathematicians and naturalists: Sir Isaac Newton and the Royal Society." In Buchwald and Cohen, eds.

Feldhay, R. (1998) "The Use and Abuse of Mathematical Entities." In Machamer, ed.

Fichant, M. (1978) "Les concepts fondamentaux de la mécanique selon Leibniz, en 1676." In G. W. Leibniz Gesellschaft and Centre national de la recherche scientifique, eds. *Leibniz à Paris: 1672–1676.* Wiesbaden: Steiner.

Figala, K. (2002) "Newton's Alchemy." In Cohen and Smith, eds.

Fogelin, R. (2001) *Berkeley and the Principles of Human Knowledge.* London: Routledge.

(2003) *A Defense of Hume on Miracles.* Princeton University Press.

Fontenelle, Le Bovier, B. (1818) *Œuvres Complètes.* G.-B. Depping, ed. Paris: A. Berlin.

Forbes, E. G., ed. (1971) *The Euler-Mayer Correspondence (1751–1755).* London: The Macmillan Press.

Franks, P. W. (2005) *All or Nothing: Systematicity, Transcendental Arguments, and Skepticism in German Idealism.* Cambridge, MA: Harvard University Press.

Freguglia, P. (2008) "Viete, Reader of Diophantus: An Analysis of Zeteticorum Libri Quinque." *Bollettino di storia delle scienze matematiche* 28: 51–95.

French, S., and J. Ladyman. (2003) "Remodelling structural realism: quantum physics and the metaphysics of structure." *Synthese* 136: 31–56.

Freudenthal, G. (1990) *Études sur/Studies on Hélène Metzger*. Leiden: Brill.

Friedman, M. (1990) "Kant and Newton: Why Gravity is Essential to Matter." In Bricker and Hughes, eds.

(1992) *Kant and the Exact Sciences*. Cambridge, MA: Harvard University Press.

(1997) "Descartes on the Real Existence of Matter." *Topoi* 16: 1–10.

(2008) "Descartes and Galileo: Copernicanism and the Metaphysical Foundations of Physics." In J. Broughton and J. Carriero, eds. *A Companion to Descartes*. Oxford: Blackwell.

Gabbey, A. (1992) "Newton's 'Mathematical Principles of Natural Philosophy': A Treatise on 'Mechanics'?" In P. M. Harman and A. E. Shapiro, eds. *The Investigation of Difficult Things*. Cambridge University Press.

(1993a) "Descartes's Physics and Descartes's Mechanics: Chicken and Egg?" In S. Voss, ed. *Essays on the Philosophy and Science of René Descartes*. Oxford University Press.

(1993b) "Between *ars* and *philosophia naturalis*: Reflections on the Historiography of Early Modern Mechanics". In J. V. Field and F. James, eds. *Renaissance and Revolution: Humanists, Scholars, Craftsmen and Natural Philosophers in Early Modern Europe*. Cambridge University Press.

Galileo, G. (1967) *Dialogue Concerning the Two Chief World Systems*. S. Drake, trans. Berkeley and Los Angeles: University of California Press.

(1974) *Discourses on Two New Sciences*. S. Drake, trans. Madison: University of Wisconsin Press.

Garber, D. (1992) *Descartes' Metaphysical Physics*. University of Chicago Press.

(2000) "A Different Descartes: Descartes and the Programme for a Mathematical Physics in his Correspondence." In S. Gaukroger, J. Schuster, and J. Sutton, eds. *Descartes' Natural Philosophy*. London and New York: Routledge.

(2009) *Leibniz: Body, Substance, Monad*. Oxford University Press.

Garrett, D. (1997) *Cognition and Commitment in Hume's Philosophy*. Oxford University Press.

Garrison, J. (1987) "Newton and the Relation of Mathematics to Natural Philosophy." *Journal of the History of Ideas* 48: 609–627.

Glass, D. A. and N. Hall. (2008) "A Brief History of the Hypothesis." *Cell* 134: 378–381.

Glymour C. (1980) *Theory and Evidence*. Princeton University Press.

Greene, R. (1712) *The Principles of Natural Philosophy*. London: University Press.

s'Gravesande, W. J. v. (1721) *Mathematical Elements of Natural Philosophy Confirmed by Experiments, or an Introduction to Sir Isaac Newton's Philosophy*. Second edition. J. T. Desaguliers, trans. London: Senex and Taylor.

Guerlac, H. (1981) *Newton on the Continent*. Ithaca: Cornell University Press.

Guicciardini, N. (1989) *The Development of Newtonian Calculus in Britain 1700–1800.* Cambridge University Press.

(1999) *Reading the* Principia. *The Debate on Newton's Mathematical Methods for Natural Philosophy from 1687 to 1736.* Cambridge University Press.

(2003) "Conceptualism and Contextualism in the Recent Historiography of Newton's *Principia.*" *Historia Mathematica* 30: 407–431.

(2004) "Geometry and Mechanics in the Preface to Newton's *Principia*: A Criticism of Descartes' *Géométrie.*" *Graduate Faculty Philosophy Journal* 25: 119–159.

(2006) "Method versus Calculus in Newton's Criticisms of Descartes and Leibniz." *Proceedings of the International Congress of Mathematicians, Madrid, Spain.* Zürich: European Mathematical Society Publishing House.

Hacking, I. (1980) "Proof and Eternal Truths: Descartes and Leibniz." In S. Gaukroger, ed. *Descartes: Philosophy, Mathematics and Physics.* Brighton: Harvester.

Hall, A. R. (1963) *From Galileo to Newton.* New York: Harper.

(1970) "'sGravesande, Willem Jacob." In C. Gillispie, ed. *Dictionary of Scientific Biography.* Vol V. New York: Scribner.

(1980) *Philosophers at War: Newton and Leibniz.* Cambridge University Press.

(1992) *Isaac Newton: Adventurer in Thought.* Oxford University Press.

Hall, M. B. (1958) *Robert Boyle and Seventeenth Century Chemistry.* Cambridge University Press.

(1962) *The Scientific Renaissance.* New York: Harper.

(1991) *Promoting Experimental Learning: experiment and the Royal Society.* Cambridge University Press.

Hall, M. B. and A. R. Hall, eds. (1965–) *The Correspondence of Henry Oldenburg.* Madison: University of Wisconsin Press.

Halley, E. (1694) "Monsieur Cassini his New and Exact Tables for the Eclipses of the First Satellite of Jupiter, reduced to the Julian Stile, and Meridian of London." *Philosophical Transactions of the Royal Society* 18: 237–256.

Hankins, T. L. (1970) *Jean d'Alembert, Science and the Enlightenment.* Oxford: Clarendon Press.

Hannequin, A. (1908) "La première philosophie de Leibnitz." In A. Hannequin, ed. *Études d'histoire des sciences et d'histoire de la philosophie.* Paris: F. Alcan.

Hanson, N. R. (1970) "Hypotheses fingo." In R. E. Butts and J. W. Davis, eds. *The Methodological Heritage of Newton.* University of Toronto Press.

Hara, K. (1965) *Étude sur la théorie des mouvements de Roberval.* Ph.D. thesis. Faculté des Lettres, University of Paris.

Harper, W. L. (1990) "Newton's Classic Deductions from Phenomena." *Proceedings of the 1990 Biennial Meeting of the Philosophy of Science Association* 2: 183–196.

(1991) "Kant on Incongruent Counterparts" In J. Van Cleve and R. E. Frederick, eds. *The Philosophy of Right and Left: Incongruent Counterparts and the Nature of Space.* Dordrecht: Kluwer.

(1998) "Measurement and Approximation: Newton's Inferences from Phenomena versus Glymour's Bootstrap Confirmation." In P. Weingartner, G. Schurz, and G. Dorn, eds. *The Role of Pragmatics in Contemporary Philosophy.* Vienna: Holder-Pinchler-Tempsky.

(1999) "The First Six Propositions in Newton's Argument for Universal Gravitation." *The St. John's Review* 45: 74–93.

(2002a) "Newton's Argument for Universal Gravitation." In Cohen and Smith, eds.

(2002b) 'Howard Stein on Isaac Newton: Beyond Hypotheses?' In Malament, ed.

(2007) "Newton's Methodology and Mercury's Perihelion before and after Einstein." *Philosophy of Science* 74: 923–942.

(2009) "Newton's Methodology." In W. Myrvold and J. Christian, editors. *Quantum Reality, Relativistic Causality, and Closing the Epistemic Circle: Essays in Honour of Abner Shimony.* Berlin: Springer.

(forthcoming) *Isaac Newton's Scientific Method: turning Data into Evidence for Universal Gravity.* Oxford University Press.

Harper, W. and G. E. Smith. (1995) "Newton's New Way of Inquiry." In J. Leplin, ed. *The Creation of Ideas in Physics: Studies for a Methodology of Theory Construction.* Dordrecht: Kluwer.

Harper, W., S. R. Valluri, and R. Mann (2002) "Jupiter's Moons and the Equivalence Principle." In V. G. Gurzadyan, R.T. Jantzen, and R. Ruffini, eds. *Proceedings of the Ninth Marcel Grossman Meeting on General Relativity.* Singapore: World Scientific.

Harrison, J. R. (1978) *The Library of Isaac Newton.* Cambridge University Press.

Harrison, P. (2002) "Voluntarism and Early Modern Science." *History of Science* 40: 63–89.

Hart, H. L. A. and T. Honoré (1959/1985) *Causation in the Law.* Oxford University Press.

Heath, T. L., trans. (1956) *The Thirteen Books of Euclid's Elements.* New York: Dover.

Heilbron, J. (1982) *Elements of Early Modern Physics.* Berkeley: University of California Press.

Heimann, P. M. and J. E. McGuire. (1971) "Newtonian Forces and Lockean Powers: Concepts of Matter in Eighteenth-Century Thought." *Historical Studies in the Physical Sciences* 3: 233–306.

Hempel, C. (1965) *Aspects of Scientific Explanation and other Essays in the Philosophy of Science.* New York: The Free Press.

(1988) "Provisos: A Problem Concerning the Inferential Function of Scientific Theories." In A. Grünbaum and W. G. Salmon, eds. *The Limitations of Deductivism.* Berkeley: University of California Press.

Henry, J. (1994a) "'Pray do not Ascribe that Notion to Me': God and Newton's Gravity." In J. E. Force and R. Popkin, eds. *The Books of Nature and Scripture: Recent Essays on Natural Philosophy, Theology, and Biblical Criticism.* Dordrecht: Kluwer.

(1994b) "Boyle and Cosmical qualities." In M. Hunter, ed. *Robert Boyle Reconsidered.* Cambridge University Press.

Herivel, J. (1965a) *The Background to Newton's* Principia. Oxford University Press.

(1965b) "Alexandre Koyré." *British Journal for the History of Science* 2: 257–259.

Hess, H-J. (1978) "Die unveröffentlichten naturwissenschaftlichen und technischen Arbeiten von G. W. Leibniz aus der Zeit seines Parisaufenthaltes. Eine Kurzcharakteristik." In Gottfried-Wilhelm-Leibniz-Gesellschaft and Centre national de la recherche scientifique (France), eds. *Leibniz à Paris: 1672–1676.* Wiesbaden: Steiner. Volume I.

Hughes, T. S., ed. (1831) *The Works of Isaac Barrow.* London: A. J. Valpy.

Huggett, N. (1999) "Why Manifold Substantivalism is Probably Not a Consequence of Classical Mechanics." *International Studies in the Philosophy of Science* 13: 17–34.

(2006) "The Regularity Account of Relational Spacetime." *Mind* 115: 41–73.

(2008) "Why the Parts of Absolute Space are Immobile." *British Journal for the Philosophy of Science*, 59: 391–407.

(2009) "Review of Physical Relativity and Understanding Spacetime." *Philosophy of Science* 76: 404–422.

Hume, D. (1757) "A Letter to Rev. Mr. Hume, Author of Douglas, A Tragedy." *The London Chronicle for 1757* (March 8–10): 239

(1998) *An Enquiry Concerning the Principles of Morals.* T. L. Beauchamp, ed. Oxford University Press.

(1999) *An Enquiry Concerning Human Understanding.* T. L. Beauchamp, ed. Oxford University Press.

(2000) *A Treatise of Human Nature.* D. F. Norton and M. J. Norton, eds. Oxford University Press.

(2009) *Dialogues Concerning Natural Religion.* J. C. A. Gaskin, ed. Oxford University Press.

Hunter, M. (2000) *Robert Boyle (1627–91): Scrupulosity and Science.* Woodbridge: Boydell Press.

Hurlbutt, III, R. H. (1965) *Hume, Newton, and the Design Argument.* Lincoln: University of Nebraska Press.

Huygens, C. (1669) "A Summary Account of the Laws of Motion." *Philosophical Transactions of the Royal Society* 4: 925–928.

(1690a) *Traité de la lumiere . . . Avec un discours de la cause de la Pesanteur.* Leiden: P. Van der Aa.

(1690b) *Discourse on the Cause of Gravity.* K. Bailey, trans., with annotations by Bailey and G. Smith. Unpublished manuscript.

(1888–1950) *Oeuvres Complètes de Christiaan Huygens publiées par la Société hollandaise des sciences.* The Hague: Martinus Nijhoff.

(1986) *Christiaan Huygens' The Pendulum Clock or Geometrical Demonstration, Concerning the Motion of Pendula as Applied to Clocks.* R. J. Blackwell, trans. Ames: Iowa State University Press.

Inwagen, P. v. (2007) "Metaphysics." *The Stanford Encyclopedia of Philosophy.* E. Zalta, ed. Fall 2007 Edition. http://plato.stanford.edu/archives/fall2007/entries/metaphysics/.

Israel, J. (2001) *Radical Enlightenment: Philosophy and the Making of Modernity, 1650–1750.* Oxford University Press.

Jalobeanu, D. and P. Anstey, eds. (2011) *Vanishing Matter and the Laws of Motion.* London: Routledge.

Jammer, M. (1954) *Concepts of Space: The History of Theories of Space in Physics.* Cambridge, MA: Harvard University Press.

(1957) *Concepts of Force: A Study in the Foundations of Dynamics.* Cambridge, MA: Harvard University Press.

(1961) *Concepts of Mass, in Classical and Modern Physics.* Cambridge, MA: Harvard University Press.

Janiak, A. (2006) "Newton's Philosophy." *Stanford Encyclopedia of Philosophy.* E. Zalta, ed. Summer 2008 edition. http://plato.stanford.edu/archives/sum2008/entries/newton-philosophy/.

(2007) "Newton and the Reality of Force." *Journal of the History of Philosophy* 45: 127–147.

(2008) *Newton as Philosopher.* Cambridge University Press.

Janssen, M. (2002) "*COI* Stories: Explanation and Evidence in the History of Science." *Perspectives on Science* 10: 457–522.

Johnson, W. E. (1964) *Logic.* Volume II. New York: Dover.

Joy, L. S. (2006) "Scientific Explanation from Formal Causes to Laws of Nature." In K. Park and L. Daston, eds. *The Cambridge History of Science,* Volume III: *Early Modern Science.* Cambridge University Press.

Kant, I. (1902–) *Gesammelete Schriften.* Berlin Akademie der Wissenschaften, eds. Berlin: Reimer.

(1968) *Kant: Selected Pre-Critical Writings and Correspondence with Beck.* G. B. Kerfred and D. E. Walford, trans. Manchester University Press.

(1992) *Theoretical Philosophy, 1755–1770.* D. Walford and R. Meerbote, eds. Cambridge University Press.

(1998) *Critique of Pure Reason.* P. Guyer and A. W. Wood, trans. Cambridge University Press.

(2002) *Theoretical Philosophy after 1781.* H. E. Allison, P. L. Heath, and G. C. Hatfield, eds. Cambridge University Press.

(2004) *Metaphysical Foundations of Natural Science.* M. Friedman, trans. Cambridge University Press.

Kargon, R. (1966) "Newton, Barrow, and the Hypothetical Physics." *Centaurus* 11: 46–56.

Kaufman, D. (2008) "Cartesian Substances, Individual Bodies, and Corruptibility." Unpublished manuscript.

Keill, J. (1758) *An Introduction to Natural Philosophy.* London: for Andrew Millar.

(1809) "On the Laws of Attraction and other Physical Principles." *The Philosophical Transactions of the Royal Society of London from their commencement, in 1665, to the year 1800; abridged, with notes and biographical illustrations.* Volume V: 417–424. Hutton, Shaw, and Pearson, eds. London: C. and R. Baldwin.

Kemp Smith, N. (1941) *The Philosophy of David Hume.* London: MacMillan.

Kirkby, J. (1734/1970) *The Usefulness of Mathematical Learning Explained and Demonstrated, being Mathematical Lectures read at the Public Schools at the University of Cambridge.* London: S. Austen /London: Cass.

Kinckhuysen, G. (1661) *Algebra Ofte Stelkonst....* Haarlem: Passchier van Wesbusch.

Knorr, W. (1986/1993) *The Ancient Tradition of Geometric Problems.* Boston: Birkhäuser/ New York: Dover.

Kaslow, A., ed. (1967) *The Changeless Order: The Physics of Space, Time and Motion.* New York: George Braziller.

Koyré, A. (1968) *Newtonian Studies.* University of Chicago Press.

(1978) *Galileo Studies.* Atlantic Highlands, NJ: Humanities Press.

Koyré, A. and I. B. Cohen. (1962) "Newton and the Leibniz-Clarke Correspondence." *Archives Internationales d'histoire des Sciences* 15: 63–126.

Kuhn, T. S. (1970a) *The Structure of Scientific Revolutions.* Second edition. University of Chicago Press.

(1970b) "Comment on Westfall's 'Uneasily Fitful Reflections'" In Palter, ed.

Ladyman, J. (1998) "What is structural realism?" *Studies in the History and Philosophy of Science* 29: 409–424.

Lakatos, I. (1978) *The Methodology of Scientific Research Programmes: Philosophical Papers.* Volume I. Cambridge University Press.

Larmore, C. (1987) "Newton's Critique of Cartesian Method." *Graduate Faculty Philosophy Journal* 12: 81–109.

Laywine, A. (1993) *Kant's Early Metaphysics and the Origins of the Critical Philosophy.* Atascadero, CA: Ridgeview Press.

Leibniz, G. W. (1849–63) *Leibnizens mathematische schriften.* C. I. Gerhardt, ed. 7 volumes. Berlin: A. Asher.

(1875–1890) *Die philosophischen Schriften.* C. I. Gerhardt, ed. 7 volumes. Berlin: Weidmann.

(1923–) *Sämtliche Schriften und Briefe.* Deutsche Akademie der Wissenschaften zu Berlin, eds. Berlin: Akademie-Verlag.

(1973) *Marginalia in Newtoni "Principia mathematica": 1687.* E. A. Fellmann, ed. Paris: Vrin.

(1976) *Philosophical Papers and Letters.* L. E. Loemker, ed. and trans. Second edition. Dordrecht: Reidel.

(1989) *Philosophical Essays.* R. Ariew and D. Garber, eds. and trans. Indianapolis: Hackett.

(2001) *The Labyrinth of the Continuum: Writings on the Continuum Problem, 1672–1686.* R. Arthur, ed. and trans. New Haven: Yale University Press.

Lennox, J. (1986) "Aristotle, Galileo and 'Mixed Sciences'." In *Reinterpreting Galileo.* W. A. Wallace, ed. Washington, DC: Catholic University of America Press.

Lewis, D. (1970) "How to Define Theoretical Terms." *The Journal of Philosophy* 67: 427–446.

(1973) *Counterfactuals.* Cambridge, MA: Harvard University Press.

Locke, J. (1693/1989) *Some Thoughts Concerning Education.* J. W. Yolton and J. S. Yolton, eds. Oxford: Clarendon Press.

(1823) *The Works of John Locke in Ten Volumes.* London: Thomas Tegg.

(1824) *The Works of John Locke, in Nine Volumes.* 12th edition. London: C. and J. Rivington *et al.*

(1892) *Locke's Conduct of the Understanding.* T. Fowler, ed. Oxford: Clarendon Press.

(1975) *An Essay Concerning Human Understanding.* P. H. Nidditch, ed. Oxford University Press.

(1990) *Drafts for the Essay Concerning Human Understanding, and Other Philosophical Writings.* P. Nidditch and G. A. J. Rogers, eds. Oxford: Clarendon Press.

Luckey, P. (1941) "Tābit b. Qurra über den geometrischen Richtigkeitsnachweis der Auflösung der quadratischen Gleichungen." *Berichte über die Verhandlungen der Sächsischen Akademie der Wissenschaften zu Leipzig, Mathematisch-physikalische Klasse* 93: 93–114.

McCann, E. (1983) "Lockean Mechanism." In A. J. Holland, ed. *Philosophy, its History and Historiography.* Dordrecht: Kluwer.

McGuire, J. E. (1966) "Body and Void and Newton's *De Mundi Systemate*: Some New Sources." *Archive for the History of Exact Sciences* 3: 206–248.

(1967) "Transmutation and immutability: Newton's doctrine of physical qualities." *Ambix* 14: 69–95.

(1968) "The Origin of Newton's Doctrine of Essential Qualities." *Centaurus* 12: 233–260.

(1970) "Atoms and the 'Analogy of Nature': Newton's Third Rule of Philosophizing." *Studies in History and Philosophy of Science* 1: 3–58.

(1972) "Boyle's Conception of Nature." *Journal of the History of Ideas* 33: 523–542.

(1977) "Neoplatonism and Active Principles: Newton and the *Corpus Hermeticum.*" In R. S. Westman and J. E. McGuire, eds. *Hermeticism and the Scientific Revolution.* Los Angeles: University of California Press.

(1978) "Existence, Actuality and Necessity: Newton on Space and Time." *Annals of Science* 35: 463–508.

(1994) "Natural Motion and Its Causes: Newton on the *Vis Insita* of Bodies." In M. L. Gill and J. G. Lennox, eds. *Self-motion: from Aristotle to Newton.* Princeton University Press.

(1995) *Tradition and Innovation: Newton's Metaphysics of Nature.* Boston: Kluwer.

McMullin, E. (1978) *Newton on Matter and Activity.* University of Notre Dame Press.

(1985) "Galilean Idealization." *Studies in History and Philosophy of Science* 16: 247–273.

(1989) "The Explanation of Distant Action: Historical Notes." In J. T. Cushing and E. McMullin, eds. *Philosophical Consequences of Quantum Theory.* University of Notre Dame Press.

(2001) "The Impact of Newton's *Principia* on the Philosophy of Science." *Philosophy of Science* 68: 279–310.

(2002) "The Origins of the Field Concept in Physics." *Physics in Perspective* 4: 13–39.

Mach, E. (1893/1960) *The Science of Mechanics: A Critical and Historical Account of Its Development.* T. J. McCormack, trans. Chicago: Open Court.

Machamer, P., ed. (1998) *The Cambridge Companion to Galileo.* Cambridge University Press.

Machin, J. (1729) *The Laws of the Moon's Motion According to Gravity*, an Appendix to Andrew Motte's translation of the *Principia* of 1729. Reissued in facsimile edition, *The Mathematical Principles of Natural Philosophy*, London: Dawsons of Pall Mall, 1968.

MacLaurin, C. (1742) *A Treatise on Fluxions.* Edinburgh: T. W. T. Ruddimans.

(1748/1968) *An Account of Sir Isaac Newton's Philosophical Discoveries.* London: A. Millar/New York: Johnson Reprints.

Maglo, K. (2003) "The Reception of Newton's Gravitational Theory by Huygens, Varignon, and Maupertuis." *Perspectives on Science* 11: 135–169.

(2007) "Force, Mathematics, and Physics in Newton's *Principia*: A New Approach to Enduring Issues." *Science in Context* 20: 571–600.

Mahoney, M. S. (1990) "Barrow's Mathematics: Between Ancients and Moderns." In M. Feingold, ed. *Before Newton: The Life and Times of Isaac Barrow.* Cambridge University Press.

Malament, D. B., ed. (2002) *Reading Natural Philosophy: Essays in the History and Philosophy of Science and Mathematics to Honor Howard Stein on his 70th Birthday.* Chicago: Open Court.

Malet, A. (1997) "Isaac Barrow on the Mathematization of Nature: Theological Voluntarism and the Rise of Geometrical Optics." *Journal of the History of Ideas* 58: 265–287.

Mancosu, P. (1996) *Philosophy of Mathematics and Mathematical Practice in the Seventeenth Century.* Oxford University Press.

Matthews, M. R. (1989) *Scientific Background to Modern Philosophy.* Indianapolis: Hackett.

Maupertuis, P. L. M. (1732) *Discours sur les Différentes Figures des Astres.* Paris: l'Imprimerie Royale.

(1756) *Œuvres.* Lyon: Chez Jean-Marie Bruyset.

Menn, S. (1998) *Descartes and Augustine.* Cambridge University Press.

Mercator, N. (1668) *Logarithmothechnia; sive Methodus construendi Logarithmos Nova, accurata & facilis. . . . Cui accedit vera quadratura hyperbolæ et inventio summæ logarithmorum.* London: M. Pitt.

(1676) *Institutionum Astronomicarum Libri Duop, De Motu Astrorum Communi & Proprio.* London: Samuel Simpson.

Mersenne, M. (1644) *Cogitata physico-mathematica....* Paris: Sumptibus A. Bertier.

Metzger, H. (1930) *Newton, Stahl, Boerhaave et la doctrine chimique.* Paris.

(1938) *Attraction universelle et religion naturelle chez quelques commentateurs anglais de Newton.* Paris.

Milton, J. (1981) "John Locke and the Nominalist Tradition." In R. Brandt, ed. *John Locke: Symposium Wolfenbuettel 1979.* New York: Walter de Gruyter.

(1994) "Locke at Oxford." In G. A. J. Rogers, ed. *Locke's Philosophy: Content and Context.* Oxford: Clarendon Press.

Mossner, E. C. (1980) *The Life of David Hume.* Oxford University Press.

Mueller, I. (1981) *Philosophy of Mathematics and Deductive Structure in Euclid's Elements.* Cambridge, MA: MIT Press.

Murray, G., Harper, W., and Wilson, C. (2011) "Huygens, Wren, Wallis and Newton on Rules of Impact and Reflection." In D. Jalobeanu and P. Anstey, eds.

Nadler, S. (1998) "Doctrines of Explanation in Late Scholasticism and in the Mechanical Philosophy." In D. Garber and M. Ayers, eds. *The Cambridge History of Seventeenth-Century Philosophy.* Cambridge University Press.

Nagel, E. (1961) *The Structure of Science: Problems in the Logic of Scientific Explanation.* New York: Harcourt, Brace & World.

Nerlich, G. (2006) "Can Parts of Space Move? On Paragraph Six of Newton's Scholium." *Erkenntnis* 62: 119–135.

Newcomb, S. (1898) *Tables of the Four Inner Planets. Astronomical Papers Prepared for the Use of the American Ephemeris and Nautical Almanac.* Vol. VI. Washington: Bureau of Equipment, Navy Department.

Newman, W. R. (2006) *Atoms and Alchemy: Chymistry and the Experimental Origins of the Scientific Revolution.* University of Chicago Press.

Norton, J. (1993) "Determination of Theory by Evidence: How Einstein Discovered General Relativity." *Synthese* 97: 1–31.

(1994) "Science and Certainty." *Synthese* 99: 3–22.

(1995) "Eliminative Induction as a Method of Discovery: How Einstein Discovered General Relativity." In J. Leplin, ed. *The Creation of Ideas in Physics.* Dordrecht: Kluwer.

Osler, M. (1994) *Divine Will and the Mechanical Philosophy: Gassendi and Descartes on Contingency and Necessity in the Created World.* Cambridge University Press.

Palter, R., ed. (1970) *The* Annus Mirabilis *of Sir Isaac Newton, 1666–1966.* Cambridge, MA: MIT Press.

Panza, M. (1997) "Classical Sources for the Concepts of Analysis and Synthesis." In M. Otte and M. Panza, eds. *Analysis and Synthesis in Mathematics.* History and Philosophy. *Boston Studies in the Philosophy of Science.* Volume 196. Dordrecht: Kluwer.

(2002) "Mathematisation of the Science of Motion and the Birth of Analytical Mechanics: A Historiographical Note." In P. Cerrai, P. Freguglia, and C. Pellegrini, eds. *The Application of Mathematics to the Sciences of Nature: Critical Moments and Aspects.* New York: Kluwer, Plenum.

(2005) *Newton et les origines de l'analyse: 1664–1665.* Paris: Blanchard.

(2007) "Euler's *Introductio in Analysin Infinitorum* and the Program of Algebraic Analysis: Quantities, Functions and Numerical Partitions." In R. Backer, ed. *Euler Reconsidered: Tercentenary Essays.* Herber City: Kendrick Press.

(2008) "What Is New and What Is Old in Viète's *Analysis Restituita* and *Algebra Nova*, and Where Do They Come From? Some Reflections on the Relations between Algebra and Analysis before Viète." *Revue d'Histoire des mathématiques* 13: 83–151.

Patrides, C. A., ed. (1980) *The Cambridge Platonists.* Cambridge University Press.

Pedersen, K. M. (1968) "Roberval's Method of Tangents." *Centaurus* 13: 151–182.

(1980) "Techniques of the Calculus, 1630–1660." In I. Grattan-Guinness, ed. *From the Calculus to Set Theory, 1630–1910.* London: Duckworth.

Pemberton, H. (1728) *A View of Sir Isaac Newton's Philosophy.* Dublin: John Hyde.

Phemister, P. (1993) "Locke, Sergeant, and Scientific Method." In T. Sorell, ed. *The Rise of Modern Philosophy: The Tension between the New and Traditional Philosophies from Machiavelli to Leibniz.* Oxford: Clarendon Press.

Piccolomini, A. (1565) *Alexandrei Piccolominei in mechanicas quaestiones Aristotelis paraphrasis . . . eiusdem commentarium de certitudine mathematicarum disciplinarum . . .* Venice: Navo, Curzio Troiano.

Poincaré, H. (1905) *Science and Hypothesis.* London: Walter Scott.

Popper, K. (2002) *Conjectures and Refutations.* London: Routledge. Originally published in 1963.

(2003) *The Logic of Scientific Discovery.* London: Routledge. Originally published in 1935.

Pourciau, B. (2006) "Newton's Interpretation of Newton's Second Law." *Archive for History of Exact Sciences* 60: 157–207.

Rees, G. (1986) "Mathematics and Francis Bacon's Natural Philosophy." *Revue Internationale de Philosophie* 40: 399–426.

Rigaud, S. J., ed. (1841) *Correspondence of Scientific Men of the Seventeenth Century.* Oxford University Press.

Rindler, W. (1977) *Essential Relativity: Special, General, and Cosmological.* New York: Springer-Verlag.

Roberval, G. P. (1693) "Observations sur la composition des mouvemens, et sur le moyen de trouver les touchantes des lignes courbes." In *Divers Ouvrages de Mathématique et de Physique par Messieurs de l'Académie Royale des Sciences.* Paris: Impr. Royale.

(1730) "Divers Ouvrage de M. Personnier de Roberval." *Mémoirs de l'Academie Royale des Sciences, depuis 1666 jusqu'au 1699* 6: 1–478.

Rogers, G. A. J. (1966) "Boyle, Locke, and Reason." *Journal of the History of Ideas* 27: 205–216.

(1978a) "Locke, Newton and the Enlightenment." *Vistas in Astronomy* 22(4): 471–476.

(1978b) "Locke's *Essay* and Newton's *Principia.*" *Journal of the History of Ideas* 39.

(1979a) "The Empiricism of Locke and Newton." In S C. Brown, ed. *Royal Institute of Philosophy Lectures*, Volume XII: *Philosophers of the Enlightenment.* Sussex and New Jersey: The Harvester Press and Humanities Press.

(1979b) "Locke, Newton, and the Cambridge Platonists on Innate Ideas." *Journal of the History of Ideas* 40: 191–205.

(1981) "Locke, Law, and the Laws of Nature." In R. Brandt, ed., *John Locke: Symposium Wolfenbuettel 1979.* New York: Walter de Gruyter.

(1982) "The System of Locke and Newton." In Bechler, ed.

Ruestow, E. G. (1973) *Physics at Seventeenth and Eighteenth-Century Leiden: Philosophy and the New Science in the University.* The Hague: Martinus Nijhoff.

Rynasiewicz, R. (1995a) " 'By Their Properties, Causes and Effects': Newton's Scholium on Time, Space, Place and Motion, I: The Text." *Studies in History and Philosophy of Science* 26: 133–153.

(1995b) " 'By their Properties, Causes and Effects': Newton's Scholium on Time, Space, Place, and Motion, II: The Context." *Studies in History and Philosophy of Science* 26: 295–321.

Sargent, R-M. (1994) "Learning from Experience: Boyle's Construction of an Experimental Philosophy." In M. Hunter, ed. *Robert Boyle Reconsidered.* Cambridge University Press.

(1995) *The Diffident Naturalist: Robert Boyle and the Philosophy of Experiment.* University of Chicago Press.

Schliesser, E. (2005a) "On the Origin of Modern Naturalism: the Significance of Berkeley's Response to a Newtonian Indispensability Argument." *Philosophica* 76: 45–66.

(2005b) "Wonder in the Face of Scientific Revolutions: Adam Smith on Newton's 'Proof' of Copernicanism." *British Journal for the History of Philosophy* 13: 697–732.

(2007). "Two Definitions of 'Cause,' Newton, and the Significance of the Humean Distinction between Natural and Philosophical Relations." *The Journal of Scottish Philosophy* 5: 83–101.

(2008) "Hume's Newtonianism and Anti-Newtonianism." *The Stanford Encyclopedia of Philosophy.* E. Zalta, ed. http://plato.stanford.edu/archives/win2008/entries/hume-newton/.

(2009) "Hume's Attack on Newton." *Enlightenment & Dissent* 25: 167–203.

(2010a) "Review of G.A.J. Rogers, Tom Sorrell and Jill Kraye (eds.), Insiders and Outsiders in Seventeenth-Century Philosophy." *Notre Dame Philosophical Reviews.* http://ndpr.nd.edu/review.cfm?id=19147.

(2010b) "Without God: Gravity as a Relational Quality of Matter in Newton." In D. Jalobeanu and P. Anstey, eds.

(2011) "Newton's Challenge to Philosophy." *HOPOS: The Journal of the International Society for the History of Philosophy of Science* 1: 101–28.

(forthcoming a) "Spinoza's Criticism of Mathematical-Physics and a Scottish Newtonian Response." *Vereeniging Het Spinozahuis.*

(forthcoming b) "Spinoza and Science." In M. Della Rocca, ed. *Oxford Handbook of Spinoza.* Oxford University Press.

(manuscript) "Philosophy Between Newton's Challenge and Spinoza's Treatment of the Socratic Problem: David Hume and Adam Smith in a Newtonian era."

Schliesser, E. and G. E. Smith. (1996) "Huygens's 1688 Report to the Directors of the Dutch East India Company on the Measurement of Longitude at Sea and Its Implications for the Non-uniformity of Gravity." *De zeventiende eeuw* 12: 198–214.

(forthcoming) "Huygens's 1688 Report to the Directors of the Dutch East Indian Company on the Measurement of Longitude at Sea and the Evidence it Offered Against Universal Gravity." *Archive for the History of the Exact Sciences.*

Schmaltz, T. M. (1999) "Spinoza on the Vacuum." *Archive fur Geschichte der Philosophie* 81: 174–205.

(2008a) *Descartes on Causation.* New York: Oxford University Press.

(2008b) "Descartes on the Extensions of Space and Time." Unpublished manuscript.

Schönfeld, M. (2000) *The Philosophy of the Young Kant.* Oxford University Press.

Schuurman, P. (2004) *Ideas, Mental Faculties and Method: The Logic of Ideas of Descartes and Locke and its reception in the Dutch Republic, 1630–1750.* Leiden: Brill.

Seidelmann, P. K., ed. (1992) *Explanatory Supplement to the Astronomical Almanac.* Mill Valley: University Science Books.

Sepkoski, D. (2005) "Nominalism and Constructivism in Seventeenth-Century Mathematical Philosophy." *Historia Mathematica* 32: 33–59.

Shank, J. B. (2004) "'There was no such thing as the 'Newtonian Revolution,' and the French initiated it:' Eighteenth-century Mechanics in France Before Maupertuis." *Early Science and Medicine* 9: 257–292.

(2008) *The Newton Wars and the Beginning of the French Enlightenment.* University of Chicago Press.

Shapin, S. (1988) "Robert Boyle and Mathematics: Reality, Representation, and Experimental Practice." *Science in Context* 2: 25–58.

Shapiro, A. E. (2002) "Newton's Optics and Atomism." In Cohen and Smith, eds.

(2004) "Newton's Experimental Philosophy." *Early Science and Medicine* 9: 185–217.

Skinner, Q. (1966) "Thomas Hobbes and His Disciples in France and England." *Comparative Studies in Society and History* 8:153–167.

Sklar, L. (1974) *Space, Time and Spacetime.* Berkeley: University of California Press.

Slowik, E. (2002) *Cartesian Spacetime.* Dordrecht: Kluwer.

Smeenk, C. and G. E. Smith. "Newton on Constrained Motion: A Commentary on Book 1, Section 10 of Newton's *Principia.*" Unpublished manuscript.

Smith G. E. (1997) "Huygens' Empirical Challenge to Universal Gravity." Unpublished manuscript notes and overheads for presentation.

(1999) "How Did Newton Discover Universal Gravity?" *The St. John's Review* 45, 32–63.

(2001a) "J. J. Thomson and the Electron: 1897–1899." In J. Z. Buchwald and A. Warwick, eds. *Histories of the Electron and the Birth of Microphysics.* Cambridge, MA: MIT Press.

(2001b) "The Newtonian Style in Book II of the *Principia.*" In Buchwald and Cohen, eds.

(2002a) "The Methodology of the Principia." In Cohen and Smith, eds.

(2002b) "From the Phenomenon of the Ellipse to an Inverse-Square Force: Why Not?" In Malament, ed.

(2005) "Was Wrong Newton Bad Newton?" In J. Z. Buchwald and A. Franklin, eds. *Wrong for the Right Reasons.* Dordrecht: Springer.

(2007a) *Turning Data into Evidence: The Isaac Newton Lectures at the Suppes Center, Stanford University.* www.stanford.edu/dept/cisst/events0506.html.

(2007b). "Newton's *Philosophiae Naturalis Principia Mathematica.*" *The Stanford Encyclopedia of Philosophy.* http://plato.stanford.edu/entries/newton-principia/.

(forthcoming a) "Pending Tests to the Contrary: The Question of Mass in Newton's Law of Gravity."

(forthcoming b) "Closing the Loop: Testing Newtonian Gravity, Then and Now."

Smith, S. R. (2007) "Continuous Bodies, Impenetrability, and Contact Interactions: The View from the Applied Mathematics of Continuum Mechanics." *The British Journal for the Philosophy of Science* 58: 503–538.

Snobelen, S. (1999) "Isaac Newton, Heretic: The Strategies of a Nicodemite." *British Journal for the History of Science* 32 381–419.

Sober, E. (2006) "Parsimony." In S. Sarkar and J. Pfeifer, eds. *The Philosophy of Science: An Encyclopedia.* New York and London: Routledge.

Spinoza, de, B. (1737) *A Treatise Partly Theological and Partly Political: Containing Some Few Discourses, to Prove the Liberty of Philosophizing . . . May Be Allowed without Any Prejudice to Piety, . . . Translated from the Latin of Spinoza.* Booksellers of London and Westminster.

(1985) *The Collected Works of Spinoza.* E. M. Curley, ed. and trans. Volume I. Princeton University Press.

Stachel, J. (1977) "Introduction." In J. Earman *et al.*, eds.

Stein, H. (1967) "Newtonian Space-Time." *Texas Quarterly* 10: 174–200.

(1970a) "Newtonian space-time." In R. Palter, ed.

(1970b) "On the Notion of Field in Newton, Maxwell, and Beyond." In R. H. Stuewer, ed. *Historical and Philosophical Perspectives of Science. Minnesota Studies in the Philosophy of Science,* Volume V. Minneapolis: University of Minnesota Press.

(1977) "Some Philosophical Prehistory of General Relativity." In J. Earman *et al.*, eds.

(1990) "On Locke, 'the Great Huygenius, and the incomparable Mr. Newton'." In P. Bricker and R. I. G. Hughes, eds. *Philosophical Perspectives on Newtonian Science.* Cambridge, MA: MIT Press.

(1991) "From the Phenomena of Motions to the Forces of Nature': Hypothesis or Deduction?" *Proceedings of the 1990 Biennial Meeting of the Philosophy of Science Association.* Vol. II.

(2002) "Newton's Metaphysics." In Cohen and Smith, eds.

"On Metaphysics and Method in Newton." Unpublished manuscript.

Steinle, F. (1992) "Was ist Masse? Newtons Begriff der Materiemenge." *Philosophia Naturalis* 29: 94–117.

Stewart, M. A. (1981) "Locke's Professional Contacts with Robert Boyle." *The Locke Newsletter* 12: 19–44.

Strawson, G. (2000) "David Hume: Objects and Power." In R. Read and K. A. Richman, eds. *The New Hume Debate.* London: Routledge.

Streete, T. (1661) *Astronomia Carolina, A New Theory of the Celestial Motions.* London: Lodowick Lloyd.

Suppes, P. (1962) "Review of Concepts of Force: A Study in the Foundations of Dynamics, by Max Jammer." *Philosophical Review* 71: 117–120.

Terrall, M. (2002) *The Man who Flattened the Earth: Maupertuis and the Sciences in the Enlightenment.* University of Chicago Press.

Thackray, A. (1970) *Atoms and Powers: An Essay on Newtonian Matter-Theory and the Development of Chemistry.* Cambridge, MA: Harvard University Press.

Torricelli, E. (1644) *Opera Geometrica.* Florence: I. de Landis & A. Massæ.

Truesdell, C. (1968) *Essays in the History of Mechanics.* Berlin: Springer Verlag.

Valluri, S. R., C. Wilson, and W. L. Harper. (1997) "Newton's Apsidal Precession Theorem and Eccentric Orbits." *Journal for the History of Astronomy* 28: 13–27.

Van Fraassen, B. C. (1970) *An Introduction to the Philosophy of Time and Space.* New York: Random House.

Viète, F. (1591) *Zeteticorum libri quinque.* Tours: apud J. Mettayer.

Voltaire, F. M. A. (1738/1967) *The Elements of Sir Isaac Newton's Philosophy.* Trans. J. Hanna. London: Stephen Austen/London: Frank Cass.

(1741) *Elémens de la philosophie de Neuton.* London.

Waff, C. (1976) *Universal Gravitation and the Motion of the Moon's Apogee: The Establishment and Reception of Newton's Inverse-Square Law, 1687–1749.* Ph.D. dissertation, History of Science, Johns Hopkins University.

(1995) "Clairaut and the Motion of the Lunar Apse: The Inverse-Square Law Undergoes a Test." In R. Taton and C. Wilson, eds. *The General History of Astronomy, vol. II, Planetary Astronomy from the Renaissance to the Rise of Astrophysics, Part B: The Eighteenth and Nineteenth Centuries.* Cambridge University Press.

Watkins, E. (2005) *Kant and the Metaphysics of Causality.* Cambridge University Press.

Westfall, R. (1958) *Science and Religion in Seventeenth-century England.* New Haven: Yale University Press.

(1967) "Hooke and the Law of Universal Gravitation: a Reappraisal of a Reappraisal." *British Journal for the History of Science* 3: 245–261.

(1971a) *Force in Newton's Physics: The Science of Dynamics in the Seventeenth Century.* New York: American Elsevier.

(1971b) *The Construction of Modern Science.* Cambridge University Press.

(1980) *Never at Rest. A Biography of Isaac Newton.* Cambridge University Press.

Whewell, W., ed. (1860) *The Mathematical Works of Isaac Barrow.* Cambridge University Press.

Whiston, W. (1696/1978) *A New Theory of the Earth.* London: R. Roberts/New York: Arno Press.

Wilson, C. (1976) "Newton and the Eötvös Experiment." In *Essays in Honor of Jacob Klein.* Annapolis: St. John's College Press.

(1978) "Horrocks, Harmonies, and the Exactitude of Kepler's Third Law." In *Science and History: Essays in Honor of Edward Rosen, Studia Copernica* 16: 235–259. Reprinted in Wilson, *Astronomy from Kepler to Newton: Historical Studies*, London: Variorum Reprints, 1989.

(1980) "Perturbations and Solar Tables from Lacaille to Delambre: the Rapprochement of Observation and Theory." *Archive for History of Exact Sciences* 22: 53–188.

(1985) "The Great Inequality of Jupiter and Saturn from Kepler to Laplace." *Archive for History of Exact Sciences* 33: 15–290.

(1989) "Predictive Astronomy in the Century after Kepler." In R. Taton and C. Wilson, eds. *The General History of Astronomy, vol. II, Planetary Astronomy from the Renaissance to the Rise of Astrophysics, Part A: Tycho Brahe to Newton.* Cambridge University Press.

(1995) "The Precession of the Equinoxes from Newton to D'Alembert and Euler." In R. Taton and C. Wilson, eds. *The General History of Astronomy, vol. II, Planetory Astronomy from the Renaissance to the Rise of Astrophysics, Part B: The Eighteenth and Nineteenth Centuries.* Cambridge University Press.

(2000) "From Kepler to Newton: Telling the Tale." In R. H. Dalitz and M. Nauenberg, eds. *The Foundations of Newtonian Scholarship.* Singapore: World Scientific.

Wing, V. (1651) *Harmonicon Coeleste, or The Celestial Harmony of the Visible World.* London: Robert Leybourn.

(1669) *Astronomia Britannica.* London: Johannis Macock.

Winkler, K. P. (2000) "The New Hume." In R. Read and K.A. Richman, eds.

(2008) "Locke's Defense of Mathematical Physics." In P. Hoffman, D. Owen, and G. Yaffe, eds. *Contemporary Perspectives on Early Modern Science: Essays in Honor of Vere Chappell.* Toronto: Broadview Press.

Wolfson, P. R. (2001) "The crooked made straight: Roberval and Newton on tangents." *The American Mathematical Monthly* 108: 206–216.

Woodward, J. (2003) *Making Things Happen: A Theory of Causal Explanation.* Oxford University Press.

Woolhouse, R. (1971) *Locke's Philosophy of Science and Knowledge.* Oxford: Blackwell.

(1994) "Locke's Theory of Knowledge." In V. Chappell, ed. *The Cambridge Companion to Locke.* New York: Cambridge University Press.

Worrall, J. (2000) "The Scope, Limits, and Distinctiveness of the Method of 'Deduction from the Phenomena': Some Lessons from Newton's 'Demonstrations' in Optics." *The British Journal for the Philosophy of Science* 51: 45–80.

Yoder, J. G. (1988) *Unrolling Time: Christiaan Huygens and the Mathematization of Nature.* Cambridge University Press.

Yolton, J. (1969) "The Science of Nature." In J. Yolton, ed. *John Locke: Problems and Perspectives.* Cambridge University Press.

(1970) *Locke and the Compass of Human Understanding.* Cambridge University Press.

INDEX